Ways of the World

Ways of the World

A History of the World's Roads and of the Vehicles That Used Them

M. G. Lay

RUTGERS UNIVERSITY PRESS
New Brunswick, New Jersey

First published in cloth throughout the world by Rutgers University Press, 1992
First published in paperback in Australia by Primavera Press, 1992

Library of Congress Cataloging-in-Publication Data

Lay, M. G. (Maxwell G.)
    Ways of the world : a history of the world's roads and of the vehicles that used them /
M. G. Lay.
        p.   cm.
    Includes bibliographical references and index.
    ISBN 0-8135-1758-3
    1. Transportation—History.   2. Roads—History.   3. Vehicles—History.   4. Bridges—
History.   I. Title.
    TA1015.L39   1992
    629.04'9—dc20   91-23148
                                                                                     CIP

British Cataloging-in-Publication information available

Every improvement of the means of locomotion benefits mankind morally and intellectually as well as materially . . . and binds together all branches of the great human family. MACAULAY, HISTORY OF ENGLAND, 1855

First let the ways be regularly brought
To artificial form, and truly wrought, . . .
That not a stone's amiss; but all complete,
All lying smooth, round, firm, and wondrous neat. . . .
Then comes a gang of heavy laden wains
Of carts and wagons, spoiling all our pains.
T. MACE, CLERK OF TRINITY COLLEGE, CAMBRIDGE, 1675

Contents

List of Illustrations

List of Tables

First, a few snapped twigs and bruised blades of grass. Soon, a path, a trail. Then, in surprisingly little time, the way becomes a road. Roads are truly exceptional among human works, in both their mutability and their longevity. So many artifacts of ancient times survive only through the deliberate efforts of archaeologists and museums, but thousands, even millions, of people today continue to use roads built many centuries ago. Sometimes, as with Roman roads, even the original surface survives. Yet, even where the pavement has long since been worn away, the early carts have become obsolete, and the original purpose of the road has been forgotten, the route remains in daily use.

Ever since ancient times, topographers, geographers, merchants, diplomats, sailors, soldiers, and travelers have known that a full and complete life demands movement. Given the human yen to know what lies over the visible horizon—and to find ways to make it easy to get there—it is surprising that standard histories of transportation pay so little attention to the world's roads and how they have come into being. Archaeologists and historians of transportation have written mostly in the context of a specific time or place and within a fairly narrowly defined culture. But we have long needed something more: a history that acknowledges both the human need for roads and their distinctive physical character. This book does that.

As a practicing civil engineer, Maxwell Lay had viewed roads primarily from an interest in a technology "which I had thought I understood and within a society that I had taken for granted." When he began looking for a good history of roads that dealt with his interests, he found nothing. Fortunately for us, he decided to write his own history of the world's roads. By giving central place to the roads themselves and their technology, rather than to social and institutional forces in history, Lay in fact tells us much more about the human use of roads than the standard histories of transportation.

Because roads outlast the people and conditions that created them and the vehicles that first traversed them, roads become a central element in social infrastructure. They possess at once great simplicity and an amazing complexity. Unlike other important mediums of transportation—canals, railroads, airways—which have required equally massive feats of construction, roads have been readily amenable to almost infinite elaboration of uses. For example, the recent decline in business travelers' use of civil aviation in the United States has been ascribed to the convenience of travel by private car: with a car you can travel when and where you like, unhindered by the airlines' circumscribed schedules.

The freedom of movement offered by roads has encouraged economic prosperity and discouraged the power of monopoly. The medieval world understood that mobility is tantamount to equality: the institution of feudalism depended on the geographic immobility of serfs. In 1957, in the seminal work *On the Road,* the Canadian-American author Jack Kerouac celebrated the joy that the beat generation felt as it set out on its own road: "God, man," he exulted, "I rode around this country as free as a bee." In society after society, the search for individual economic, political, and cultural freedom has meant literally taking to the road. In this book, Maxwell Lay tells us how those *Ways of the World* were shaped and how the history of the road becomes a paean of freedom.

JAMES E. VANCE, JR.

Acknowledgments

The introduction acknowledges how heavily I have drawn on the work of others. But this is a minor debt compared with those which I now discuss. First I must record the assistance that I have received from almost every librarian I have dealt with—what a profession of service they render. I offer them all my belated thanks.

I also have pleasure in acknowledging both the financial support for the initial preparation of this book, which was provided by the Rees Jeffreys Foundation in England, and the helpful inputs of my friends and colleagues.

Otherwise, and with two exceptions, writing this text has been a solitary and lonely task performed as interludes in my daily labors as a bureaucrat in the Victorian Roads Corporation. While Cass provided me with patient company and many timely reminders of higher obligations, my sole and heartfelt personal acknowledgment belongs to my dear wife, Margaret, who aided the initiation and progress of the text, tolerated my obsession with it, supported my commitment to its publication, but nevertheless accurately defined it as more a mistress than a manuscript.

Ways *of the* World

Introduction

The present state of things is the consequence of the
past; and it is natural to enquire into the sources of
the good we enjoy and the evils we suffer.
BEN JONSON, CA. 1600

*T*his book has a number of objectives. The first is to provide a comprehensive history of roads and bridges and the vehicles that have used them. Roads, and all those associated ways, have been an ever-present part of our society. Their influence makes them an excellent medium through which to observe the changing human condition. As a daily participant in the management of a road system my sensitivities are perhaps heightened to the lessons we can learn from our roads and the folks who use them. I say "folks" because I am struck by the remarks of one of Priestly's characters in *Good Companions:* "I nivver knew there were so many folks wandering about." A good history gives us the wonderful opportunity to make our contemplative observations over time as well as over space. Thus, the chance to write this book and develop some new perspectives was an irresistible temptation.

A second objective is to properly document that story. I first felt this need in 1985 when looking for a good history of the world's roads to quote in my *Handbook of Road Technology.* What I found were typically sectional, even biased, books concentrating on a particular aspect of roads or on a particular region. In the end I found myself writing a thirty-four-page chapter, which subsequently formed the starting point for this book.

As I progressed with the task I became increasingly aware of the strength of Teilhard de Chardin's famous comment, "Nothing is comprehensible, except through its history." I was seeing with new eyes a technology that I had thought I understood and a society that I had taken for granted. The book changed from a mechanical task into an attempt to record and share some of those understandings and insights.

Nevertheless, this is a technical history. Since most previous writers on the subject have not had a strong engineering background, I make no excuses for attempting to fill a few vacuums and correct the occasional past misconception (particularly after reading histories on the subject that required awareness of neo-Marxist philosophy and of the writings of Habermas). This book falls squarely into just one of C. P. Snow's two cultures. I left its writing even more committed to the above stance and with a heightened respect for the talents of early road and bridge builders.

My third objective is to explore the development of roads as part of a larger societal system. To achieve this I have taken both a synoptic world view and a systems view—with the road at the center of that system. I have been fascinated by the broader implications of the interplays between social and technological change and the development of roads. I have explored how roadmaking has been affected by such factors as the need to travel, the ability to travel, and the changing technologies of travel and of construction. In a broader sense I have also examined the interplays between society and its transportation system, particularly insofar as those interplays are reflected in roads and vehicles.

An interaction that I could have pursued further is the socioeconomic role of human settlement. However, this topic seemed so large as to risk subsuming the theme of the book, and so I have instead provided contact points and guideposts to assist the reader in exploring settlement issues further. The best of these guides is surely James Vance's wonderful book *The Scene of Man* ([1977] 1986).

Finally, my hobby interest in information systems led me to my fourth objective: to provide a good reference list and literature survey of the world's English-language literature on the history of roads. The best existing list would be Robert Forbes's comprehensive but somewhat dated catalogs, and some previous work in this field suffers noticeably from poor referencing. I found my material by computer searching, citation searching, browsing, rifling old card catalogs, sneaking undetected into library stockrooms and stackrooms, and gossiping. Citation searching was the most useful tool, but I must admit that serendipity was more fruitful than I would have predicted. I am happy with my references, but would be delighted to be told of more.

It follows that this is not a history based on primary sources. It is a secondary compilation founded on the hard work of others. I make no apology for my synthetic role. I have seen my task as that of coalescer of individual contributions into a coherent whole and then as an informed observer of that whole. Nevertheless, the text does contain some original contributions to the history of roads. For example, the discussion of asphalt in chapter 7 certainly represents the first coherent and correct history of that important material.

To achieve my objectives I have written at a number of levels, using a structure that sequentially examines the development of pathways; looks at the associated rise of wheeled vehicles and draft animals; studies the consequential development of surfaced roads; sidesteps to examine the motives for road building; plots the rise of self-powered vehicles and explores their influence and application; examines the surfaces that arose to meet their needs; describes the somewhat separate history of bridges; and, finally, reviews developments in this century and looks fleetingly at the future. This structure does not always result in a strictly chronological history. To assist the reader seeking order out of my chaos, the book includes a detailed chronology of events and a comprehensive index.

You will find no short, sharp conclusions to this history. But I trust that it leaves you with some pride in the inheritance we have all received—the wonderful ways of the world.

Names and Units

Place names used in the text are current names rather than those in use at the time under discussion. This allows easier location in modern atlases. *National Geographic* maps are used as a guide, although Chinese names are taken from current official Chinese maps. Chinese names are given in China's preferred Pinyin spelling—conversion tables to older anglicizing are available.

Biblical quotations are from the Revised Standard Version unless otherwise noted.

For international consistency in measurements, the universally adopted Système international d'unités, or International System of Units (SI, i.e., metric system) is used. The units are as follows:

> For length the basic unit is the meter. A meter (m) is 1.09 yards, and a kilometer (km) is 1,000 meters and equals 0.62 mile. The least common unit for the lay reader may be the megameter, Mm, which is equivalent to a million meters, or 1,000 kilometers, or about the distance between Washington, D.C., and Chicago. A Mm is therefore 620 miles. A millimeter (mm) is a thousandth of a meter. There are 25.4 mm to the inch. A terameter (Tm) is a million megameters, or about seven times the distance to the sun.
>
> For speed, 100 km/h is equivalent to 62 mph
>
> The metric ton, which is called a tonne, is used for mass. A tonne (t) is 1,000 kilograms and equals 2,205 pounds, i.e., 1.10 American short tons and 0.98 British long tons. It is therefore very close to the conventional ton.
>
> The kilowatt (kW) is used rather than the horsepower. A kW is 1.34 horsepower.
>
> The hertz (Hz) is used rather than RPM. The Hz is RPM/60.
>
> The megapascal (MPa) is used rather than psi. A MPa is 135 psi.

All measurements are rounded off to the appropriate level of accuracy for the circumstances, rather than given as an accurate conversion from the original measurements.

The
First
Ways

Chapter 1

There be three kinds of wayes, whereof you shall
reade in our ancient bookes.
EDWARD COKE, *LAWS OF ENGLAND*, 1628

*T*his story begins in a countryside still recovering from lengthy glaciation and as yet untouched by human ingenuity. It is a countryside populated by many species, who by now depend on movement over the ground for their very survival. But nature has not made that movement easy: the terrain is difficult, the obstacles enormous, and the ground cover impenetrable. The species have two survival options: either to continue to adapt to the world as they find it, or to modify that world and make their own way within it. This chapter explores the taking of that second option.

Animal Paths

The first pathways to cross the countryside were created by animals pushing aside vegetation and pounding the earth with their feet. Relatively recent evidence of this sequence can be found in an 1802 report to the British House of Lords by the famous civil engineer Thomas Telford. Discussing the north of Scotland he noted that "previous to the Year 1742, the Roads were merely the Tracks of Black Cattle and Horses." Similarly, a Kentucky settler noted in his journal in 1765, "We came to a large road which the buffaloes have beaten spacious enough for a wagon to go abreast." A major American commentator subsequently observed that "the buffalo because of his sagacious selection of the most sure and direct courses, has influenced the routes of trade and travel of the white race as much, possibly, as he influenced the course of red men in earlier days. [The buffalo blazed] the course of many of our roads" (Hulbert [1904] 1971).

Such animal ways arose to accommodate large-scale migratory movements and to service the need for water and salt. For example, the Natchez Trace was created by buffalo heading north to salt licks at Nashville, Tennessee (Rose 1951b).

The animal path explanation for the origin of our first ways has not been universally accepted, with some arguing that animals do not follow consistent paths (Forbes 1934; Roe 1929; Roe 1939). The reality depends on the local terrain and vegetation. Difficult terrain or dense vegetation in fertile areas would certainly have required narrow and specific animal ways. Subsequently, animal husbandry and then the cultivation of crops would have reinforced the need for ways to either permit animals to access pasture land or prevent them from doing so (Taylor 1979). On the other hand, in open territory the ways would have been much broader, although the North American experience clearly shows that they still provided convenient paths for many travelers.

The wheeled vehicles that helped to settle the vast, flat prairies of the New World used only the simplest of broad tracks. These routes subsequently opened up the American West and carried over 300,000 people into a new life. One of the first was the Santa Fe Trail, which linked a major Mexican highway—one of the Camino Reales to be discussed in chapter 3—to Santa Fe, New Mexico, and thence to the Missouri River towns of St. Joseph, Independence, and Franklin. Parts were probably used by the Spanish explorer Francisco de Coronado in 1541, and the trail was a well-used trading route from the time of Mexican independence and Missouri statehood in 1821 to the coming of the railroad in 1880.

Perhaps the most famous was the 3.2 Mm Oregon Trail, which took travelers from the Missouri River towns to Oregon City, Oregon. Blazed in 1813, it operated as a wagon trail between 1836 and the coming of the railroad in 1869. The California Trail also began from the same staging points, deviating from the Oregon Trail beyond Wyoming's South Pass through the Continental Divide.

Even in more established areas, communities frequently maintained droveways to provide for the seasonal movement of animal herds. Many still exist in Spain as Canadas Reales, carrying sheep between their summer and winter pastures. The Reales are commonly about 75 m wide and can be hundreds of kilometers in length. In Britain, major droveways linked Scotland and Wales to the London markets. Indeed, many came to be called "Welsh roads" (Addison 1980). They have been described as "no more than broad tracks, suitable only for the beasts that made them," and they often used the older greenways to be discussed below (Wilkin 1979). Their usage peaked early in the nineteenth century, after which time the railways began to provide a better alternative; many droveways then reverted to greenways (Taylor 1979). Driftways were short droveways going, for example, from farm to water.

Routes created solely for the needs of animals or to accommodate mass human migrations did not have the strength to survive the quite different and sustained economic pressures of a powerful and developing society. Hence, the original ways have been either lost or archived and only short lengths remain as random parts of today's economy-driven networks. Thus, although many of our first human ways undoubtedly had an animal path origin, the same cannot be said about most of today's road systems.

Human Intervention

Linguistically, the words *path* and *pad* originally indicated earth beaten by the foot, and the word *way* came from the concept of moving and traveling. Indeed, *way* has the same Sanskrit root (*vah*) as the words *wagon* and *vehicle,* and reappears as *waie* and *weg* in French and German. This ancient word *way* will often be more useful in our story than will the more recent *street* and that relative newcomer *road*.

By 10,000 B.C. pathways were being used by human travelers in temperate zones, and retreating glaciation then permitted their extension to the remainder of today's inhabited world. Agricultural communities began to arise in about 9000 B.C. A general improvement and stabilization in the weather in 8000 B.C. led to consistent patterns of animal migration and to significant and widespread human travel (Taylor 1979).

Although animals could push and trample, they could not actively construct. People, on the other hand, could construct ways but needed both incentive and organizational support. The manufactured path was thus a consequence of increased social and economic pressures exerted by a growing civilization. It usually arose in either more difficult terrains or more fertile lands than the open prairie.

Indeed, the first human pathways led to campsites, to food, and to water. In time the travel needs became more than local, and extended pathways used fords, mountain passes, routes through swamps, and bypasses of dangerous areas. The subsequent growth of communities created the first major need for organized travel, the catalysts being the requirements of trade and the collection of superimposed taxation.

The extent of these paths can be imagined by noting that early European explorers in Africa reported having been able to cross the continent from east to west without ever leaving a village-to-village footpath (Gregory 1931). Likewise, when Europeans arrived in North America they found the land traversed by a myriad of narrow footpaths up to 600 mm wide and intended for single-file human use (Labatut and Lane 1950). The most famous of these was possibly the Iroquois/Mohawk Trail, which was a 600 km beaten path between Buffalo and Albany in New York State and which is now New York Route 5 (Rose 1950c). American government documents in 1808 were still referring to roads that did not follow Indian trails as artificial roads (Gallatin 1808).

The Australian Aborigines in the nineteenth century provided evidence that early societies used their pathways for intertribal trade in such items as flints, dyes, and narcotics and other special plants. Early European explorers in Australia noted how the relatively primitive and nomadic aboriginal tribes had actively improved their pathways by such techniques as removing stones and piling them into wayside heaps (Lay 1984). The intervention of humans by the deliberate improvement of animal ways was a major step in our transport development.

The next stage of this discussion will concentrate largely on the situation in England, because only there have major remnants of the prehistoric path system been shielded from the worst ravages of war and thus survived the intervening millennia (Forbes 1934). The situation in prehistoric England has been described by Hippisley-Cox:

> The triangular pattern of high land surrounding the village of Avebury, in Wiltshire, forms the common meeting place of the range of hills that divide the upper Thames from the Severn, from the Kennett, and from the small southern rivers. Before any drainage of the country had been attempted, all communication had of necessity to be made along these watersheds, the valleys being little better than bogs and morass. In the down country, where agriculture has not destroyed them, these trackways may still be traced as

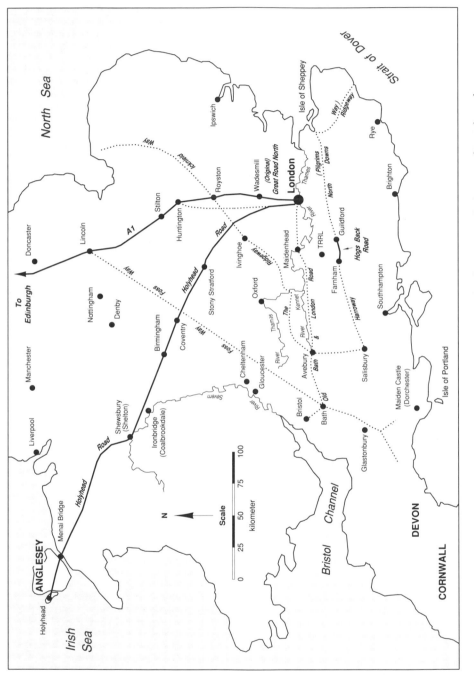

Figure 1.1. Map of Wales and southern England, showing key locations in the history of British roads.

broad green roads, showing evidence of ancient travel. Their turf, from long trampling, is fine and darker in color than on the surrounding land. In their closer soil innumerable daisies turn the old trackways white during their early summer, while here and there long lines of thistles mark the journeys of many pack horses. (Scott-Giles 1946)

Four major pre-Roman routes converged on Avebury, the important neolithic center located in the English chalk country (fig. 1.1), a terrain naturally well disposed to waymaking. The pathways were sometimes called greenways, with the term *green road* still used in modern England to denote an unmade public right of way.

As the preceding quotation implies, many of the early pathways rose quickly out of waterlogged valleys to follow high contours and ridges, which were firmer, less densely vegetated, and safer. These ridgeways were also protected from the destructive effect of running water, the major destroyer of roadmaking efforts. Hillside paths suitable for wet weather use and for wheeled vehicles tended to follow the sunnier southern side of the ridge and were sometimes called summerways (Kennerell 1958). Ridgeways date from between 5000 and 2000 B.C. and were probably based on even older pathways; prior to 4000 B.C. sea levels had been higher, and much of the fertile flatland in use after 3000 B.C. had previously been underwater. In such times the ridgeways would have been closer to the various fertile settlements and therefore would have served a more immediate need.

A number of ridgeways still exist as unimproved paths in the English countryside. One such is the Ridgeway, or Rudge, now a popular 140 km walking path from Avebury to Ivinghoe in Buckinghamshire, where it links with the somewhat newer Icknield Way (Taylor 1979; Burden 1985).[1]

The Hog's Back Road, on the current A31 route between the relatively low-lying towns of Farnham and Guildford, is an example of a modern road following an old ridgeway rather than taking a more direct, valley-based option (Belloc 1923). The overall route is part of the North Downs Way (see fig. 1.1), which probably developed to service the chert trade from the Isle of Portland. Another part is a portion of the famous Pilgrims' Way.

The name Harroway used for a segment of the North Downs Way between Guildford and Salisbury possibly meant "shrine way," being derived from the Old English term "hearing-weg" (Taylor 1979). However, some suggest that the name meant "army way" and, as such, was applied to a number of long-distance ways (Knightly and Cyprien 1985). One famous army way was the 260 km prehistoric Haervejen.[2] The Haervejen links Scandinavia and mainland Europe via Viborg and Vojens in Denmark and southward to Schleswig in Germany, where (from 808 A.D.) an earthen wall called the Dannevirke provided Europe with protection from the Vikings, and vice versa (Matthiessen 1961).

By 3000 B.C. ridgeways were being used as trade routes for salt, tin, bronze, rushes, flint, and chert. Paths leading directly to markets became known as portways. The entire European network was extensive and comprehensive, with Avebury and Salisbury forming the hub of a well-connected ridgeway system (Bodey 1971). The Wyche Cutting on a saltway near Malvern is an example of the major construction effort expended on some of these routes.

Ridgeways were also common in North America prior to European settlement—one of the most famous was the Natchez Trace. Subsequently, colonial highways often followed such ways, making them an "American institution" (Labatut and Lane 1950).

As foot and hoof travel over the millennia wore away the path surfaces, many became quite deep and were called hollow ways or holloways. Some in the Devon limestone are now 3 m deep trenches, suggesting use over six millennia (Rush 1960). Early in the nineteenth century the Reverend John Marriot described these hollow ways by a curious analogy:

> Then the banks are so high, to left and to right,
> That they shut up the beauties around them from sight; . . .
> Thinks I to myself, half inspired by the rain,
> Sure marriage is much like a Devonshire lane.

White, in his famous *Natural History of Selborne* published in 1788, graphically described local hollow ways about 5 m deep in the soft slopes surrounding Selborne. In addition, some hollow ways were deliberately constructed, either to ease travel, as with the millstone ways of Derbyshire (Hey 1980), or for defensive purposes (Addison 1980).

In North America, hollow ways were frequently rutted by moccasined Indian feet to depths of 300 mm, and 1,500 mm troughs were occasionally encountered in New England (Shank 1974).

The creation of major lowland ways required a degree of engineering skill and organization and began to develop in about 4000 B.C. Some manufactured paths were built at that time around Glastonbury in Somerset. These include the Sweet Path, the world's oldest road, which has spent most of its life safely buried in a peat bog (Coles and Coles 1986).

The size of Queen Bodicea's Celtic fleet of about four hundred chariots, with their narrow 40 mm wide iron tires, suggests that the pre-Roman English pathways must have possessed some quite hard surfaces (Newcomen 1951).[3] Indeed, good local roads helped the Britons defeat Julius Caesar and his Roman legions when they first landed at Kent in 55 B.C. (Syme 1952). Writing of that invasion, Caesar noted that the Celtic general Cassibelaunus "disbanded most of his forces, keeping only some four thousand charioteers. . . . He sent his charioteers out of the woods by every road and track" (Wiseman and Wiseman 1980). The wicker-sided chariots predate Cassibelaunus and Bodicea; the Greek traveler Pytheas of Marseilles described chariots he saw in Britain in 330 B.C. Indeed, chariots had been in use in the more temperate climes of the Middle East for at least a millennia. The significance of their presence in the wetter and more difficult British countryside is that it implies a well-developed system of carriageways. Similarly, the speed with which Caesar moved his legions in Gaul also suggests a well-established path system in ancient France.

The English countryside provides many examples of old pathways that have survived over the millennia as well-used independent routes. Seemingly pointless kinks in these ways are often reminders of some long-gone obstacle, perhaps as impermanent as a fallen tree (Gregory 1931). Some ways used the strips of land least suited for agriculture. Others derive from Saxon times when the practice was to leave uncultivated strips between adjoining properties. Roads subsequently followed the corners of the property boundaries (Addison 1980). In addition, local practice—and subsequent laws—relating to unenclosed land permitted travelers to pick and choose the easiest way on any particular day, thus leading to further meanderings. Chesterton offered a famous alternative explanation when he wrote in 1914:

> Before the Romans came to Rye or out to Severn strode,
> The rolling English drunkard made the rolling English road.
> A reeling road, a rolling road, that rambles round the Shire.[4]

In China, kinks were sometimes deliberately introduced to prevent the roads and bridges from being used by fast-flying evil spirits.

A number of the Roman roads followed preexisting millennia-old pathways rather than specially created alignments, and most of the new roads of the eighteenth and nineteenth centuries also followed old pathways.[5] Many other ways remain today as independent foot and bridle paths in the English countryside, now used mainly for pleasure and local private trips. Good reviews of the early English path system are given in Wilkinson (1934) and Taylor (1979).

Roads that developed from such interconnected local pathways could not be expected to produce a rational regional road network. What did arise has been described by Belloc (1923) as "haphazardly established roads, long neglecting opportunities which would have been obvious to the eye of the most cursory and moderately intelligent survey." During a presidential trip to New England in 1789, George Washington noted in his diary, "The roads . . . are amazingly crooked to suit the convenience of everyman's fields and the directions you receive from people are equally blind and ignorant."

This seeming haphazardness is still observable in many of the inner streets of Old World towns and in local roads connecting adjacent small villages. Despite their individual randomness, such roads do eventually form an interconnected system, as with the African village paths mentioned above. The influence of pathways can also be seen in the urban New World. For example, the street system of Newark, New Jersey, is based on preexisting Indian trails, and Detroit's main street follows the Indian Saginaw Trail (Labatut and Lane 1950).

Occasionally a different pattern of road development occurred in nonurban areas of the New World, where land was often subdivided or sectioned into rigid rectangular patterns, sight unseen and with no recognition of local topography. This resulted in rectangular-grid road layouts with "more roads having excessively steep grades in Iowa than in Switzerland" (FHA 1976) and, in South Australia, to roads "being marked indiscriminately over impossible acclivities or precipitous ravines" (Lay 1984).

At Thomas Jefferson's initiative, in thirty of the American states, largely those beyond the thirteen original colonies, land was divided by the Land Ordinance Act of 1785 and subsequent acts into parishes six miles square. Each square was then divided into 36 square-mile units. The squares were then quartered into farms. Each farmer was then required to donate 33-foot strips to create a 66-foot right of way for the roads, which followed the property boundaries at a one-mile spacing (FHA 1976). The road width was based on the 66-foot Gunter's chain used to survey the land and was said to accommodate the turning circle of a horse and wagon team. This was the geometrical inheritance of the famous Homestead Act of 1862, which opened up much of the West.

Many of these ways on which the world has so much relied have inevitably become major threads in the fabric of the countryside. The interpretation of the pattern of the threads has varied from time to time and from country to country. For example, roads in the United States have been described as "basic social and cultural expressions, artifacts of [a] national obsession with mobility and change, with the horizon, with the frontier. . . . The creation of collective design. They reflect unconscious as well as conscious patterns of politics, economics and culture" (Patton 1986).

Serving the Town

Human settlements occurred naturally around sources of food, water, and security. The world's first major town was Jericho, which began in about 8000 B.C. on raised ground near the springs of Elisha. Carbon dating indicates that hunters had been using the spring since about 9000 B.C.

The further needs of agriculture and animal husbandry led to the establishment of many permanent settlements by 5000 B.C. Urban societies appeared in Mesopotamia around 4000 B.C. and, not coincidentally, villages with street paving also date from about this time. By 3000 B.C. there were well-kept ways connecting towns on the Indian subcontinent (Vaknalli 1980).

Towns and then cities arose for four main reasons: to meet social needs, for joint defense, to minimize travel, and to facilitate trade and manufacture. By coming together in cities, people collectively gained strength and freedom.[6] Given the above general factors, towns were typically sited at river crossings, defensible locations, ports, crossroads, the navigable limits of rivers, sites possessing dominant features (such as low hills), and strategic peninsulas (Dickinson 1961). Thus, London was located at the first possible upstream crossing of the Thames. Paris was on the route north from Orléans and was located at a river crossing that used the Île de la Cité. Located below the congruence of the Seine and the Marne and above the congruence of the Seine and the Oise, this route avoided crossing a second major river.

Many castle cities, such as Edinburgh, are examples of towns located at defensible locations. Indeed, the availability of a raised area suitable for a castle, fortress, or temple was often a prime criterion for site selection. Athens with its Acropolis is perhaps the best-known example. Port cities are numerous and obvious, and towns of both the Old and the New World such as Rome and Melbourne were frequently established at the navigable limits of rivers. Istanbul is an example of a peninsula city.

As time passed, trade rather than defense became the dominant factor in city location. The economics of transport strongly favored the movement of freight on water rather than on land. Consequently, productive development occurred adjacent to harbors, rivers, canals, and railways, and little activity occurred in the intervening areas (Hutchins 1977). Within the cities, warehouses and factories similarly clustered along the waterfronts.

Street systems in the earliest towns had many of the characteristics of contemporary North African cities. Mumford (1961) comments that "in the earliest cities, as the excavations at Ur show, the street as an open, articulated means of circulation was exceptional: The narrow tortuous alley, well shaded from the torrid sun, was the usual channel of traffic, better adapted to the climate than the wide thoroughfare." Such cities were a web of narrow streets, all below 2.5 m in width and barely permitting passage of a pack animal, let alone a wheeled vehicle. The randomness of the network was a useful tool in defending the city against invaders who had penetrated the outer wall (Mumford 1961). Another commentator characterized these towns as "an irregular entanglement of streets, houses and blocks" (Dickinson 1961).

Geometrically ordered arrangements of streets did exist. Straight streets were employed in the ancient Middle Eastern cities of Assur and Nineveh (see fig. 3.1) and in Harappan cities such as Mohenjo-Davoir in Pakistan. In about 1900 B.C. Egyptian workers building a pyramid for Sesostris II were housed near El Faiyûm in a new town called Kahun, which was laid out with straight streets 3–4 m in width. In about 615 B.C. King Nabopolassar built a long, straight processional road in Babylon to honor the god Marduk. Such grand processional avenues, or kingsways, were to reappear in various cities throughout history (Haverfield 1913). Herodotus described the streets of Babylon in 450 B.C. as laid out in ordered straight lines.[7]

Hippodamus, a Greek contemporary of Pericles and follower of Pythagoras, was the first recognized town planner. His career began when he supervised the reconstruction of his hometown of Miletus between 440 B.C. and 408 B.C., following its destruction by the Persians. Hippodamus subsequently created a number of new towns, including Priene, the Athenian port of Piraievs, and Thurii in southern Italy.[8] His planning approach was not responsive to the locale; rather, it was rectilinear and geometric and reliant on a checkerboard regularity and rectangularity to lead inevitably to a gridlike street network. Hippodamus saw this process as a triumph of reason over the wanton riot of nature. He thus formalized the practice of rigidly applying absolute principles while paying scant attention to local topography—a tradition boldly followed by many subsequent town planners. The responsibility is not entirely with Hippodamus. In the first place his method merely codified existing practice; for example, grid arrangements were used from about 2000 B.C. in Mohenjo-Davoir and from about 1500 B.C. in Knossus. Second, the grid simplified and encouraged land subdivision, development, and speculation, and these forces were always stronger and more imminent than were the needs of good living and smooth traffic circulation.

Aristotle was among those who acknowledged that Hippodamus had introduced Greek cities to both the use of wide, straight streets and the proper grouping of buildings. Nevertheless, it was Hippodamus' underlying grid structure that came to have the most lasting impact. Known as Milesian planning in recognition of its origin in Miletus, its influence can still be seen in cities formed by early colonization in Asia Minor, Greece, and Italy. Adopted by the Romans and extensively documented by Vitruvius, Milesian planning became the basis for new military towns in medieval Europe and for colonial towns established by European nations in the New World. Spanish colonial planning policy for new towns, defined in the Laws of the Indies issued in A.D. 1573, was based directly on the Vitruvius description of Milesian planning and imposed the grid on all new towns. St. Augustine in Florida and Sonoma in California are two American cities that fell under its direct influence.[9] Similarly, the grid was applied by British military officers as the town planning standard for early Australia (Lay 1984).

Milesian planning was well used by William Penn in his development of Philadelphia, as a matter of commercial course rather than of forced colonization. The effect was widespread: the 65 × 110 m street grid adopted in downtown Chicago was not that different from the 35 × 115 m grid of ancient Pompeii. Through all this time, Milesian planning bravely maintained its tradition of ignoring local topography, as in the alignment of the streets on the hills of San Francisco.

Street layouts within a city can thus be categorized as either evolved or imposed. The evolved patterns relate to the local topography and usually contain elements of the old entangled street systems, of a larger grid-based star pattern, and of linear strip development; the latter two reflect the crossroads origins of many towns. Imposed layouts are either Milesian or radial. The radial system of roads reflects modern traffic flows, but it has evolved in a natural way in only a few cities (Dickinson 1961).

To the ordered Milesian planning of the Greeks, the Romans from about 500 B.C. added two new dimensions. First, they required their cities to be systematically walled and for that wall to follow a rectangular layout and be a primary feature of the town. The city wall was certainly not a new Roman idea; the earliest known walled city is Uruk, which was built on the Euphrates in southern Mesopotamia (see fig. 3.1) prior to 3000 B.C. Second, and following Etruscan religious practice, the Romans required the main streets of the grid to run north–south and east–west. This incidentally created cross streets at the town center. In most towns, these were also the rural highways and so all traffic met at the center of the city, which became the prime location for political, administrative, and religious buildings and for colonnaded forums. In medieval times the layout of the rectangular walled city with the crossroads dividing the town into four quarters was seen to accord with basic Christian symbolism (Jackson 1980). Even today the symbolism continues, with most modern streets aligned to the major compass directions.

As Roman practice developed, additional requirements were added. These included a consideration of prevailing wind directions to reduce the effect of foul smells from cesspits and the more deliberate development of a forum as the city center.

In London, which the Romans founded in A.D. 50, the original east–west streets and crossroads still underlie the current street structure in that part of the City centered around the Bank of England. The governor's palace was at the site of Cannon Street Station, and the temple was just a few hundred meters south at the rear. The forum was a further few hundred meters away near London Bridge. The strategic influence this core had on the development of London can be seen from the eight roads that today radiate out from the Bank of England intersection.

The Chinese also adopted a rectangular grid, but with mystical overtones. Using the old capital of Xian as a precedent, they laid out the Imperial City in Beijing according to a divine plan reflecting a rectangular universe. The four prime compass directions were used to align the streets and these also represented the four seasons. Visitors entered from the south, which was also the summer, red, bird, and holy direction. In about A.D. 700 the rectangular grid and geometric philosophy of Xian was transplanted to Japan and to such cities as the old capitals of Nara and Kyoto.

Roman town size was restricted to about 700 × 500 m to ensure that all points were within hearing distance, particularly of the defensive walls (Mumford 1961). Plato gave a human thrust to this criterion by suggesting that the size of his ideal city would be such that all citizens could be addressed by a single voice. A more practical criterion was the need to restrict the length of the walls, which were expensive to build and to maintain.

City walls were a dominant feature of many European towns until quite recent times. Indeed, Paris built its last set of walls in 1840–1844, and these walls withstood a siege in 1870–1871 (Toynbee 1970). Nevertheless, the walls of Vienna, Berlin, and Rome were removed between 1850 and 1870. During the First World War the German advance in August 1914 was delayed for a week by the walls of Liège. Subsequently, one of the three remaining sets of walls around Paris was a key element in planning the defense of that city. Their actual performance was less impressive, and they were demolished soon after the war to render another form of public service. For obvious defensive reasons, a belt of clear land had been required outside such walls, and the conjunction of land and wall was to provide the route for many of today's circumferential roads.

Even large cities such as Babylon and Rome, which had peak populations of over half a million, were contained within an area of 14 sq km or less and had an effective radius of under 2 km. Indeed, the peak sizes reached by many cities, such as Baghdad in about A.D. 900, suggests that a population of 900,000 was the largest that could be practically accommodated within a walking city. Population densities usually peaked at about 600 people/hectare (ha). In addition, a walking city could rarely sustain more than 50,000 people within a 2 km radius without developing multiple centers.

The centuries following the Fall of Rome saw a dramatic decline in European urban life. The needs of the gods were supplanted by the needs of humans (Vance [1977] 1986). By A.D. 1000 Rome's population had dropped to 35,000, and most towns were to remain much smaller with populations of 10,000 or less until well after A.D. 1500. Population densities dropped to about 100 people/ha and town areas to under a square kilometer (Clark 1977).

Even in the seventeenth century, few points in a city were more than a kilometer from the city center. A commentator of 1635 gave a description of a typical prosperous town when he wrote of Exeter, "The City is invirond about with a wall, about [three kilometers] in compasse, with five gates, and some watch towers, and on the out parts thereof she is guarded about with pleasant walkes and diverse Bowling Grounds. . . . the Buildings and Streets are faire, especially her high street from East to West Gate" (Bridenbaugh 1968).

London did not expand beyond its Roman walls until 1574, its population rocketing from 50,000 in 1500 to 190,000 in 1600 (Chandler and Fox 1974). Not until 1580 were the consequences of that expansion realized and the first regulations introduced to inhibit the urban sprawl. Paris overflowed its city walls in the seventeenth century, and both cities reached their critical size of 800,000 in about 1750 as the Industrial Revolution gained momentum. Population densities once again approached 600 people/ha (Clark 1977).

Even in the period of rapid urban population growth during the Industrial Revolution, many cities contained that growth within their old walls, thus creating great crowding and congestion. The natural developmental consequence within the walking city was to cram everything into the smallest possible space (Schaeffer and Sclar 1975). Most people lived within a fifteen-minute walk of their work, and streets were predominantly for pedestrian movement. As late as 1870, new building in Liverpool and Birmingham was occurring within 4 km of the city centers (McKay 1976). Population densities in London and Paris peaked at over 700 people/ha. In New York it reached 1,350 people/ha by 1900.

In general, all this led to European towns at the end of the nineteenth century that were "badly overcrowded, plagued with poor, unhealthy, and relatively expensive housing, especially for the lower classes in the old buildings of the central quarters" (McKay 1976). This issue is pursued further in chapter 9.

Early street systems rarely survived through to the present day. In discussing the various provincial towns established by the Romans, Haverfield (1913) commented:

> The modern successors of these towns have rarely kept the network of their ancient streets in recognizable detail. . . . The paths and passages by which men once moved have vanished. . . . There is hardly one modern town in all the provinces of the Roman Empire which still uses any considerable part of its ancient street plan. . . . Rome herself, the Eternal City, hardly uses one street today which was used in the Roman Empire.

The pattern of change was aided by the impermanency of many structures, leaving them vulnerable both to the elements and to people. Apart from flood, fire, and war, there were also the large-scale demolitions of such visionaries as Baron Haussmann and the more incremental actions of many city officials. Between 1855 and 1889 London Council demolished 14 ha of buildings per year to make way for road widenings and extensions, often in the name of improved health and social well-being (Edwards 1898).

As cities developed from towns, pressures on land use also caused a change in street patterns. In towns, individually owned pieces of land were usually narrow rectangles, in order to allow the maximum number of properties to abut the street, each with its own garden. As the towns became cities, the need for private gardens diminished and the desirability of street frontages increased. Properties therefore became wider and shallower and required a closer street spacing.

One development from the early Greek approach, with its 3.5 m wide streets, was the creation of a somewhat wider main street, which then usually became the shopping avenue and pseudo marketplace. Such streets can also be found in older cities such as Damascus and Jerusalem (Mumford 1961). The Acts of the Apostles (9:11) refers to the most famous of these—"the street called straight" in Damascus—whose alternative name was, indeed, the Long Bazaar. They are mirrored in current strip developments. Market squares often developed as a consequence of the street markets causing an initial widening of the right of way, usually at an intersection (Dickinson 1961). Such squares did not become common in European cities until about the eleventh century.

Like the Greeks, the Romans were not much given to the creation of public open space within their cities, and their famous forums were closely tied to official or commercial purposes. In London the complex of great parks comprising St. James's Park, Green Park, Palace Gardens, Hyde Park, and Kensington Gardens arose when Henry VIII confiscated the market gardens of the religious orders and turned them into deer parks for royal hunting. Indeed, the common provision of urban open space is a recent innovation and derives from the great urban architecture developed in Spain during its period of colonial wealth from the fourteenth to seventeenth centuries. In effect, Spain democratized the old forum to produce the plaza.

The network of wide streets found in modern cities was unknown until the nineteenth century; only processional roads provided even the seeds of such a system. Baron Haussmann began a new era in the second half of that century when he created the broad boulevards of Paris in response to military needs. Generally, however, wide streets and urban parks only began to occur in cities after urban transport arose late in the nineteenth century and consequently reduced the demand for and price of inner urban land. The transport aspects of recent city development are pursued further in chapter 9. The broader aspects of city growth are outside the scope of this book and are more than adequately covered elsewhere (e.g., Vance [1977] 1986).

Summary

Our first ways were created by animals and subsequently improved and coordinated by human travelers. Nevertheless, their local and specific origins meant that they usually presented a haphazard pattern when viewed on a broader and more systematic scale. They were also paths for walking, just as the cities that developed during the same period were predicated on walking as the prime means of urban transport.

By 2000 B.C. many parts of the world contained relatively advanced civilizations in both a technical and a commercial sense. Agriculture and large permanent settlements were common, and trade between settlements was frequent rather than exceptional. Religion, war, and administrative dominance also became part of the social structure. Simple walking paths were often inadequate to supply the new travel needs.

Not surprisingly, a number of transport developments arose to meet the travel demands caused by the onset of civilization. These developments beyond walking will be explored in the next chapter. It will be seen that they greatly influenced the manner in which the various ways were used over the subsequent millennia. In some instances the old footways remained viable, but in many cases new ways were required to accommodate the new travel modes. Locally, these new ways had to be wider and better surfaced than were the footways. Strategically, they had to follow better grades and lead to new destinations. The cities were to respond much more slowly than the countryside to these new transport developments, partly because of the slow pace at which land use changes can occur in a closely settled city and partly because foot travel adequately served most of the transport needs of the contained city. The situation changed only in very recent times when cities began to expand outward. Chapter 9 explores the subsequent interactions between city size and transport technology.

The
Demands
of Transport

Chapter 2

The chariots shall rage in the streets,
They shall jostle one against the other in the broad
  ways:
They shall seem like torches,
They shall run like the lightning.
                    NAHUM 2:4, CA. 620 B.C.

*T*he previous chapter discussed the various needs which gave rise to the demand for travel and explored the many ways that were developed to help satisfy that demand. Of course, in addition to a way, there must also be a means. This chapter examines both the travel technologies used to provide the means and the interactive requirements that those developing technologies placed on the existing ways.

Without the Wheel

The first of the new travel technologies was not the wheel but the provision of power. Indeed, after the creation and widespread application of footpaths, the world's second major transport development was the use of animals, initially as beasts of burden and subsequently for pulling plows, sleds, carts, wagons, and carriages.[1]

The domestication of large animals probably occurred about 7000 B.C., initially to provide humans with a secure source of food. Although the use of animals to provide transport power was a convenient secondary development, for most of its history the world's roadway system has operated with domesticated animals as its sole source of motive power. Cattle, onagers, donkeys, asses, dogs, goats, horses, mules, camels, elephants, buffaloes, llamas, reindeer, yaks, and, of course, humans are some of the better known species that have found useful transport employment.[2]

Personal Travel

Until the nineteenth century most people rarely traveled more than a day's walk from their home, and the spacing of many settlements grew around this distance (Bagwell 1974). At the best, walkers could only manage between 15 and 40 km in a day, depending on need and burden, whereas a marching army under good conditions would expect to travel at least 80 km. The very occasional long-distance trip was predominantly made on foot or horseback; until the fifteenth century, wheeled vehicles were rarely used for personal travel.

For speed, initial reliance was placed on fleet-footed runners, epitomized by the young Greek who in 490 B.C. brought the news of Greek victory at the Bay of Marathon to Athens some 37 km distant, and then fell dead. Indeed, the Greeks were able to send messages at 200 km/day by using runners in a relay system. Similar speeds were reached by North American Indians running along the Iroquois Trail (Rose 1950c). The foot-travel record was probably held by the Incas, who developed a system that was able to transmit messages at 400 km/day by using fast runners over 2.5 km stages, suggesting that the runners were achieving about 15 km/h (i.e., 6-minute miles) and running stages by day and night.

Footposts became common in seventeenth-century Europe, and it was believed that a good "footman" could outdistance a horse after seven or eight days. Irish footmen were particularly well regarded, traveling up to 120 km/day in emergencies (Crofts 1967).

Horse riding probably began in Russia in about 3000 B.C. and was introduced into Mesopotamia in 2000 B.C. It gained popularity at a slow pace and only began to appear with some frequency after 1000 B.C. The horse was initially used primarily for war rather than for transport—the Assyrians introduced the first cavalry force in 900 B.C. However, in the absence of stirrup or saddle, riders were forced to cling to their horse through the pressure of their knees. Little leverage could be gained for propelling weaponry, so the chariot remained the preferred fighting machine. Metal bits were in use by this time and, by 500 B.C., both bits and horses had improved to such an extent that the horse began to supplant charioteering as a means of both elite personal travel and warfare (Bokonyi 1980). Throughout this riding period the horse remained a quite expensive animal, and most riders traveled on the much cheaper donkey and mule.

The next stages in the development of horse riding were the stirrup and the saddle. Neither technology spread rapidly. The stirrup was probably developed in India in about 200 B.C. and allowed horse and rider to act as one (White 1962). Stirrups arrived in Europe via the barbarians in about A.D. 700, well after the Romans, and sounded the death knell to the chariot so prevalent in Roman land transport. Saddles were not in use in Europe until about A.D. 200. Horseshoes, also uncommon in Roman times, became widespread in about A.D. 700.

A single horse and rider could routinely manage about 60 to 80 km/day. To extend this capacity meant using horses in stages, and this in turn required the establishment of post houses and stables, under the control of a postmaster, where tired horses could be rested and fresh ones obtained. Using teams of horses and riders, the Persians of King Cyrus the Great in about 550 B.C. achieved 250 km/day on their exclusive royal roadway.[3]

The Romans and the Chinese also used this method, with post houses at spacings of about 20 km and overnight lodging houses at 60 km or so. These distances became common town spacings as settlements grew around both the post houses and the lodging houses. Average speeds in the systems were about 10 km/h or 200 km/day for continuous travel. The official speed record in Roman times was about 300 km/day, established by Emperor Tiberius between Germany and Lyon (Borth 1969). The Chinese record was 430 km/day (Needham 1971). The Roman post house system began to decline as the empire aged and as the servicing of the system changed from one of laissez-faire private enterprise to an unpopular series of obligations and regulations that only added to the concurrent road-maintenance burden (Postan and Rich 1952).[4]

Nevertheless, long-distance communications continued after the Romans. News of King Richard the Lion-Hearted's capture in Austria in 1192 traveled across Europe at about 300 km/week or 40 km/day, whereas ordinary travel was averaging 200 km/week (Haskins 1929). By the end of the twelfth century, a messenger system had developed to service the needs of students moving in and out of Paris. It was operated by the University of Paris between 1314 and 1598 and was the incentive for Louis XI to establish an official system—Couriers de France—in 1464. The post houses established for the Couriers were still about 20 km apart, and four stages (80 km) were traveled in a typical day.

Finally, more than a millennium after the demise of the Roman post system, a master of posts was appointed in England in 1509 (Anderson 1932). In 1603, news of Queen Elizabeth's death was carried by post horses over almost 700 km in what was then considered to be the exceptional speed of 200 km/day—so exceptional, in fact, that the rider, Sir Robert Cary, was rewarded by James I with a peerage (Margetson 1967). At the other extreme, Thomas Macaulay noted that news of the queen's death did not reach some parts of Devon until after the long period of mourning was over. In 1834 Robert Peel returned from Rome to London to take over the British prime ministership and averaged 120 km/day, predominantly along old Roman routes (Upton 1975).

In 1860 the famed pony express in the United States carried its mail over about 3,000 km, from Missouri to California, approximately following the Oregon Trail. The riders traveled at 300 km/day, changing mounts at post houses spaced every 10 to 30 kilometers. The pony express operated for only eighteen months before it was made redundant by the transcontinental telegraph (Findley 1980).

Humans have also been used as teams of bearers to carry other humans. Commonly, the bearers shoulder two longitudinal shafts on which the passenger is supported. The most notable examples of the technique are the Elizabethan sedan chair, the Roman litter, and the Chinese palanquin. The sedan chair is an elaborate enclosed chair carried between two longitudinal shafts and was named after the French town of Sedan where it was first made. Although no faster than walking or coach travel, it was more comfortable and allowed the occupant to totally escape the filth of the street. Chairs were quite popular, and some 400 were operating in London in 1750. They went out of fashion in the early 1800s, and their last use in Europe was in Vienna where, as tragsessel, they were on hire until 1888. Chairs remained common in the East until well into the twentieth century. About 900 were operating in Hong Kong in 1916, and many were still in use there and in mainland

T LXV

Figure 2.1. Ancient litter (*top*) and a palanquin, with human bearers. The obvious artistic license taken in this and many subsequent drawings is typical and indicates the care that must be taken in interpreting prerealist drawings. From Ginzrot (1817)—Ginzrot in particular misled many scholars.

China during the 1950s. The Japanese used a tiny house suspended from a single pole, which was called a Norimon.

The litter and palanquin (fig. 2.1) were couchlike or even bedlike, and the palanquin was covered and curtained. In 1600 Queen Elizabeth was carried through London at above head height in an elegant open litter cum chair (Belloc 1926). For practical reasons, litters were sometimes carried by horses or mules, rather than by humans.

The Movement of Freight

When freight first had to be moved, human hands, shoulders, hips, and heads were all gainfully employed. When the capacity of the unaided human was exceeded, the solid stick would have been the obvious tool to use, first to transfer the load to the shoulders and then to allow it to be shared as a yoke between two people. For less coherent loads, the technology would have expanded to include wicker baskets hung from the shoulders by rope or carried on the head, more often than not on the female head. Such people-powered freight techniques are still in quite effective use today in parts of Asia and Africa. African porters could carry 25 kilograms (kg) for 25 km/day (Holmstrom 1934). For shorter distances, loads of about half body weight are common, and peak loads over very short distances can exceed 175 kg (Balogun et al. 1986). The Chinese used laborers carrying slings and bamboo poles to move loads of up to a tonne distributed at about 25 kg per bearer.

When the loads to be carried demanded greater strength or power than could be supplied by humans, the humans innovated by using their domesticated feed animals as beasts of burden. The wicker baskets were transferred from human shoulders to the backs of cattle to produce the first pack animals.

Pack transport—or *summage*—took a step forward in about 3500 B.C. when the domesticated donkey came out of Africa.[5] It was admirably suited for the task at hand. From that time forward the pack animal has been an unobtrusive but vital part of our transport operations. There are written records from 2000 B.C. of organized pack animal convoys operating in the Middle East.

The early packhorse could carry less than 50 kg in two pannier baskets. By the end of the Middle Ages, breeding and loading improvements meant that a packhorse could carry a load of about 120 kg, or a third of its body weight, for up to 25 km. Similarly, donkeys could carry about 75 kg, mules 100 kg, and camels 175 kg. Despite this improvement, the role of the packhorse was clearly limited by its load capacity. To overcome this restriction, long strings, or *drifts,* of nine to fifty packhorses tied tail to nose worked many regular freight routes, with regular packhorse services operating between London and the north from the fifteenth to the nineteenth century (fig. 2.2; Bird 1969).

Road maintenance efforts that did occur during this period were often devoted to packhorse paths beside the road itself. The narrow width meant that it was often practical to raise the path above the mud and pave it with flagstones. Over time, some paths came to resemble embankments. By the seventeenth century, packhorse operations had grown to such an extent that there were many stone-paved paths with their own bridges, particularly in the north of England where an extensive system existed (Crofts 1967).[6] Most of the associated packhorse bridges were built

Figure 2.2. The packhorse convoy. The collar of bells seen on some of the horses warned other convoys of their approach to proclaim right of way on narrow paths—a frequent cause of dispute and argument—and was also believed to regulate the movement of all the horses. *By L. Huard, from Smiles (1874).*

between 1660 and 1740. Their entrances were often guarded by posts to deter their use by carts, and their parapets were low enough to permit the panniers to protrude over each side (Hey 1980).

In his famous seventeenth-century legal text, Edward Coke saw packhorse ways playing a major role in a distinct hierarchy of roads, declaring that "there be three kinds of wayes whereof you shall reade in our ancient bookes. First a foot way . . . and this was the first way. The second is a foot way and horse way . . . and this is vulgarly called a packe and prime way because it is both a foot way, which was the first or prime way, and a packe or drift way also. The third . . . contains the other two, and also a cartway" (1628).

Packhorses were used by many armies. The British army used them extensively in the wars it waged with the French and the Indians in North America. The preexisting Indian trails were well able to accommodate strings of packhorses, and the mode subsequently became a major commercial operation during the eighteenth century, lasting until better roads and wagons arrived at the beginning of the nineteenth century (Shank 1974). Packhorse transport reached its European zenith in about 1750, after which the canal system began to eat into the trade (Wilkinson 1934). Numbers dropped even further with the advent of the railways, although some were still working in Yorkshire in 1914 (Addison 1980; see also Hey 1980).

Given the simplicity of pack animal operations and the relative complexity of the technologies of harnessing, cart building, and roadmaking, it is not surprising that through much of human history the pack animal walking on narrow footways and carrying its load in containers simply slung from its back has been a major means of land transport. In addition, the relative economics of pack transport often improved as the distance to be moved increased.

Nevertheless, the major movement of freight in the last millennium has been by cartage rather than by summage, due mainly to the pack animal's low 50 kg load capacity. This meant that about twelve packhorses were needed to carry the same load as a single horse and cart (Postan and Rich 1952). Many freight needs were left unsatisfied. Agrarian society had to move produce from fields to storehouses and processing areas and, later, to market. This demand could not be met by simply carrying the load on an animal's back. A breakthrough in freight technology was required.

The potential for a breakthrough arose in about 5000 B.C. when the castration of domesticated cattle was found to produce an excellent power source in the ox, which could be made to haul horizontal forces between four and ten times greater than the vertical forces that it could carry on its back. This development was probably driven more by the needs of agriculture than of transport. Indeed, the first hauled device was probably the plow, beginning as a hooked branch or log that oxen pulled through the earth. For both power and ease of harnessing, oxen worked in pairs, connected to either side of the plow log by a wooden crossbar yoke. Oxen pull from a hump at the back of the neck and have prominent shoulders, so they were relatively easy to harness in this fashion.

It would not have been long before the cattle-harnessing technology developed for plowing led to the thought that the same harness and crossbar yoke coupled to two dragged logs, rather than to one plow, would provide a platform for load carrying. Two such devices are the travois and the slide car, in both of which the front of the load platform is carried on the animal's back and the rear slides along the ground. In the travois, or *V*, the hauler—human or beast—drags the forked branch of a tree or two separate branches or poles with their far ends tied together. The freight spans between the two branches near midlength. Such devices have been widely used, by North American Indians for example, and are still in use in rugged parts of Scotland. In the slide car the two poles are parallel.

The next development was the hauled, sliding sled, which was a flat platform that was dragged along the ground. Compared with the travois and slide car, the sled required a more elaborate construction and a new type of harness. However, if it operated over smooth surfaces with a friction coefficient of under 0.10, it could carry a greater load than could the travois, slide car, or packhorse (fig. 2.3). The sled is still used for freight transport in various parts of the world. The word *sledge,* incidentally, covers both the dry-land sled and the sleigh used on snow and ice. Sleighs require very little haulage force and therefore need a simpler technology and less power, as reflected in the common use of dog teams. There is evidence of sleighs in use in 6000 B.C. and sleds in 5000 B.C.

Thus, by about 5000 B.C. domesticated cattle had become the original workhorses of the road. The use of the ox for dragging and the donkey and horse for carrying and then for riding provided the first quantum leap in human travel times and capacities. Because the ridden horse, horse litter, loaded packhorse, and harnessed, dragging ox required larger horizontal and vertical clearances than did people on foot, some path development in the form of bridle paths and trackways

A—Sledge with box placed on it.   B—Sledge with sacks placed on it.   C—Stick.
D—Dogs with pack-saddles.   E—Pig-skin sacks tied to a rope.

Figure 2.3. Agricola's view of some sixteenth-century freight-movement methods. Note the sled in the foreground. *Drawing from and with the permission of the Science Museum, London.*

was necessary to take advantage of these new sources of motive power. The next chapter discusses the ways that arose to meet these needs.

Inventing and Using the Wheel

Development was at a stalemate. Agriculture and travel were now possible, but all were locked into a human dimension. To expand out of that scale, a lever was needed by which the world could be moved, load by load and place to place. That lever was to be the wheel, which, after the path and the domesticated animal, became the world's third major transport invention.

Invention

The initial stage of the development of the wheel for transport was probably associated with the dragged sled and the travois. On level ground, small sled wheels or a single wheel between the pole ends of the travois would have usefully reduced the considerable dragging friction. However, whether these transport issues were the demands that led to the invention of the first wheel remains a matter of conjecture. Early wheels were also used for pottery, and perhaps the original motivation for the invention was production rather than motion. Nevertheless, it does appear that the wheel was invented in Mesopotamia, the birthplace of civilization, in about 5000 B.C. The first wheels were of wood and so required the prior development of woodcutting technology. Knowledge of the initial stages is somewhat sketchy—the oldest known vehicle wheel comes from the southern Russian steppes and dates from about 3000 B.C. The uniqueness of the invention is evident in that, until Columbus reached the Americas in 1492, none of the relatively advanced civilizations there had developed practical wheeled vehicles. Nor was the wheel developed indigenously in Southeast Asia, southern Africa, or Australia (Piggott 1968; Piggott 1983).

An important next stage in transport development was the use of an axle to join two wheels together and thus give increased stability and load capacity. Some, without any archaeological evidence, have suggested that perhaps the above stages were reversed and the wheel developed from logs used as rollers, hypothesizing that this began when the undersides of sleds were notched to hold rollers in place. To reduce the weight of the rollers, any unnecessary timber was removed, thus giving a single unit of two wheels connected by an axle. Certainly a number of societies have used low sledlike wagons with small solid wheels (fig. 2.4). These trolleylike devices were commonly called trucks and truckle carts.

The single-wheel-first theory is supported by the fact that many unearthed early wheels are about a meter in diameter and constructed from either a single

Figure 2.4. Truckle cart. *From an 1886 painting by Care, from and with the permission of the National Museum of Wales (Welsh Folk Museum).*

wooden plank or from three pieces of plank joined by dowels to produce a tripartite wheel. This indicates that wheelmaking techniques were directed at single-wheel rather than roller construction (Lee 1947). Single-piece wheels were formed from planks split longitudinally from the trunks of oak trees (Piggott 1983).

Most early wheels operated on fixed, nonrotating axles. Common lubricants between wheel and axle were animal fats, leather, and vegetable oils. To be effective, metal contact surfaces were desirable, but this requirement had to wait for subsequent advances in metallurgical technology.

Initial Application

By about 3000 B.C. a variety of vehicles in ancient Mesopotamia and northern Iran had begun to make practical use of the wheel. The first were two-wheeled A-framed carts, which probably developed from the travois. The two-wheeled chariot with a single harness pole—more in the style of the ox-drawn plow than of the travois—was developed in Sumer in about 2800 B.C. It was drawn, not by oxen, but by donkeylike onagers, which continued to perform that role until the advent of the horse in the next millennium. Cumbersome four-wheeled wagons followed in about 2500 B.C. (Piggott 1968). Their weight meant that only oxen could provide the necessary haulage power. They thus came into prominence in regions where crop cultivation was common.

The harnessing of the ox can be improved by carrying the forward vertical component of the weight being hauled on the animal's back rather than on its shoulder. This change increases haulage capacity by about 50 percent, but has eluded many civilizations. It was irrelevant with the plow, but became increasingly critical in transport applications.

One problem with these newfound vehicles was that the wooden running surfaces of their wheels had a very short life. To overcome this problem, leather tire coverings were introduced in about 2500 B.C. and protruding copper nails in around 2000 B.C.

Vehicle ownership was initially a matter of some distinction and status, and the vehicles were sometimes buried with their distinguished owners. Indeed, part of our knowledge of the vehicles of the time comes from the discovery of well-preserved examples of such burial items (Piggott 1983).

Times changed, and by 2000 B.C. wheeled vehicles had become common throughout the Middle East and had arrived in Europe. This led to the Celts inventing the spoked wheel, creating its wooden rim from a series of C-shaped segments called *fellies*, which were formed from bending hot timber. Each felly would typically carry two of the eight or twelve spokes.

The spoked wheel represented an important step in the change from ponderous, heavy vehicles to light, fast ones. The change occurred because war had become more sophisticated and higher-value wares were being traded over longer distances (Bokonyi 1980). The associated quest for better chariots introduced a whole new set of construction technologies (Casson 1974). The resulting light chariot soon became the conveyance of warring kings and princes.

In concert with the development of lighter, faster wheeled vehicles, modern horses with primitive hauling harnesses first appeared in 2000 B.C., and the horse soon became established as an animal for hauling as well as for riding. In particular, the combination of the light chariot and the horse in about 2000 B.C. provided a popular but expensive military vehicle. Powered by a pair of horses on either side of a single shaft, the chariot traveled at high speeds and avoided the steering problems associated with the heavier carts and wagons. From a military viewpoint, from 1700 B.C. the light chariot allowed the barbarians to successfully attack the Middle Eastern empires. Generally, it represented the world's first significant increase in transport speeds. The jump from 4 km/h in an oxcart and 6 km/h on foot to 30 km/h in a chariot was not to be bettered until the coming of the steam engine some three and a half millennia later.

Joshua (11:4) described the many wooden chariots assembled against him in about 1225 B.C. The changing technology is evident in that, whereas he defeated his attackers by burning their chariots, two hundred years later the somewhat more successful Canaanite general Sisera was using nine hundred iron chariots (Judges 4:3).

Iron making was established by 1500 B.C. and, following the use of iron for building vehicle bodies, iron tires were introduced in about 700 B.C. They were made of heavy iron segments nailed, bolted, or riveted to the wooden wheel. In a major advance, the Celts developed the shrink fitting of iron tires in about 400 B.C.[7] In this method the iron tire was made smaller than the wheel, expanded by heat, and then cold shrunk into place. This smooth tire not only directly improved rolling resistance, but also—together with the availability of iron fasteners—greatly simplified the production of spoked wheels. Celtic pre-Roman spoked chariot wheels unearthed in Britain have iron tires between 30 and 45 mm wide and a diameter of between 750 and 1,200 mm.[8] Metals were also used for bearings, pins, harnesses, and horseshoes.

Despite these dramatic developments, the chariot was soon to be superseded by developments in horse riding. Nevertheless, the technology it created lived on long after the chariot disappeared. One such end product, a light Celtic wagon built in about 100 B.C., is shown in figure 2.5. The wagon has accurately turned oak hubs containing cast-and-turned bronze collars, iron tires with flanged edges, many bronze ornamentations, strap suspension, and a partly steerable front axle (Klindt-Jensen 1949).

Nonsteerable wagon axles created major problems, particularly when the wheels became caught in the deep ruts and sharp curves common in many roads. The development of generally useful four-wheeled vehicles therefore required steerability to be developed as the third stage in the invention sequence. This was achieved in about 500 B.C. (some commentators put the date as early as 1500 B.C.) with the production of an axle capable of swiveling about a vertical axis (Lee 1947; Piggott 1983). Such vehicles can be readily detected in accurate drawings because the front wheels had to be small enough in diameter to pass under the floor of the vehicle (figs. 2.5 and 2.6 show wagons with only limited steerability). The technology did not spread rapidly. There were only a few steerable wagons in fourteenth-century England, and they were not commonplace until the seventeenth century (Boyer 1960).[9]

The initial vehicles all had a single hauling shaft because they were based on yoked-oxen precedents; however, the paired-shaft system is more suitable for horses. A relief from the palace of King Ashurnasirpal II at Nimrud (now in the British Museum) would appear to show chariots with pairs of shafts in about 850 B.C. The technique was extensively used in China in 400 B.C. and was widespread on Roman

Figure 2.5. Celtic wagon built in about 100 B.C., found in a Danish bog in 1881, and now reassembled in the Nationalmuseet in Copenhagen. Called the Dejbjerg wagons after the town in Jutland where they were found, they were probably made in middle Europe. The original vehicle had strap suspension, which is not visible in the restoration. *Photo by the kind permission of the Danish Nationalmuseet.*

roads by A.D. 200. Harnessing animals in file became common by about 100 B.C. and dramatically increased the size of the payloads that could be hauled. However, it was not widely used by the Romans, thus restricting them to two effective animals per heavy vehicle and severely limiting the haulage capability of the vehicle. On the other hand, the Romans did introduce the concept of using only one animal between a pair of shafts on a vehicle.

Roman harnesses tended to choke animals during a hard haul and did not permit the animal to slow the vehicle on a downhill slope (Borth 1969). Hauling requires a much more sophisticated technology for the horse than for the ox, because the horse pulls from forward of the shoulder. Under load, the breast band and neck strap of a yoke tend to press on the horse's windpipe, causing choking and suffocation. An effective horse harness therefore needs a carefully structured, padded collar resting on the horse's shoulder in order to prevent harness pressure on the windpipe. It also works best with a pair of shafts, rather than a single shaft. The introduction of such a harnessing arrangement increased the haulage capacity of the horse fourfold (White 1962). Nevertheless, its application was not widespread, preventing many communities from using the horse to its fullest (Casson 1974). It did not arrive in Europe from Central Asia until A.D. 750, well after the Romans. All this had major transport implications. A pair of horses that should have hauled three tonne could only manage half a tonne with Roman carts and harnesses.

Indeed, Roman vehicle technology was relatively primitive and drew heavily on Celtic practice. They had copied the Celtic technique of using leather straps to

***Figure* 2.6**. Reconstructed Roman passenger vehicle in the Cologne Romisch-Germanisches Museum. Note the strap suspension and swiveling front axle of this relatively advanced vehicle. *Photo with the kind permission of the Rheinisches Bildarchiv, Cologne.*

suspend the passenger compartment from four posts (fig. 2.6). The method then vanished, to be reintroduced in A.D. 900 by the Slavs of eastern Europe, using chains rather than straps. The technology led to the term *suspension,* which now describes a device that usually does not suspend. The demands of high-speed travel on leather coach straps later gave rise to the expression "hell for leather." The "Cologne" Roman vehicle in figure 2.6 shows a relatively advanced technology and a strong Celtic influence.

After the Romans

After the Romans, only small pockets of wagon technology continued to flourish. Carts were largely restricted to farm and local travel and were only occasionally used for long-distance travel (Gregory 1931; Jackman 1916).

Horses became more common after the tenth century. Many factors influenced the decision as to whether horses or oxen were to be used for haulage. Both had about the same haulage capacity, but horses could make 30 km cart trips each eight-hour day, whereas oxen could only travel half that distance and were more difficult to organize. A horse thus produced about twice the power of an ox. Put another way, the ox produces about the same tractive pull as the horse but at only about half the speed (at 2 km/h rather than 4 km/h), so its power output is halved. Oxen had the advantage that they could keep going over more days, required less water, were easier to feed and harness, were more able to manage difficult terrain, and were less likely to be bogged. Their hooves were more durable than those of an unshod horse; however, this advantage began to disappear with the development of the horseshoe in A.D. 700.

In the eleventh and twelfth centuries carts rather than wagons still predominated in Europe. In 1168 Thomas à Becket used carts for his journey of exile through France, and King John's last military campaign in England in 1216 made use of carts and packhorses. By the thirteenth century, as the Crusades drew to an end and as the improved harnessing technology spread, both carts and wagons were in common service (Boyer 1960). The fourteenth century was a time of major land transport development and vehicles became widespread. Tipperary farm tenants were required to bring their landlord's wheat to his barns using his wagons and carts (O'Keeffe 1973). Many estates went much further and began shipping their produce to town markets in carts driven by the farmers. This had a profound effect on the development of the countryside, changing farming from a subsistence lifestyle to a market-oriented industry. Professional carriers also became commonplace, providing a remarkably economical service which only added about 10 percent to the cost of a typical commodity for hauls of 80 km.

By 1450, freight was being moved by cart from Southampton to places as far north as Coventry. Nevertheless, most citizens of the fifteenth century still regarded wheeled vehicles as external interlopers (Postan and Rich 1952). When Queen Elizabeth traveled to Warwick in 1572, her baggage was conveyed in 600 carts. By 1599 regular cart-based freight services were operating between London and the Ipswich cloth industry. Similar scheduled services expanded rapidly over the next forty years (Bodey 1971). Nevertheless, the technology was only selectively available. Carts were not introduced into the more remote parts of Devon, Wales, and Scotland until the nineteenth century (Smiles 1874; Bird 1969). In such areas, freight movement by dragged sleds remained common as long as roads remained poor. Within urban communities much freight was moved by wheelbarrow, which also had some use for intercity transport.

The growing use of cart and wagon was not met with universal acclaim. In 1669 Courteney Poole urged his colleagues in the English Parliament to ban all carts and wagons because they "discouraged navigation" (Hartmann 1927). The vehicles also severely damaged many roads, provoking strong administrative reactions and ingenious technical rejoinders.

The large four-wheeled wagon was primarily a German development. In the mid eighteenth century, German settlers on the Conestoga Creek in Pennsylvania developed the famous Conestoga wagons. They were distinctively painted with a bright blue body, red trim, and a white canopy and weighed a little over a tonne.

The wagons were hauled by four to six horses traveling at about 3 km/h and had a capacity of between three and six tonne, depending on the surface. A useful feature was a boat-shaped floor, which prevented cargo from being displaced during a rough ride. A later version called the Prairie Schooner had a flat bottom and lower sides.

By the 1850s the loads that could be carried on conventional vehicles ranged from 2 tonne on two-wheeled carts to 8 tonne on four-wheeled wagons. For up to eight hours of travel at 5 km/h, a properly harnessed horse produced a pulling force of one-tenth of its weight when traveling over good foothold. Thus a half-tonne horse produced a pulling force of $0.5/10 = 0.05$ tonne, which was equivalent to the output of five men (Coane, Coane, and Coane 1908; Cron 1976b). A poor surface has a coefficient of friction of about 0.05 (see table 3.1), so a typical load capacity for a range of highway conditions was $0.05 \times 20 = 1$ tonne per horse. For long distances the load capacity was closer to half a tonne per horse, and this was also the value used by the British army for battlefield conditions. Roman vehicles and horses had a load capacity of only a third of a tonne, even over relatively good surfaces.

Over a day, a horse produced a constant energy output, no matter how it was worked (Rose 1952a). The well-known horsepower unit was first calculated by James Watt and was based on a strong three-quarter-tonne horse traveling for a short time at 3.7 km/h being able to pull a cart with a tractive force of one-tenth of its weight ($0.1 \times 0.75 \times 9.8 = 735$ newtons), which gives a power output of $750 \times 3.7/3.6 = 746$ watts. Over a full day, the same horse can only produce about 500 watts.

Its low load capacity made freight transport by wagon inordinately expensive for long-distance hauls. For much of history, the ship has been by far the most effective means of moving freight. Even when Rome was at its peak and at the hub of its enormous road system, it preferentially received its food supplies by ship (Toynbee 1970). In the twelfth century, Frederick I of the Holy Roman Empire passed an edict declaring the Rhine to be the king's highway (Forbes 1938). One estimate from 1818 was that it cost as much to haul a tonne of payload 50 km by road as it did to move it across the Atlantic by ship (Taylor 1951). Relative costs to move a tonne 1 km in the nineteenth century were downriver barges, 1; canal barges, 5; rail, 10; road, 30.

Overall, the main arteries of trade were the rivers and seas up to 1750, the canals between 1750 and 1830, and the railways between 1830 and 1914. The availability of these alternative transportation modes greatly diminished the need for road development in many areas. The pinnacle of horse freight usage came with the burgeoning of the canal system. When hauling a canal barge, each horse could pull 50 tonne of freight, or about twenty-five times the effectiveness of a horse and cart on good level ground, although at only half the speed. Indeed, one of the main reasons for the development of canals was the manifest freight inadequacies of the roads (Bagwell 1974).

Horse-drawn delivery vehicles with iron tires were still in use in many of the major cities of the developed world in the late 1940s, and an Australian vehicle census in 1945 counted some fifteen thousand operating horse-drawn vehicles. In many countries much freight is still carried in hard-tired animal-drawn carts and wagons. In 1986 India had some 15 million animal-drawn vehicles, only 7 percent of which had rubber tires.

Determining the Width

The use of wheeled vehicles gave rise to a whole new set of pathway needs, because the vehicles were wider and heavier than a human or a ridden horse or a beast of burden. The first Roman legal code in 450 B.C. specified a width of 300 mm for footpaths, 900 mm for bridle paths, and 1200 mm for a carriageway suitable for single vehicles (Cron 1976c). The width of vehicles was dictated by a number of interacting factors:

1. Any great increase in width over that of the preexisting pathways would have required an excessive amount of new path construction or of old path widening.
2. There was a limit to the distances that the timbers used for the axle and the vehicle floor could span.
3. There was a limit to the pressures that the path surface could withstand at the path-wheel contact area; the wider the vehicle, the larger those pressures became.
4. There was a limit to the loads primitively harnessed animals could haul.
5. There was a common need for paths to be wide enough to accommodate pairs of harnessed animals on either side of a single shaft. Vehicle dimensions naturally took advantage of the resulting lateral clearances.

Given these factors, a fairly standard vehicle width of a little under 2 m arose and has remained unchanged over the millennia, to the extent that it still exists in today's motor vehicles. Interesting evidence of the constancy of this dimension can be seen in measurements of wheel-rut spacings on Cretan roads from around 2000 B.C. and on later Greek and Roman roads. The ruts indicate a transverse wheel spacing, or *gauge,* of 1.4 m, which is consistent with a vehicle width of a little under 2 m, given a vehicle body overhang on the side of each wheel of about 250 mm (fig. 2.7; Pritchett 1980; see also Lee 1947 and Cron 1975b). Vehicles found as burial companions in graves in southern Russia and dated at about 2000 B.C. show a wheel spacing of 1.3 m, and most unearthed early vehicles lie in the range of 1.4 to 1.6 m (Piggott 1983).

Wheel rutting tended to impose a de facto standard on wheel widths in a similar way to the modern railway track. Rutting was common, even with stone pavements, and may have initiated with the dragging action of the travois (Pritchett 1980). The raised region between ruts sometimes provided convenient stepping-stones for pedestrians wishing to avoid the road's normal mud and slush. The technique was further developed in Pompeii, where raised crossings or stepping-stones were provided between the wheel paths on the Via Stabiana. The technique forcibly imposed a standard wheel spacing on all vehicles using the area (Geddes 1940). To ignore the standard was to court travel disaster, and Van Loon suggests that one of the reasons Napoleon's invasion of Russia failed was that the wheel spacing of the French cannon was incompatible with the dimensions of Russian road ruts.

A number of Greek and Maltese roads prior to 1000 B.C., and subsequent Roman roads, were built with ruts deliberately carved into the natural rock surface.

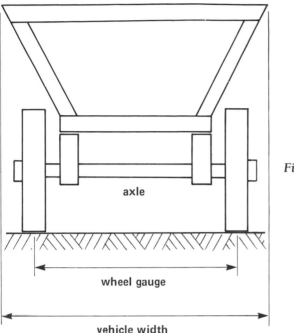

axle

Figure 2.7. Cart nomenclature.

wheel gauge

vehicle width

In Greece this particularly occurred on precipitous ledges and curves, because most wagons of the time were not steerable. This was a self-perpetuating system: the ruts would have required relatively large-diameter wheels, which would have made the provision of steering even more difficult. One public purpose of the ruts was to prevent accidents to vehicles on the way to the Olympic games—any mishaps on this journey were regarded as a rather ill omen. The system was continued in softer ground by paving the base of the wheel ruts with flat stones. One of the best known of the rut roads ran about 20 km from Athens to Eleusis. In Malta, ruts carved in coralline sandstone were also used in open ground to guide vehicles bringing soil and water to barren mountain terraces. The initial impetus for the ruts probably came from the grooves made by sleds and slide cars dragged over the surfaces in prewheel days. With the advent of the wheel, many of the ruts were deliberately enhanced (Zammit 1928; Forbes 1953).

The older Greek roads had rut spacings of 1.3 m, but this dimension tended toward 1.4 m on the newer roads and in Malta. The ruts were typically 60 mm wide and 250 mm deep in Greece and 400 mm wide (tapering down to 100 at the base) by 100 mm deep in Malta. Except for a double-track route from Athens to the marble quarries at Mount Pentelicus, the Greek rut roads had only one pair of ruts. However, the Maltese system had up to ten pairs of associated ruts, with sidings to permit passing (Zammit 1928; Casson 1974). This predominantly one-way system provided some problems for oncoming traffic. The classic case is the Theban legend in which failure to agree on who had priority on the Cleft Way from Thebes to Delphi led to the fatal quarrel between Oedipus and King Laius and thus to history's most famous patricide.

The rut technique was not forgotten over the millennia. In 1935, ruts to guide motorists were incorporated into the deck of the Queensboro Bridge in New York. They were not successful.

There are reports of Babylonian roads of about 1000 B.C. built of 400 × 400 × 150 mm flagstones laid as two tracks with a centerline spacing of 1.5 m, which would accommodate wheel gauges of 1.2 to 1.8 m (Lee 1947). Flagstones found at Maiden Castle near Dorchester in England and dated at about 250 B.C. also show 1.4 m wheel ruts. Maiden Castle is at the intersection of the Fosse Way (see fig. 1.1) and a tinway from Cornwall. The carts that made the ruts were probably being used for hauling tin to the coast (Newcomen 1951).

In 221 B.C. Chinese emperor Shi standardized the guage of chariot wheels at a "double pace," which appears to have been about 1.5 m. He then very logically determined standard road widths on the basis of that gauge. Over the millennia, such opportunities have rarely been taken by road builders, who usually struggle to keep pace with the more rapidly changing vehicle technology (Needham 1971).

The wheel gauge of Roman vehicles varied from 1.15 to 1.45 m. For instance, the ruts in the Roman road shown in figure 3.5 have a gauge of about 1.2 m (Rose 1952a). Thus, the typical 5.5 m wide Roman road permitted two vehicles to pass and, incidentally, a legion of soldiers to march in their customary formation of six abreast. Perhaps this tradition was also determined by the de facto wheel-gauge standard?

A Norse law of A.D. 950 required a road to be as wide as the height of the thumbs of a mounted man with his arm extended upwards (Cron 1976a). In 1280 Judge Philippe Beaumanoir's Coutumes de Beauvaisis, which codified French law, distinguished five types of ways: the *sentier,* or path, which was about a meter wide; the *carriere* for carts, which was about 2.5 m wide; a way that permitted carts to pass and was about 5 m wide; a way that permitted carts and cattle and most other things to pass and was about 10 m wide; Roman roads, which could be 20 m wide (Mesqui 1980).

In 1135 Henry I decreed that English roads should be wide enough for two carts or carriages to pass each other or for six armed knights to ride side by side. A third requirement was that "rustics standing one on each side of the way could make their goads meet" (Stenton 1951). The decree was clearly ineffective; in a 1555 act, legislators found it necessary to require cartways to be 2.4 m wide and horse cause-ways 900 mm wide. A further English act of 1691 again urged that all roads be widened to 2.4 m. The urging for ways to be widened to 2.4 m continued with a 1773 act, also demanding 2.4 m for cartways (Bodey 1971).

Legislators often turned their attention to controlling the vehicle rather than the infrastructure, thus usually transferring any costs to someone else's shoulders. In 1753 a British act limited wheel gauges to 1.65 m (Jackman 1916). A typical gauge for nineteenth-century animal-drawn vehicles was 1.4 to 1.5 m, and a common railway gauge was and still is 56.5 inches, which converts to 1.44 m. George Stephenson later copied this 1.4 m inside measurement from the gauge of coal wagons, whose train tracks had a 1.5 m overall width. On the other hand, Isambard Brunel selected his 2.1 m broad gauge so that his rail wagons could carry the 2.0 m wide road vehicles (Rose 1952a).

Thus, this historical liking for 1.4 to 1.5 m as a wheel gauge and under 2 m for vehicle widths led to common widths of 2.5 to 4 m for one-way roads and 4 to

8 m for two-way roads. Such widths would have been much more demanding to develop or construct than the half-meter required for footpaths or the meter needed for bridle paths and cattleways. On the other hand, Kautilya's Arthasastra produced during India's Mauryan Empire in about 320 B.C. recommended 3.7 m widths for elephant ways.

By the nineteenth century, society's carriageway horizons had begun to widen. Napoleonic law required major French roads to have a 7 m paved width. In 1806 the U.S. Congress specified a 6 m paved width when it authorized the construction of the National Road to the West. The British Highway Act of 1835 required cart-ways to be 6 m wide (Hadfield 1934), and in 1845 New York State required 6.5 m. However, after the dark ages of modern roads between 1850 and 1900, the standard had to be dropped, with another American state then accepting 4.2 m (Cron 1976c). In 1916 the U.S. federal road agency considered 5 m to be an adequate two-way width (Cron 1976a).

Modern pavement widths for two-way roads range from 4.5 to 9 m. Pavements under 6.5 m wide are commonly considered to be narrow and prone to problems of safety and of pavement deterioration, the latter due to the outer wheels of vehicles causing structural damage to the pavement edge. For larger roads, lane widths are usually between 2.5 and 4 m in order to accommodate a fairly universal vehicle width of just under 3 m (Lay 1990).

It is one matter to require a certain width of pavement, but quite another matter to have sufficient width of land to accommodate not only that pavement but also all the associated needs for adjacent roadside verges. This total width is called the right of way. There was initially no need to specify its boundaries because the king's highway concept (discussed in the next chapter) meant that the road's practical presence could not be disputed. Following Roman practice of keeping an arrow's flight clear on either side of the roadway, the English Statute of Winchester in 1285 required the sides of roads leading to market towns to be cleared for 60 m from the road. Gradually, a practical need emerged to define the right of way. In 1705 French regulations required a width of 18 m for major routes and 9 m for town-to-town roads. Likewise, New World land subdivision practice frequently required owners to provide a 20 m right of way. All the above paved widths would easily fit within such a land reservation.

Wheel Loads

The total load that a cart or wagon can carry is of vital interest to the hauler operating the vehicle. However, for the road manager the key question is load and the pressure that each individual wheel applies to the pavement. In addition, to reduce both the weight of the wheel and its rolling resistance over good surfaces, the hauler needs a narrow wheel with a narrow tire, whereas the road manager requires a wide tire to protect the pavement. Thus, the joint questions of the maximum load to be carried by wheels and the required width of tires have long been key points of debate in road management.[10]

Traditionally, when load limits have been widely promulgated, they have been just as widely flouted. The earliest recorded load limits date from 50 B.C., when the Romans restricted loads to about 250 kg (Lee 1947). Five hundred years later in A.D.

TABLE 2.1
1809 Wheel Load Limits for Four-wheeled Vehicles

Tire width (mm)	Load per wheel (t)	Contact pressure (MPa)	Number of horses
50	1.0	4.0	4 (stagecoach)
75	0.9	1.6	4
150	1.1	0.5	6
275	1.3	0.6	6 (conical wheel)

SOURCE: Data derived from Law and Clark 1907, which, in turn, was based on regulations made under the then-current United Kingdom Act.

438, their Theodosian Code set the limits at 750 kg on an ox-drawn wagon, 500 kg on a horse-drawn wagon, and 100 kg on a cart. These low loads are consistent with our earlier comments on the relative inefficiency of Roman freight vehicles. In 1622 James I of England prohibited loads greater than one tonne on any vehicle operating during the winter (Bodey 1971). The maximum permissible load on English roads was raised to 6 tonne in 1765, reflecting a significant improvement in the condition of English roads.

Six tonne, or about 1.5 tonne per wagon wheel, was a practical upper limit for many years. The wheel load limits in use in 1809 are given in table 2.1. The illogic of these load limits giving an increase in contact pressure as the width decreased was recognized but ignored. At the beginning of the twentieth century, many laws still existed requiring a millimeter of tire width for each 10 to 18 kg carried. A typical formula was (Coane, Coane, and Coane 1908):

$$\text{width in mm} = (\text{mass carried in kg}) \times C / \sqrt{\text{diameter in mm}}$$

where the values of $C = 15$ for earth, 10 for macadam, and 2 for paved roads approximated the strength of those pavements.

However, at this time gasoline-powered trucks with solid rubber tires were launching a new, destructive attack on road pavements. The problem was most severe in the United States when surplus trucks from the First World War caused particular havoc. Guidance on protecting the road was received from a major Bureau of Public Roads research program, which showed the great advantage of using pneumatic rather than solid tires and recommended a higher maximum wheel load of about 4.5 tonne if pneumatic tires were used, due to their greater area of contact with the pavement. Although European practice was to adopt a somewhat higher value, wheel loads themselves have remained relatively constant since those decisions in the early 1920s. The major increases in gross truck loads to around 50 tonne have been the result of adding more wheels and more axles to trucks, rather than of raising wheel loads. This type of reaction is an almost inevitable result of having a road infrastructure that changes far more slowly—perhaps at fifty-year intervals—than the associated vehicle technology.

Direct policing of load regulations was not easy because quantitative load levels were not obvious to the eye and their direct control required some means by which they could be ascertained. The initial method involved winching the vehicle to be weighed off the ground and determining its weight by a system of steelyard levers and scales. Public weighing facilities were introduced into Dublin in 1555 "to eschew the loss to excessive and untrue tolls" (O'Keeffe 1973). In 1602 a toll on carts weighing over a tonne was introduced in Kent. In 1741, British turnpikes were permitted by law to charge extra tolls on loads over 3 tonne, although noblemen were exempt from the toll. The excess toll was to be applied to road maintenance. The tax was strongly but fruitlessly opposed, with opponents pointing out that it would merely encourage more small carts.

In 1744 John Wyatt invented the modern weighbridge platform and hence made weighing vehicles far more feasible. Not coincidentally, at the same time, weighing devices were legalized and the rights to operate them were let annually, usually by public auction (Bird 1969; Jackman 1916). The moves were clearly successful as in 1751 a further law made such facilities compulsory on turnpikes within 50 km of London (Dyos and Aldcroft 1969).

The Roman Theodosian Code also restricted the number of animals that could be used to haul a vehicle because a horsepower approach was far easier to police than was a load limit. In 1508 this method was adopted by the city council of Paris, banning wagons drawn by more than two horses. In 1629 an English act prohibited more than five horses from drawing any vehicle; after forty years the horsepower limit was relaxed from five to seven (Gregory 1931).

A similar alternative was to limit the number of wheels. For instance, from 1622 to 1661 an English statute of James I decreed that only two-wheeled vehicles could be used on English roads. The main consequence of the two-wheel law was to produce grossly overloaded two-wheeled carts (Crofts 1967).

The problem with iron-shod wheels was often one of wear due to frequent usage, rather than overload due to single-load levels. This continues today, and the road manager's concern with traffic within legal load limits is usually with the wheels of large numbers of these vehicles causing either rutting or structural fatigue failure.[11]

Iron tires have always been favored by haulers because they not only permit tires and wheels to be narrowed, but also last far longer than the timber alternative. However, road maintainers view them far less favorably because they have a much more abrasive effect on the road surface and produce higher contact stresses. The problem was often worsened by driving iron nails with prominent heads radially into the tire's running surface to provide better surface traction.

In medieval times iron tires were sufficiently common and damaging for a number of towns to prohibit their use within the town limits. One measure in Beverley in the north of England was introduced in 1418, based on a regulation of 1367.[12] Subsequently, many cities had occasion to ban the entry of iron-wheeled vehicles.

Another technique was to regulate tire width. An English act of 1662 required tires to be at least 100 mm wide. This act was soon suspended and then repealed in 1670 when it was found that the new wide wheels would not fit into many of the country's well-established ruts. The rut problem must have diminished, for the width limit was raised to 225 mm in the 1753 Broad Wheels Act, which also required tires to be 450 mm wide for very heavy loads (Bodey 1971). Protests led to an amendment two years later that permitted the use of 150 mm wide tires for wagons pulled by fewer than seven horses. A subtle variation introduced in a 1765 act imposed lower penalties on wagons whose fore and aft wheels were staggered laterally so that they ran in different but adjacent wheel paths.

The first General Highway Act of 1767 and an amending act of 1773 permitted vehicles with 225 mm wide tires to be drawn by up to seven horses and those with 400 mm wide tires to be drawn by any number of horses (Jackman 1916). These 400 mm wheels were built so as to flatten any surface over which they passed and were generally only found on turnpike roads, where the act suggested they should receive five years of toll-free operation. An 1816 act made it clear that stagecoaches could use much narrower tires. However, the turnpikes of the eighteenth and nineteenth century imposed higher tolls on narrow-tired vehicles in an attempt to discourage their use.

A major effect of these regulations was the use of 400 mm wide tires to produce broad-wheeled freight wagons. These inefficient vehicles required teams of a dozen or so horses pulling loads of about 700 kg each, although horses could normally be expected to pull loads of 1,000 kg over hard surfaces. Such vehicles were commonplace until the final abolition of many of the tire-oriented regulations in the Highway Act of 1835 (Bird 1969).

A closely related and equally technically ineffective eighteenth-century development to circumvent wheel-width regulations was the use of very large conical wheels that rolled and slid on the roadway. They were clearly little more than narrow wheels legalistically disguised as broad ones to gain the broad-wheel concessions (Deacon 1810). The result merely demonstrated the distorting effect of wheel-control legislation, for their tapering meant that much of the contact surface had to be dragged along the road. A sometimes more constructive adaptation was the use of dished wheels with the rim outside the plane of the hub. This provided lateral strength and obviated the need for heavy spokes—"sometimes" because the adaptation was also used to evade the wheel-width laws by keeping the actual contact width small.

On poor surfaces wide tires were used to permit easier passage of the vehicle, provided that enough motive power was available. Indeed, a few late eighteenth-century heavy wagons were built with great rollers rather than conventional wheels. However, it has not usually been possible to make rolling surfaces wide enough and find large enough power sources to make this system work, and so operators have had to find other ways to overcome poor pavement running surfaces. A more direct solution was to use very large diameter wheels; for example, 1.6 m diameter wheels were common in 1800, but their use was somewhat restricted by the need to keep the diameter of the wheels on the turning axle below the height of the wagon floor. Even larger wheels were used for the same purpose on the famous nineteenth-century steam traction engines (Lane 1935).

It can be seen from the preceding discussion that the basic principle followed over most of our transport history has been to make the vehicle suit the road, with little attempt at adapting the road or the system to the vehicle. Sidney and Beatrice Webb in their comprehensive review ([1913] 1963) referred to events in the eighteenth and nineteenth century as an

> interminable series of enactments, amendments, and repeals—successive knots of amateur legislators laying down stringent rules as to the breadth of the wheel; the form of the rim; the use of iron tires and lead nails; the height of the wheel; the position of the felly, the spokes and the axle; the space between each pair of wheels and the respective line of draught between back wheels and front.

> The wheeled carriage was an intruder on the highway, a disturber of the existing order, a cause of damage—in short an active nuisance to the road-way—to be suppressed in its most noxious forms and where inevitable to be regulated and restricted as much as possible.[13]

The correct solution, of course, was not to penalize the development of more efficient vehicles but to improve the inadequate pavement surface. The pursuit of this solution will be discussed in later chapters.

Summary

The discussion in this chapter has concentrated on roads, but the change from beast of burden to wider and heavier carts, wagons, and carriages also had a major impact on the effectiveness of fords and on the construction of bridges. Similarly, once the ways were capable of carrying the first generation of wheeled vehicles, the door was open for the normal technological leapfrog to occur and for a whole new class of vehicles to develop. This next quantum jump is discussed in chapter 5.

We have seen that strong animals hauling relatively simple wheeled vehicles gave the world a greatly expanded and extremely useful ability to transport people and freight over significant distances. Indeed, four millennia were to pass before it was possible to significantly improve on the performance provided by domesticated animals. This ability was put to many uses, with trade, military conquest, and administrative control ranking very high on the list. At each stage of this transport development there was an increasing demand for more efficient and effective ways, with respect to their extent, their capacity, and their condition. The response to these demands is discussed in the next chapter. We will see that it has traditionally been a difficult response, because those who construct and maintain the ways are usually quite separate from those who operate the vehicles that use them.

Some of the early transport modes, such as the pack animal, remain in use today, for they are relatively undemanding on infrastructure and thus appropriate where development and resources are low. Indeed, we can still find examples of most of the early transport modes in operation somewhere in today's world. Nevertheless, most have been steppingstones in the evolution of today's new transport systems. A few of the consequences of these early technologies, such as wheel gauges and road widths, still have a major influence on contemporary life, despite the many changes that have occurred in the intervening millennia.

One strong lesson that comes from the observation of that evolution is that the common responses to it of either overregulating to protect an existing transport mode, or of expecting the infrastructure to be immovable and instead requiring vehicles to adapt to suit it are both particularly inappropriate. As we have seen, history shows that transport must be regarded as an interactive system, with each component reacting to changes in all parts of the system.

Roadways

*Whereby all the said high-waies shall or may be
made more sound and strong, pleasant and comfortable for all manner of way-faring, journeying or
travelling with carts and carriages.*

THOMAS PROCTER, *A WORTHY WORKE*, 1610

Wheeled vehicles hauled by draft animals were in use by about
3000 B.C. The preceding chapter examined their impact on the road system, and this
chapter explores the response of that system to those ever-increasing impacts.

The basic roadmaking dilemma is that natural material soft enough to be
formed into a smooth, well-graded surface is rarely strong enough to bear the weight
of a loaded wheeled vehicle, particularly when the material is wet. On the other
hand, rock strong enough to carry wheel loads under all moisture conditions is
rarely able to be easily formed into a suitable surface for traveling. Nevertheless, by
the time of the wheel it would have been possible to manufacture an adequate road
surface, given the time and resources. In effect, the remainder of this book explores
the technical and administrative techniques used by inventive people to aid in the
accomplishment of this task.

In the Beginning

The first indications of manufactured roads are of stone-paved streets in Ur in
the Middle East in about 4000 B.C. (fig. 3.1; Kirby et al. 1956), of corduroy roads in
Glastonbury in England in 4000 B.C., and of brick paving in India in 3000 B.C.
(Kennerell 1958).[1] The previous chapter discussed cart tracks constructed in Malta
between 2000 and 1500 B.C.

By 2000 B.C. the availability of metal tools meant that many villages and towns
were able to shape stones into flagstones for paving local streets and paths. At about
this time the Minoans constructed the world's oldest extant stone road. It ran for

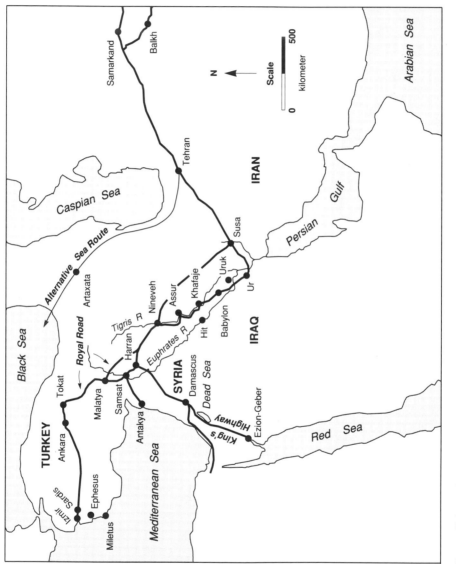

Figure 3.1. Early roads in the Middle East, the cradle of civilization. Mesopotamia—the land between the rivers—occupied the flat alluvial area between the Tigris and the Euphrates.

some 50 km from their capital of Knossus, through the mountains of Crete and the town of Gortyna, to a southern seaport at Leben that serviced east–west Mediterranean trade. The road was a significant piece of construction, with elaborate longitudinal sidedrains, a 200 mm thick pavement made of sandstone pieces bound together by a clay-gypsum mortar, and a 4 m wide surface of basaltic flagstones flanked by mortared pieces of limestone in the manner of a modern road shoulder (Forbes 1934; Lee 1947). In many ways it was technically superior to the subsequent Roman roads, relying on structure rather than on size. There is also evidence of some construction using broken stone in the modern style on the island in about 500 B.C. (Merdinger 1952). Indeed, Minoan techniques were not to be improved upon for over three millennia.

Asian Roads

The first major arterial road was the 2.5 Mm route used by Assyria to connect its capital of Susa, via Nineveh, to Sardis in western Turkey and thus to the nearby Mediterranean ports of Izmir and Ephesus (see fig. 3.1). It was a route rather than a manufactured road, was built for foot rather than for wheeled traffic, and relied on ferries rather than on bridges. Nevertheless, the route was of great importance because it joined Assyria, after its conquest of Mesopotamia in about 1200 B.C., with the many trading activities based on the Mediterranean (Schreiber 1961).

The Assyrian route also connected to a network of similar routes, including one from Harran to Damascus, which linked in turn with the King's Highway that ran as a track along the east bank of the Jordan to Ezion-Geber on the Red Sea, the location of copper mines and King Solomon's mines since about 3500 B.C. Moses had attempted to use the southern portion of this last road in leading the people of Israel home from their Egyptian exile. Although he assured the local king that "we will go along the King's Highway, we shall not turn aside to the right hand or to the left," his passage was prevented (Num. 20:17). Later, a road from Egypt came up the coast, turned inland, and passed south of Nazareth, meeting the King's Highway south of the Sea of Galilee. The Via Maris, a famous Roman road, was a development of this system.

The first recorded road builders were the engineer pioneer corps, or *ummani,* of the Assyrian kings. Their work in 1100 B.C. in constructing a well-aligned mountain road for King Tiglath-pileser was documented for posterity in the record that they "hewed a way with bronze pick-axes."

In 670 B.C. Assyrian King Esarhaddon ordered that roads should be laid out throughout his kingdom to facilitate trade and commerce. The theme was continued by Chaldean King Nebuchadnezzar, who boasted of his conquest of Lebanon in about 600 B.C.: "I have cut through steep mountains, I have split the rocks, I have made a way through and built straight roads for [exporting] the cedars." In 586 B.C. the biblical people of Judah were exiled by Nebuchadnezzar to Babylon, crushed and hopeless. The prophet Isaiah proclaimed that God would set the people free, and in doing so he would "make a highway across the mountains, and prepare a road for my people to travel, from the north and the west, and from Aswan in the South." The theme of a road for the repatriated people was repeated by Isaiah a number of times, ending with the instruction, "Prepare a highway: clear it of stones" (see 49:11, 49:12, 11:16, 19:23, 62:10).

Isaiah was influenced in taking up this theme by his knowledge of the works of King Nebuchadnezzar and of King Cyrus II. In particular, Cyrus had built roads in Jordan in about 550 B.C. and was famous for the speed at which he moved his armies. In 538 B.C. he was responsible for the return of the Israelites to their homeland. Cyrus was followed by one undistinguished king and then by King Darius I, popularly known as Darius the Great, who reigned from 522 B.C. to 486 B.C. with his administrative capital at Susa.

Darius extended the network of roads established by Cyrus and put in place an elaborate post system to permit the efficient administration of his empire. Roads were essential to Darius for two reasons. First, his empire depended greatly on tributes from distant provinces and subjugated kingdoms. Second, he did not have the network of seaways that served the trading needs of the rival Mediterranean empires.

Darius's major work was the famous Royal Road, which was built in about 500 B.C. It broadly duplicated the well-used Susa to Izmir road of the Assyrians, but bypassed many significant towns in order to reduce travel times (see fig. 3.1). The route was used by quite large wagons, but priority was given to royal messengers both to prevent them from being impeded by common travelers and to protect the relatively poor-quality surface of packed earth, described by Xenophon in his *Cyropaedia* as a "beaten track" (Schreiber 1961). Herodotus noted in the fifth book of his *Histories* that the 2.7 Mm Royal Road contained 111 post houses and could be traveled in ninety days, although the Royal messengers took only ten to twenty days. Esther (8:9–10) tells how Xerxes I in about 470 B.C. employed "mounted couriers riding on swift horses that were used in the King's service [to send an edict to the] princes of the provinces from India to Ethiopia."

The Persian road system continued to expand and was adopted and adapted by Alexander the Great and his Thracian engineers. The expansion came to an abrupt halt with the death of Alexander in 323 B.C. Eleven years later the Romans began building the first of their major highways.

A number of major roads—or, more realistically, tracks—radiated out from the Assyrian administrative and trading hub on the Persian Gulf. They linked Europe, China, and India in a transport hub whose effect on the world has been without equal.

Undoubtedly the most famous and influential of the spokes radiating from the hub was the Silk Road (fig. 3.2), a collection of caravan routes dating from about 300 B.C., when they were first used to bring jade to China from the jade carvers of the old and mysterious town of Hotan. The Silk Road was extended to meet the Middle Eastern system in about 200 B.C., and by 100 B.C. it had become an active trade route between China and the Mediterranean.

The common route to the East passed through Iran and by Samarkand, Fergana, and Os, then through a mountain pass to the oasis market town of Kashi, where the road split. Depending on circumstances, a traveler might take the better provisioned but bandit-plagued route south of the Tian Mountains and north of the Taklimakan Desert via the oases at Aksu, Kuqa, Turpan, and Hami. In the early years, travelers generally preferred the more arduous but safer route south of the Taklimakan Desert and north of the Altun Mountains, through Shache, Yecheng, Pishan, Hotan which was the beginning of the most demanding part of the trip, Yutian, Minfeng, Qiemo, Ruoqiang, and then via the Yang Pass at 5,200 m to the ancient city of Dunhuang.[2] The road joined again at Anxi, went down the Hexi Corridor,

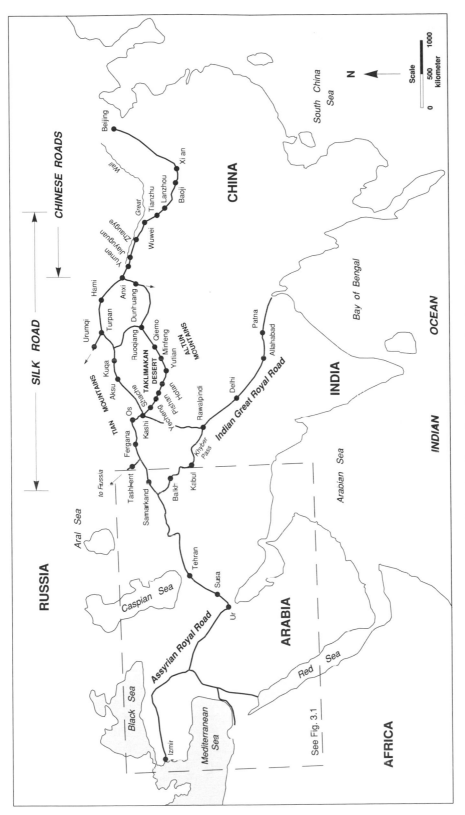

Figure 3.2. The Silk Road and its two associated royal roads.

passed through the Great Wall at Yumen and Jiayuguan, then to Zhaugye, left the Great Wall near Wuwei, via Tianzhu, met the Huang (or Yellow) River at Lanzhou, connected with the Chinese road system at Baoji, and ended at Xian, the capital of China at the time the road was opened.[3]

The southern route, incidentally, picked up trade from Qinghai, and the northern route similarly linked with trade from Urumqi and northwest to Russia via Alma-Ata. An alternative route from Russia passed north of the Caspian, went through Tashkent, and then met the Silk Road west of Fergana.

In Chinese the western portion of the Silk Road is called Tian Shan Nan Lu, which means "the road south of the celestial mountains," and the eastern portion is called the Imperial Highway in the vicinity of Xian. The name Silk Road is attributed to the famous nineteenth-century German geographer Ferdinand von Richthofen in 1877. Traders from the East and the West usually went only as far as the Stone Tower near Os, where they exchanged cargoes. Indeed, most merchants only traveled along small portions of the road. Nevertheless, from A.D. 70 onward Buddhism spread to China from Taxila near modern Rawalpindi to Kashi and then along the road. Chinese Buddhist monks, beginning with Fa Xian in A.D. 399, used the road to travel to India. The first Westerner to travel beyond Kashi was probably a Syrian Nestorian priest called Olopun in A.D. 635.

Genghis Khan made good use of the Silk Road in the early years of the thirteenth century. After reopening it to aid his conquest of China, he then improved the passage and established an elaborate system of post stations and couriers on the northern route in order to exercise control over his newly conquered Chinese territory from the distant Mongolian steppes. His men were fine road builders, and elsewhere they built bridges, cut passes, and cleared passages through thick forests. They used some of their routes to move 9 m wide mobile "houses" on wheels set 6 m apart.

Perhaps the Silk Road's most famous traveler was Marco Polo, who took advantage of the secure travel provided by the Mongol empire and used the southern route on his trip from Venice to Beijing between 1271 and 1275. The road was in use for over two millennia (a wonderful evocation of the Silk Road is given in Martyn 1987) and there are reports of traffic wearing parts near Samarkand into hollow ways up to 4 m deep (Schreiber 1961). The Silk Road reached its peak in the fourteenth century and then declined, initially due to such causes as the coming of Turkish Islam to the area, the fall of the Mongols, and the drying up of many of the oases. The decline was complete when an alternative sea route—the Spice Route—was established in 1497 by Vasco da Gama.

Without question, the Silk Road played a major role in transferring culture and technology between East and West. From China passed silk, jade, spices, ginger, tea, peach and pear trees, methods for fruit cultivation, porcelain, iron-making technology, papermaking and printing, gunpowder, and well-drilling methods. Into China came glass, grapes, wool, cotton, ivory, jewels, large horses, and ideas such as new forms of music, dancing, painting, and other artistic endeavors, Buddhism, and Islam. Mathematics, astronomy, medicine, and the calendar were other beneficiaries of the route. And, of course, there were the conventional trade items of salt, flint, amber, copper, and gold. The expensive nature of the trip naturally limited any trade to items of high value and small volume.

An extensive road system existed in ancient China, including a few paved roads and some major early bridges (Gregory 1931). The system was created in the Western Zhou dynasty, which ruled from 1066 to 771 B.C. Traces of roads of this time from Qishan to Feng and Hao are still visible. Where possible, rural roads followed an alignment as straight as the path of an arrow, and trees were planted along the way. In the capital city of Xian there were nine roads, each 9 m wide and on a grid pattern, 7 m wide circular roads, and 5 m wide local roads. Importantly, the system included the designation of officials responsible for road maintenance.

Many trade routes were established in about 500 B.C. during the Warring States period (Jiang 1986). At this time the network length was already about 7 Mm. However, Chinese road development mainly occurred after 200 B.C. during the Qin and Han dynasties, particularly in the twelve-year reign of the tyrant first Emperor Shi. A successor to Shi wrote, "He also ordered the building of post roads all over the Empire—so that all was made accessible" (Needham 1971). These 15 m wide highways were made thick and firm at the edge and tamped with metal rammers. Trees were planted at 10 m intervals.

Shi's "Empire of All under Heaven" introduced an elaborate and innovative bureaucratic system. To aid its operation the inner lanes of the major roads were reserved for the emperor and his high officials. The Shi empire, however, is best remembered, not for its bureaucracy or its roads, but for the creation of the Great Wall of China.

According to Joseph Needham (1971), the Chinese road system reached its peak about A.D. 20 and then declined for a millennium. Later Chinese sources (Jiang 1986) put the peak over 700 years later in the Tang dynasty. The maximum length of the network was about 40 Mm, compared with 80 Mm for the Roman system. Chinese roads, incidentally, provide the earliest known names of engineers associated with roadmaking: Tu Mao, Yu Xu, and the first woman road engineer, Shan C.K.S. (Needham 1971).

A second major caravan trade route (shown in fig. 3.2) linked the Middle Eastern hub via Balkh, Kabul, and the Khyber Pass to the extensive Indian road system at Rawalpindi (Schreiber 1961; CRRI 1963). There had been active trade in Afghan lapis lazuli. Another route used by the Chinese joined Kashi with Rawalpindi.

Indian road construction was quite advanced by 2000 B.C., with growing use of brick paved streets, subsurface drainage, and bitumen as a mortar between the bricks (Hindley 1971; Forbes 1934). The Aryan Rigveda, which dates from about 1500 B.C. and is probably the world's oldest surviving literature, mentions Indian Great Roads, or Mahapathas (Vaknalli 1980). India at the time was an extensive maritime trader with Africa and Arabia. By 300 B.C. the Indians under King Chandragupta and the Mauryan empire had linked the road from Susa at Rawalpindi to a well-constructed 3 Mm Great Royal Road running south–east through Delhi and near Allahabad to the then capital of Patna and the mouth of the Ganges, with a branch servicing the important salt trade on the west coast. The king created a special ministry to manage the road. Shade trees were planted—partly for religious reasons—wells dug, and post houses and ferries provided (Schreiber 1961; CRRI 1963). A good summary of early Indian roadmaking is given in Forbes (1934), and an extensive review of the Indian road network over the years is given in Deloche (1968).

Bitumen Roads

The roadmaking technologies so far encountered have largely been very appropriate and very local ones—beaten earth and stone slabs and trenches in soft rock. Just once or twice did road makers import some lateral technology into their process. The most obvious and lasting example is their occasional use of bitumen, an ancient material with some very modern progeny. It is therefore useful to devote some space now to beginning to tell its own long and contorted story. For the moment, it is enough to define bitumen as the viscous, "heavy" component of petroleum. It is a strong and durable glue. Asphalt is a mixture of bitumen, sand, and stone. Far more detailed definitions and discussions of word origins are given in chapter 7.

Just as the first ways were probably created by animals, there is also evidence of animal paths leading to petroleum springs and of animals rolling in oil to saturate their coats (Duncan 1928).

Bitumen was used in the Middle East from at least 3000 B.C. for cementing inlaid jewelry, waterproofing tanks, mortaring bricks, and grouting mosaic tiles. By about 2500 B.C. it was being used at various temple sites to provide an impervious surface from which water and/or blood could be easily collected. Nebuchadnezzar's famous Hanging Gardens of Babylon were waterproofed with bitumen. The commonest practical uses of bitumen over time were as mortar and as a waterproofing (Forbes 1936; Watson 1950; Abraham 1960).

Bitumen has always been relatively plentiful in the region, particularly in the valley of Siddim at the southern end of the Dead Sea and near the town of Hit on the Euphrates (see fig. 3.1). Indeed, the Greeks and Romans called the Dead Sea the Lake of Asphaltites after the lumps of bitumen and asphalt found floating on its surface. Bitumen from the region was exported to early Egypt as a mortar and as caulking for ships. It had other uses: as a magic potion to ward off the Babylonian she-devil Labartu; as a punishment under Assyrian law, which had it poured hot over an offender's head (Forbes 1936); and as just retribution—for example, "the valley of Siddim was full of bitumen pits: and as the kings of Sodom and Gomorrah fled, some fell into them" (Gen. 14:10).

Ancient baskets of bitumen packed for export down the Euphrates have been found at Hit, and its bitumen springs have been used by boat builders over the millennia (Woolley 1954).[4] In a Biblical context, bitumen was used for waterproofing Noah's Ark—"Make yourself an ark . . . and cover it inside and out with pitch"—for building the Tower of Babel in Babylon in 500 B.C., which used "brick for stone and bitumen for mortar"; and for Moses' basket of bulrushes, which was "daubed . . . with bitumen and pitch" (Gen. 6:14 and 11:3; Exod. 2:3).

Roadmaking using burnt clay or mud bricks was common in the urban areas of the Middle East, given the general absence of suitable stone. Excavations at Khafaje, Babylon, and Ur (see fig. 3.1) uncovered pavements dated at about 2000 B.C. or earlier with layers of brick bonded with a bitumen mortar. An Assyrian temple built in Assur in 700 B.C. contains a 3.5 m wide and 1 km long processional road made of two or more layers of bricks in a bituminous mortar, all covered by a surface course of volcanic breccia and limestone slabs. A similar road was found at Tell Asmar near Baghdad (Abraham 1960; Simon 1970; Woolley 1954).

Bitumen is mentioned in inscriptions at Nineveh and Babylon, including a note by King Nebuchadnezzar that his father, Nabopolassar, in about 615 B.C. had made in Babylon a 6.5 m wide and 1.5 km long processional road, using three or more layers of burnt brick "glistening with asphalt and burnt brick. . . . Placed above the bitumen and burnt brick [was] a mighty superstructure of shining limestone." The surface was paved with meter square limestone slabs, one of which carried the inscription, "I have paved the Babel Street with slabs from Shadu for the Procession of the Great God Marduk." Babel Street was also known as the Processional Way and as "Aiburshabu, The Street on which no enemy ever trod" (de Camp 1960; Earle 1974). The bricks and slabs were mortared with bitumen, and when brick roads had a new surge of popularity at the beginning of the twentieth century, bitumen was again the common mortar.[5] Nebuchadnezzar, incidentally, also mixed bitumen with naptha to fuel the furnace in which he attempted to incinerate Shadrach, Meshach, and Abednego (Dan. 3:23).

There were many other bitumen sources. Aristotle, Plutarch, and Pliny described deposits around the Adriatic, and a number of early religious groups made regular visits to fire temples based on natural gas or bitumen and located around the Caspian Sea and in Pakistan, India, and Burma (Frazer 1914). However, with the exception of India, the use of bitumen for roadmaking does not appear to have been adopted by early civilizations; they usually preferred simple clay mortars, which were more readily available but far less effective. Bitumen is now one of the commonest of roadmaking materials, and its history and use in more recent times are described in chapter 7.

European Roads

The swampy Somerset Levels near Glastonbury in England (see fig. 1.1)—through biological good fortune—contain the world's oldest extant manufactured way, the recently named Sweet Track, which was built in about 3300 B.C. over some 2 km and linked settlements in the Polden Hills with an island in the swamp (Coles 1984). The Sweet Track uses longitudinal logs to support oblique crossed pegs, which form a V every meter or so. Planks placed between the V's create a walking path. It is estimated that the path could have been built in a day, once the wood was on site. Indeed, a number of well-preserved timber ways exist over about 1,100 sq km of the Levels. There are paths built of brushwood and of woven panels between 3000 and 2000 B.C. and the more conventionally timbered Abbot's Way, which was built in about 2000 B.C. (Coles 1986). Similar log roads from the same era have been found in swamps throughout Europe, such as Holland at Valtherbrug, in Switzerland, and in Hungary.

Europe in the second millennium before Christ saw the development of simple trade ways such as the Heraclean Way running between Italy and Spain via Marseilles and nearby Heraclea. Heraclea was close to present-day Avignon (Chevallier 1976). Such ways were used for the relatively casual movement of flints from Denmark, freestone from Belgium, salt from Austria, lead and tin from England, and amber from northern Europe. Amber is a hardened resin, which was sought after and hoarded for its beauty, medicinal uses, and magical connotations—which possibly derived from its electrical properties. England, for example, traded its metals and metal products for the much revered amber.

By about 1500 B.C., many of the ways had linked together into an extensive trading network known—in eastern and central Europe—as the amber roads after their main initial freight (Postan and Rich 1952). One amber road ran south from near Hamburg to Marseilles, branching via either Cologne, Metz, and Lyon or via Frankfurt and Basel. Another amber road ran south from near Hamburg to Venice via the Brenner Pass. A fourth came from prolific amber sources on the Lithuanian Baltic coast, followed the rivers of eastern Europe to Kraków, and hence to Carnuntum near Vienna (see fig. 3.3). A few parts of the amber roads remain today as corduroy roads built of oak logs (Schreiber 1961). Tacitus told how the amber traders traveled everywhere in safety, even in wartime, under the protection of their emblem of the sacred boar.

Greek roadmaking peaked during the Mycenaean Age in about 1000 B.C. and earlier. The rugged countryside and the abundance of sea routes meant that roadmaking was not a major activity. The rocky roads were mainly for nonwheeled traffic, but there is evidence of some deliberate construction of wheel ruts and single-lane carriageways.

The Greeks gave Hermes, primarily the messenger of the gods, a second role as patron of roads and protector of travelers. Shrines to Hermes were initially created by travelers placing a stone by the wayside as they passed through a crossroad.[6] Subsequently, the shrines consisted of elaborately carved structures onto which passing travelers poured sacred oil. The Romans continued this pattern with roadside shrines devoted to Mercury, Hermes' Roman equivalent. The Greek roads were themselves often consecrated to gods and decorated with monuments and temples. They were lined with tombs in the vicinity of towns.

European crossroads frequently possessed magic, mystery, and fear. For instance, Hermes had competition from the goddess Hecate for travelers' attention. She presided over magic and spells and could be seen dancing with her soul-seeking ghosts and hellhounds at crossroads on moonlight nights. On the last day of each month, travelers placated her entourage by leaving meals and black pets at appropriate crossroads. Later, crossroads in eastern Europe were well-known lurking spots for vampires. Another European crossroad custom was to bury suicide victims there, staked through the heart to prevent subsequent walking. And, of course, crossroads were frequently chosen as sites for hanging and then burying criminals. "Dirty work at the crossroads" was thus a phrase of some significance.

Roman Roads—A Pinnacle Is Reached

With the possible exception of the Chinese, the roadmaking efforts described above pale into insignificance beside the roads, bridges, and tunnels of the Romans. These were a remarkable achievement, providing travel times across Europe, Asia Minor, and northern Africa that were not to be appreciably bettered until the coming of the train two millennia later.

As were the Persians before them, the Romans were very conscious of the military, economic, and administrative advantages of a good road system. Each expansion of the empire was consolidated by an associated expansion of its road network. The process allowed the Romans "to establish and maintain the most durable empire in European history" (Casson 1974). To the Romans, their empire covered the entire civilized world, so a natural corollary was that "all roads lead to Rome." In an act indicative of the key role that the roads played within the empire, Julius Caesar granted himself the old title Curatores Viarum, Director of the Great Roads.

The motivational issues behind Roman roads are explored a little further in the next chapter. In the remainder of this discussion the emphasis will be on their engineering basis, the key to which was a recognition that the fundamentals of sound road construction were good drainage, good material, and good workmanship.

The Romans learned some of their roadmaking from the Greeks (lime-cement and masonry), Etruscans (cement), Carthaginians (pavements), and Egyptians (surveying). To produce a stronger and more durable mortar, they used lime-cement from southern Italy in about 300 B.C. and natural cements in the form of pozzolan, a volcanic dust from Pozzuoli, in 200 B.C. The Romans then went a significant stage further by adding gravel to the mortar to make a concrete. This was a major Roman engineering innovation, which, after their departure in A.D. 500, was to be lost for over a millennium. Indeed, their excellence in pursuing engineering fundamentals was substantially aided by their innovative use of cement and concrete.

In the millennia before the Romans, the Minoans and then the Carthaginians introduced the concept of manufacturing an artificial road pavement, rather than merely improving the existing surface. More recently, the Etruscans had paved the streets of Vetulonia near Grosseto by 400 B.C., and so there was a local paving tradition on which the Romans could draw (most of Rome was paved by 170 B.C.). The main source on the input of Carthaginian pavement technology is Saint Isidorus, who was born in Spain of a Carthaginian family in A.D. 540. He wrote in his *Etymologiae* that pre-Roman Carthaginians "consolidated their roads with stones and flints, knit together with sand, as though it were fastened by masonry to the surface of the earth." At best his knowledge was secondhand.

Rome began building arterial roads in 334 B.C. when it constructed the Via Latina in order to provide an effective administrative link with its newly acquired colony of Calvi, which is near Capua and 200 km from Rome. The colony quickly grew in importance and, commencing in 312 B.C., Rome built the first of its major roads from its Capena gate in the ancient Servian wall via the Appia gate in the Aurelian wall, past the Albani Hills and then to Capua, following the approximate line of an earlier earthen trackway to the east of the Via Latina. It was named the Appian Way—Via Appia—after the censor Appius Claudius Caecus, who was responsible for this stage of its construction.

The first stage of the Appian Way had a gravel surface and was about 6 m wide, permitting two carriages abreast. Stone paving was added between 295 and 123 B.C. By 244 B.C. the way had been extended south to the port of Brindisi to service ships traveling to Asia. Writing in about 50 B.C. in his *Universal History* the historian Diodorus Siculus noted that making the way had been so expensive that it had exhausted the public purse. It was first called the Queen of Roads by the famous Roman poet Publius Statius in his *Silvae* in about A.D. 70. Further reconstruction was undertaken by Hadrian around A.D. 130, and the way was still in routine use in the sixth century. Significant lengths of the Appian Way survive today in their original form.

Such major routes were called *viae publicae* and, if paved with stone, *viae munitae*. Even these were predominantly used by foot and hoof traffic and only incidentally by wheeled commercial vehicles. To be called *via*, a way had to be about 5 m wide and thus able to permit two vehicles to pass. To handle heavy traffic on

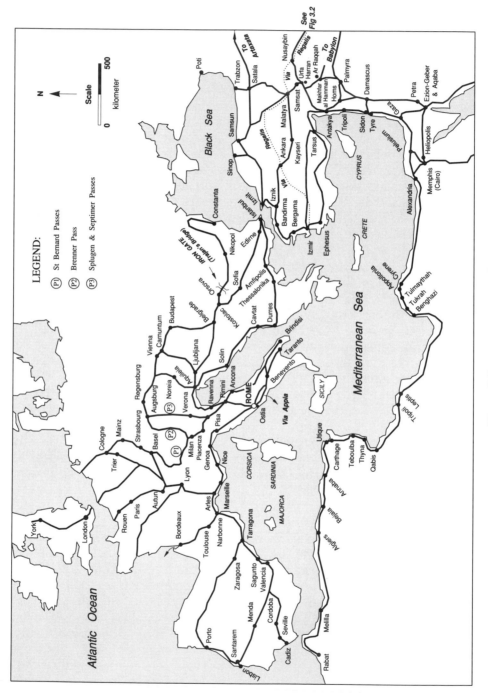

Figure 3.3. Major Roman roads.

the Via Portuensis between Rome and its port of Ostia, the Romans introduced the world's first dual (or divided) carriageway road. The median strip between the two carriageways was brick paved and catered to foot traffic.

The straight and smoothly surfaced Roman roads were more akin to a modern rail network than to the preexisting pathways and trackways (Hindley 1971), although this comparison ignores the somewhat steeper gradients of up to 15 percent found on many Roman roads. The system steadily developed, and at its peak in A.D. 200 involved 80 Mm of first-class roads—they completely ringed the Mediterranean and covered Asia Minor and Europe southwest of a line joining Istanbul, Vienna, Cologne, and Edinburgh (fig. 3.3). Perhaps the greatest of the roads in terms of construction achievement was the 2 Mm Via Nerva, which crossed North Africa from the Straits of Gibraltar via Carthage to the Egyptian capital of Alexandria. By comparison, the Chinese system grew to a peak of 40 Mm in about A.D. 100. Needham (1971) believes that the two systems were comparable when the land area served is considered.

In Britain alone the Romans constructed 12 to 15 Mm of good quality road during three hundred years of active construction following their successful invasion in A.D. 43; that is, they built about a kilometer a week on average and a kilometer every two days at their peak (Bagshawe 1979). Eight of their nine major roads radiated out from London, with the most important road to the west passing within a kilometer of the current location of the British Transport and Road Research Laboratory at Crowthorne, where parts are still used as local roads known as the Devil's Highway (Margary 1955).

For most of the Roman era, the army was responsible for road construction and maintenance, although the work force was usually composed of slaves and convicts. Roads were often funded by public subscription and by dignitaries wishing to have their names carved on mileposts. Caesar also raised funds for roadmaking by selling slaves and melting statues made in his honor. Toward the end of the Roman Empire, construction ceased and the heavy burden of maintenance was placed on private citizens. Indeed, maintenance funding was a matter of continuing and increasing contention. Emperor Augustus decided that the funds were to be obtained from the country through which the road passed. However, the tax was very unpopular and imperial subsidies were often needed.

Major Roman roads were always constructed on a natural formation excavated to firm material or strengthened with wooden piles, and then formed to shape. An essential feature of the roads outside rocky regions was that they were bordered on both sides by carefully constructed longitudinal drains. Where the roads served a military function, the roadside was cleared of vegetation for 60 m on either side to prevent attack by hidden archers.

The next step was the construction of the *agger,* an earthen structure raised up to a meter above ground level and usually built from material excavated during construction of the longitudinal drains. The agger was typically about 15 m wide, made up of two 5 m wide edgestrips on either side of a 5 m pavement. In this respect the cross section is very similar to a modern road design; the edgestrips are equivalent to modern road shoulders (Lay 1990). The agger contributed greatly to moisture control in the pavement and, not accidentally, made the roadway a readily defended position.

The agger was topped with a sand leveling course. The pavement structure on

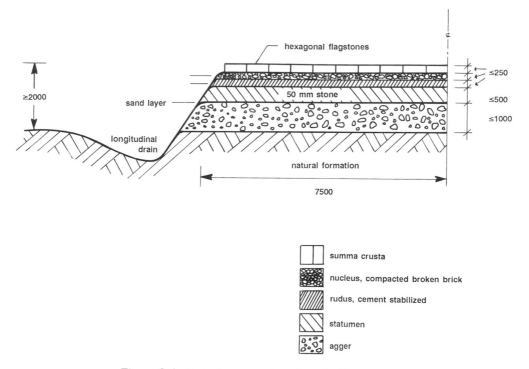

Figure 3.4. Typical cross section of a major Roman road.

top of the sand varied greatly. In the more elaborate constructions, such as that shown in figure 3.4, the structure could be another meter thick, leading to a surface up to 2 m above the surrounding countryside.[7] However, such heights were not common. Indeed, a variety of pavement structures were used, depending on local conditions.

As illustrated in figure 3.4, the main structure began with a *statumen* course up to 500 mm thick, depending on soil conditions, and composed of carefully placed flat stones at least 50 mm in size.[8] This was followed by a cement-stabilized or mortared *rudus* course up to 250 mm thick composed of smaller stones. Next came another mortared layer, or *nucleus* course, up to 250 mm thick, this time of even smaller broken stones, brick, or similar material, well compacted into place. The statumen, nucleus, and rudus courses were often omitted. Finally, if the traffic was expected to be heavy there would be a *summa crusta,* or *pavimentum,* surface course of large 600 mm by 250 mm thick, carefully fitted hexagonal flagstones. The pavement was sloped from one side to the other (*cross fall*) to aid drainage. Roman roads were relatively massive structures by modern standards, as a comparison between figure 3.4 and the later figure 3.8b will show. Indeed, they have often been critically described as "horizontal walls." Certainly, the engineering skill lay in their construction, rather than in their design.

The road structure so obtained was extremely strong and stiff and lasted up to a hundred years before major reconstruction was needed. However, the flagstone summa crusta surfaces were prone to movement, due to traffic and changing temperature and moisture levels, and required continual maintenance if they were to provide a suitable surface for wheeled vehicles.

Typical Roman road alignments consisted of long, straight lengths of road with direction changes located at saddles in high ridges or other convenient sighting points. The Romans preferred sidehill rather than ridge or valley-floor routes and thus frequently followed the older ridgeways and summerways. Cuttings were avoided when possible, as the road in figure 3.5 climbing across steep contours shows. In addition, not all Roman roads followed the generous cross-sectional geometry described above; the width was usually reduced in difficult terrain, as figure 3.5 also illustrates.

Wider pavement widths were used on curves to accommodate the large turning circles needed by relatively primitive Roman wagons, which rarely used the swiveling, steerable foreaxles adopted by the more mobile barbarians. Indeed, the Romans predominantly used carts rather than wagons for routine travel, restricting the latter to either heavy haulage or to ceremonial use. Perhaps each culture adapted its vehicles to suit its infrastructure.

There was thus some technical rigidity in the Roman approach to vehicles, road alignment, and the interaction between the two. In addition, horses of Roman times were small and lacked the hauling power of their modern counterparts, and the Romans often restricted wheel loads to 250 kg. All this generally detracted from the usefulness of Roman vehicles, and their freight movements were predominantly by water rather than by road. Indeed, Casson's (1974) view of road as opposed to sea travel in Roman times was that

> to travel by land was more time-consuming than by water and infinitely more tiring but . . . there were compensations. For one, storms were rarely a matter of life or death. For another, the season of the year did not make a difference . . . only periods of heavy snowfall brought it to a complete halt. A trip by land might involve more baggage and the like, the traveler probably had to have more changes of clothing, as well as special wear adapted to the rigors of the road.

Roads in Decline

The Roman Empire began contracting from about A.D. 300, and its decline became an obvious reality in 401 when Alaric's Visigoths launched the first successful invasion of Roman Italy and in 410 when they became the first army in eight hundred years to capture Rome. Rome now had neither the need nor the resources to maintain a far-flung administrative machine and its supportive road system.

The Old Order Crumbles

As the empire slowly crumbled, the various replacement regimes saw no merit in the marvelous road system that formed part of their booty. A number of factors contributed to this attitude.

1. The new communities did not have the resources needed to maintain the roads and bridges they had inherited. Furthermore, they had little need for

Figure 3.5. A steep, but smoothly graded, narrow Roman road in the hilly Bozbergs near Effingen in Switzerland's Aargau canton. Note the well-developed wheel ruts at a gauge of about 1.2 m. *Photo with the kind permission of the Aargau Kantonsarchaeologie.*

long-distance travel, no desire for urban life, and no central administration to be serviced by a road network (Whitelock 1952).

2. A common defense strategy was to restrict plunderers, invaders, and envious neighbors by deliberately closing any through roads (O'Keeffe 1973). The philosophy was widespread; for example, between 1610 and 1868 it was illegal to build any bridge that would link Japanese feudal cities.

3. Many of the new communities were made up of subsistence farmers who lived in river valley settlements well removed from the upland roads of the Romans.

4. There was widespread belief that riding in wheeled vehicles was beneath male dignity (Borth 1969). This was particularly so for knights. Recall how Lancelot would only enter a carriage when it was obvious that there was no other way to save Queen Guinevere.

5. Wheeled vehicles were rare, most personal land travel had reverted back to foot, high-value light freight was carried on the backs of unshod pack animals, and heavy freight stayed on the water. Foot traffic preferred to travel on firm ground rather than on hard stone surfaces, and so travel needs were better met by bridle paths and droveways (Syme 1952).

6. The stone from the many disused Roman roads was quarried for domestic and farm use.

Given this situation, it is not surprising that local communities often used the Roman roads more as property or parish boundaries than as a means of travel. For example, Watling Street was nominated in the A.D. 878 Treaty of Wedmore as the boundary between King Alfred's Anglo-Saxon England and the Danelaw regions of the northeast.[9]

Consequently, both the Roman road system and the skills needed to build, maintain, and operate it, decayed and then largely disappeared for over a millennium.[10] The system did not deteriorate as much in the eastern parts of the old empire, where the Byzantines maintained a commitment, at least in principle, to road maintenance and continued to operate a post system until the eleventh century (Lopez 1956). In the western half, only a few parts remained in service. There are records of Roman roads in use in France in A.D. 650 and in Germany and England in the eleventh century. King Harold II sped to repel the Norman invaders of England by marching his army at 80 km/day along the Roman roads from York to London and thence to Hastings.

However, the origins of the Roman roads were soon forgotten, and when Geoffrey of Monmouth wrote his *History of the British Kings* in 1147 he ascribed the construction of Watling Street, Fosse Way, and Icknield Way to a legendary English king called Berlinus. In continental Europe a similar belief often held that the Roman roads were constructed by Queen Brunhilda of Austrasia, who died in A.D. 614. However, the Roman roads were clearly not forgotten in official France. Philip II reconstructed one in 1222, and in 1280 Judge Philippe Beaumanoir defined Roman roads in law as the fifth and greatest class of French paths and roads.

The Anglo-Saxons called the Roman roads they inherited "streets," which came from the Latin word *strata* and meant "constructed" in both languages. In the Danelaw regions, the Scandinavian influence favored the word *gate,* and hence *gait* for movement. Since paved roads were only to be found in towns, *street* came to be used for a town road and no longer has its original arterial meaning. Nevertheless, its formal impact was not strong, and early English records and legal documents used either the ancient word *way* or the Latin word *via*.

The first glimmering of European road development during the millennium after the Romans was an effort by Charlemagne (771–814) to construct roads in France and place them under central control. He also placed some emphasis on reconstructing key Roman bridges. Charlemagne was the first titular head of the Holy Roman Empire and had a vision of a network of roads maintaining his Franks as the dominant members of the new empire. Nevertheless, his prime practical objectives were to ease the journeys undertaken by his imperial court and to aid his numerous military expeditions. To do all this, he appointed officials with the sole task of organizing road construction and—in the face of insufficient forced labor from his conquests—he placed an obligation on all citizens to work on road or bridge maintenance. This was a form of the *corvée* system to be discussed in the next chapter. Some roads were built for religious purposes, such as the road Saint William of Toulouse built in 806 to provide access to his monastery in the Cévennes.

In A.D. 850 the streets of Córdoba, then the wealthy showpiece of Saracen Spain and by far the largest city in Europe, were extensively paved by its fourth Moorish caliph, Abdorrahman II.[11] Abdorrahman's roadmaking activities did not extend to intercity roads because, with the caliphates controlling much of the Mediterranean region, most freight was moved by sea. Land transport was largely restricted to pack animals using narrow footpaths.

The Norsemen were also dominant during this period; indeed, the most significant new road was their ninth-century Varangian Road, which ran south from the Baltic to Istanbul and there connected with the Silk Road trade routes to the east (see fig. 3.1). Varangian meant Viking. The road was also called the "Water Road from the Varangians to the Greeks." The key towns on this ninth-century road were Kiev and Novgorod (fig. 3.6; Rose 1952a). Travel was probably more by river (Dnieper, Lovatt, Volkhov) than by land (Jordan 1985). Indeed, in his famous account of a journey from Baghdad through Russia to Sweden in A.D. 922, the Arab Ahmad Ibn Fadlan, traveling the eastern Caspian Sea and Volga route through Russia itself, spent twenty-two days in a Viking riverboat and forty-seven days on horseback (Chrichton 1976). The Arabs were adept travelers, and an Arab geographer of the time, Ibn Khurdadhbih, produced a comprehensive *Book of Roads and Provinces*. A key aspect of the route was that it allowed Europe to avoid the Moorish domination of the Mediterranean.

An important road relic of the times is preserved at Novgorod, which had extensively paved its streets with 4 m wide timber pavements, using flat-sided boards placed on logs to give a good traffic surface. Some twenty-eight layers of logs have been counted in one such road discovered during excavations at the intersection of the former Main and Serf streets. Each layer was added as the road settled into the soft ground and must have been placed at about twenty-year intervals over the centuries following the opening of the Varangian Road (Mongait 1961; Jordan 1985).

A Revival in Travel

European communities at the beginning of the second millennium after Christ were quite self-contained, leading one commentator to note that "the towns of the Middle Ages were like islands. . . . Their relations were chiefly with other towns"

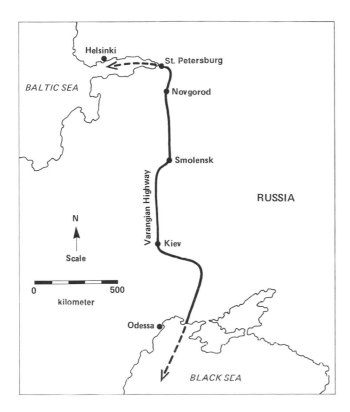

Figure 3.6. Route of the ninth-century Varangian Highway. As explained in the text, much of the highway was a boat route using such rivers as the Volkhov and the Dnieper.

(Haskins 1929). The lives of ordinary folk were governed by the requirements of their manor. Travel was a local activity based around the home and village, and long-distance travel was very much the exception rather than the rule. Special laws called *piepowder*—from the French *pied poudre,* meaning dusty feet—arose to regulate long-distance travelers, who were seen as special people outside the laws of any particular manor. Travelers were also "regarded as unfortunates worthy of protection and assistance" (Hartmann 1927). Indeed, the word *travel* comes from the Latin *travail,* meaning "work" or "hardship."

Nevertheless, there was a gradual increase in travel as power once again began to centralize (Rose 1952a). Kings and popes needed to maintain regular contact with their outlying lords and bishops. One of the first pressures for a revival in roads came when the rise of religious fervor after 1000 led to demands for more churches and cathedrals and, consequently, for building stone to be brought from distant quarries (Leighton 1972). At the end of the eleventh century, religious fervor also led Godfrey of Bouillon to build a road from Izmit to Iznik to aid the First Crusade (see fig. 3.3). In England, the coming of the Normans in 1066 had a major impact on travel between London and the southern seaports.

Some roads were kept open to allow travel to the weekly markets, to the annual round of town and country fairs, and to meet the needs of religious administration and local pilgrimage. Markets and fairs were a Roman innovation in Europe; they disappeared during the sixth and seventh centuries and were then revived by Charlemagne at the end of the eighth century. They gradually grew in importance and had widespread official recognition, sanction, and protection. Most trade between countries and communities occurred at such fairs and, during the eleventh and twelfth centuries, many of the previously small country fairs grew into major trading events. Peddlers and packhorses serviced the communities unable to attend.

The linking together of town to adjacent town gradually produced an overall road network. The consequences of this randomness can be seen in England, where the major roads certainly did not focus on London, as evidenced by the heavy use made of the Fosse and Icknield ways (see fig. 1.1; Stenton 1936).

Henry I built one of the few new major English roads of the millennium when military needs in Wales forced the construction of a road over Wenlock Edge in Shropshire in 1102. The Welsh problem continued, and in 1277 and 1283 the English widened and improved the key roads and passes into Wales for military reasons (Stenton 1951).

Windmills for processing grain began to appear in Europe at the end of the twelfth century. Since their particular requirements meant that they frequently had to be located away from established villages yet still be accessible by carts and wagons, they required significant new local road construction.

The revival of traffic became more noticeable by the thirteenth century and was due to five major factors (Forbes 1938; White 1962):

1. From about 1100, plows began to be pulled by horses rather than by oxen. This gradually led to the wider availability and better harnessing of sturdy horses and hence to their increased use for hauling a growing number of carts and wagons. Their iron-shod hooves and cart wheels had a devastating effect on the many inadequate road surfaces they encountered.
2. There was a movement, aided by the first factor, from hamlets to villages and subsequently to large towns and cities.
3. The new towns and cities began to take a more dominant role as a strong central legal and administrative system reappeared. The consequences of this move are explored further in the next chapter.
4. The Crusades had created a resurgence in international travel and awareness.
5. Government policy began to encourage the expansion of trade and manufacture.

The eleventh-century acceptance of the tenet that every dwelling and workshop in a town should have direct street access resulted in a focus on local road and street construction (Jackson 1980). Nevertheless, the first paving of a Parisian street occurred for quite different reasons. In 1184 the twenty-year-old Philip II, while looking out the window of his tower at the Louvre, commented to his physician that the stink that offended him was caused by carts stirring up the putrid street mud. Indeed, a common saying about glues was that they "stuck like the mud of Paris." Philip decided to act and issued an edict requiring all the streets and marketplaces of the town to be paved with strong and durable stone. Only the two key crossroads were actually constructed as a result of the edict, but story has it that he considered his efforts so successful that he formally changed the name of the city from Lutetia to Paris (Moaligou 1982; Rose 1952a). Lutetia was jokingly said to mean "the muddy town"—a pun on the Latin word *lutum,* meaning "mud." The name Paris—the home of the Parisi people—had been in wide use since A.D. 300, but Philip had also reorganized the city into a single unit and so his formal naming had some significance. In addition, there had been a dramatic jump in population from 20,000 in A.D. 1000 to 110,000 in A.D. 1200.

As a gauge of the depth of the common urban street problem, in 1190 a storm blew four pillars from St. Mary-le-Bow Church in Cheapside, London, onto the street. The 8 m long columns sank into the quagmire of the road, leaving only a meter protruding above the surface. Indeed, writers often described such roads as *quags*.

It was normal to empty the contents of chamber pots into the streets. Parisian practice was to require the emptier to call, "Watch out for the water" three times before tossing. However, in Edinburgh the obligation was on the traveler to shout, "Don't throw" to warn of his presence (Rose 1952a). In ancient Rome the call was "stand out of the way." The practice was banned completely in Paris in 1395 and again in 1502. To circumvent being detected, Parisians developed window boxes of flowers to camouflage their outpourings. Subsequently, these too were banned.

Within the increasingly large and prosperous towns, there arose a growing demand for street paving. Most major cities had some paved areas by the end of the thirteenth century, and paving became a reputable artisan activity. The first known use of the word *pavior* was in 1282. In 1302, London records note that four men called *paviors* were "sworn to make the pavements throughout . . . the city . . . in the manner most commodious for the public and according to the ordinance of old approved" (Malcolm 1934). They would have come from the ranks of the stonemasons and formed one of the early artisan guilds. The trade of paveur was first mentioned in 1397 by Charles VI of France in letters patent (Moaligou 1982).

In 1479 a London document referred to paviors as belonging to the "pavyours" guild (Malcolm 1934). Apprentices are mentioned from 1502. The trade was well organized, with master sworn paviors and companion paviors (juniors). One of the requirements of the London craft was that no pavior could employ more than two laborers. German paviors became famous for the one-legged stools on which they sat to perform their task. The profession was not without its blemishes. Some members created business by surreptitiously destroying existing pavements, a tactic that was sometimes abated by making it a capital offense.

The first formal road construction and maintenance contracts were possibly introduced in Paris in about 1570. Certainly in 1575 a contract was signed between the city of Paris and Master Paviors du Clondit and Eveque to repair the surface of a road from Porte Saint Martin to beyond Yblon Bridge. As a matter of some practical necessity, the low-lying streets of Amsterdam were all paved by 1650.

Right of Way

The post-Roman roads were not so much usable pavements as corridors in which the right of public passage applied. This tradition existed through much of Europe and was found in the laws of such immediate successors to the Romans as the Visigoths, Burgundians, and Bavarians (Lopez 1956). Its subsequent development and codification will be pursued herein using English law—from which many countries derive precedent—as our example.

The first great English law text was Ranulf de Glanville's *Tractatus,* which was written in the late twelfth century and which declared that the king's rights on a highway were as absolute as were his rights in the royal demesne (Hall 1965). Glanville said that there was an unjustifiable encroachment on the property of the king (a *purpresture*) when anyone was guilty of "obstructing public ways. . . . When anyone is convicted by a jury of having made such a purpresture, the whole of the fee which he holds of the lord king shall be at the king's mercy, and he shall restore his encroachment." A related concept was that of the king's peace, which initially applied to "royal ways" and subsequently to all highways. Its effect was that any attack on a traveler on such a way was treated as an attack on the king himself (Stenton 1951). Such a democratic practice was a curious anachronism in a feudal society in which many of a citizen's other rights were highly circumscribed. Perhaps the citizen's right of passage rode piggyback on the administrative need for open access to all parts of the country.

In the first systematic treatise on English law, Henry de Bracton wrote in the thirteenth century that the concept of the king's highway was "a sacred thing, and he who has occupied any part thereof by exceeding the boundaries and limits of his land is said to have made an encroachment on the King himself." Although a citizen could own land in freehold, effectively the king could own an overriding right of way, which was available to all members of the public who wished to use it (Beuscher 1956). Any obstruction to a citizen using one of the king's highways was seen by the courts as an act against the king himself. This principle of the subservience of the rights of landowners to the right-of-way needs of travelers was the foundation for the access-control limitations placed on land abutting many of today's highways. It also developed into a formal requirement for abutting landowners to keep the highway unobstructed and was reiterated as recently as 1773 in a British statute (Netherton 1963).

Gradually, the term *king's highway* came to be applied, not only to regal and military roads, but to all roads leading to ports and to markets and towns, and later became synonymous with all public highways. In his famous seventeenth-century legal text, Edward Coke (1628) described "the King's Highway for all men" as a class of roads separate from streets, which he defined as belonging to a city or town or "between neighbours and neighbours." Note that Coke's legal fame rests on his defense of the dominance of the common law over the competing claims of royal prerogative. The concept of right of way persisted strongly and consciously in English law, even after 1688 when Parliament assumed the legislative supremacy previously exercised by the king. Indeed, it was reinforced in a statute of 1691.

The practical situation from at least the twelfth to the seventeenth century was well described by the Webbs in 1913:

> What existed, in fact was not a road but what we might also term an easement—a right of way, enjoyed by the public at large from village to village, along a certain customary course, which, if much frequented, became a beaten track, . . . it was the good passage that constituted the highway—so that if the beaten track [became impassable] the King's subjects might diverge from it, in their right of passage.

In England this right to diverge had been codified in the Statute of Winchester, signed by Edward I in 1285. The countryside was largely unenclosed, so, to protect their field from being tramped on, adjoining farmers had an incentive to keep the original road clear (Bodey 1971). Perhaps more important, once the public gained the right of passage—if only by custom—it was then protected by law. Because travelers would pick their path, looking for firm and unbroken ground, the resulting rights of way came to be of generous width.

The practice still held in 1675, with one writer describing how much ground "is now spoiled and trampled down in all wide roads, where coaches and carts take liberty to pick and choose for their best advantages . . . [they] utterly confound the road in all wide places" (Smiles 1874). It continued to be supported in late eighteenth-century legal judgments (Netherton 1963), and in the early nineteenth century, Abraham Rees's *Cyclopaedia* advised its readers that "if a highway going through a field is out of repair, travellers may justify going out of the track, though there be corn sowed . . . [and] may break down the inclosure, and go over the land adjoining, until a sufficient way be made."

The concept is also embodied in the *Oxford Dictionary*'s wider definition of a road as "the entire way devoted to public travel." Figure 3.7 shows that the width of the traveled way could readily be six lanes. This was about 20 m—or 66 feet or one chain—and did become a common road reservation width. On the other hand, many urban citizens felt that they had a right for their businesses and homes to encroach onto the road space, and it was common for towns to have full-time inspectors checking for such violations.

A review of conditions in medieval times notes "the vitality of the juridical concept of the 'public highway'" (Lopez 1956). This concept was subsequently reinforced by a 1773 statute, which also introduced the first provisions allowing a landowner to be compensated for land taken during road construction. The philosophy of right of passage was reiterated in a famous 1812 ruling by British Lord Chief Justice Edward Ellenborough, who said that "every unauthorised obstruction of a highway to the annoyance of the King's subjects is a nuisance" (Sessions 1976). The phrase was to be well used; at midcentury a subsequent lord chief justice declared: "every use of Her Majesty's highway to the annoyance of Her Majesty's subjects is an abomination. The Queen's highway is not to be used as a stableyard."

The terms *obstruction* and *nuisance* have been interpreted very widely by the courts over the centuries, thus enhancing the concept of the right of passage (Lock, Gelling, and Colquhoun-Kerr 1982). The view has persisted to the current day. For example, in a contemporary (1958–1959) Australian judgment, Justice Windeyer declared that "a highway is a way open to all the King's subjects." The concept of free passage still permeates English law, with the term *highway* commonly embracing all portions of land over which any of the public may lawfully pass but may not lawfully stop. In this legal sense, even a footway may be a highway if it is open to all the public to pass upon it (Lock, Gelling, and Colquhoun-Kerr 1982). A related second road right, which arose in the Germanic Holy Roman Empire, was that of *grundruhr*. By this right, abutting landlords could claim ownership of anything that fell to or stopped on the road (Forbes 1938).

Demand without Supply

Road conditions remained poor. In a famous incident in 1414, the schismatic Pope John XXIII became inextricably bogged on his ill-fated way to the Council of Constance. He suspected darker forces than a saturated pavement and cried out in despair that his predicament was the devil's work. At the end of the fifteenth century, monasteries were in decline or disarray, religious travel had declined markedly, and the religious component of road maintenance had disappeared. It was, however, the beginning of a period of economic growth as increased trade and travel created a new set of transport needs. Growing affluence was seen in an increase in gentlemanly horse riding.

The language was also beginning to change to reflect *movement* rather than *place*. The word *road* came to be used in place of Anglo-Saxon words like *street, way, lane,* and *path*. *Road* probably came from the verb "to ride" and implied a route along which riding, or at least reasonable progress, was possible. The wider use of *street* occurred up to about 1550, and the term *road* came into the literature between 1590 and 1610.

The increased travel demand coupled with decreased maintenance plunged the roads into a sorry state.[12] There are many reports of people drowning in potholes and roadside drains and of horses sinking to their bellies in mud. The hooves of animal herds and, in particular, the various activities of pigs caused major pavement damage. A great deal of deliberate pavement damage also occurred, for the roadway was used to provide good garden soil; rammed clay for building; storage areas for rubbish, muck, coal, sand, or gravel; sites for bonfires; space for pitching market booths; pits for sawing wood; and the opportunity for building encroachment (Borth 1969; Copeland 1968; Salusbury 1948). Abraham Rees's early nineteenth-century *Cyclopaedia* describes street soil as "a most valuable manure—eagerly sought for the purposes of the farm." In the towns, "butchers used the streets as slaughterhouses, and into them threw the offal from their shops; fish dealers poured forth their fishy water . . . [but] the refuse from house and stable prevented the water from flowing freely" (Jackman 1916). Many local folk actively opposed road improvements, because delayed travelers were good sources 'of income. As suggested earlier, however, there was some pressure to keep the roads clear in case travelers exerted their right of passage over adjoining land.

Discussing the year 1685 in his *History of England,* Thomas Macaulay commented:

> On the best lines of communication the ruts were deep, the descents precipi-
> tous and the ways often such as it was hardly possible to distinguish, in the
> dusk, from the unenclosed heath and fen. . . . It was only in fine weather that
> the whole breadth of the road was available for wheeled vehicles. Often only
> a narrow track of firm ground rose above the quagmire—coaches stuck fast
> until a team of cattle could be procured to tug them out of the slough.

Regarding the Holyhead Road (see fig. 1.1), which Thomas Telford was later to rebuild, Macaulay recorded that travel progress was achieved by having carriages disassembled and the components carried on the backs of "stout Welsh peasants." In 1703 a coach carrying the Habsburg emperor Charles VI overturned twelve times during an 80 km journey from London to Petworth. It remained common for wealthy people to hire strong men to walk alongside their coaches to keep them upright and push them through quagmires (Bird 1969). Certainly, many roads were impassable in wet weather.

One commentator writing of roads in 1720 said that they were "more like a retreat of wild beasts and reptiles, than the footsteps of man." London's Mile End

Road was described in 1756 as "a stagnant lake of deep mud from Whitechapel to Stratford" (Webb and Webb [1913] 1963). In the winter of 1745, the British government was nearly overthrown, mainly because of the bad state of the roads (Kirkaldy and Evans 1915). There are stories of loads of timber taking a year to move from an English forest to the coast and of the transport costs then being 90 percent of the cost of the timber (Forbes 1938).

Until well into the eighteenth century it was common to construct and maintain roads by using bundles of brush called *fascines* to provide a coherent mat on which a road surface could be placed and to fill ruts. This was sound practice in poor or swampy ground, where the root system of native vegetation was also used. A French technique involved deliberately planting young willow trees along the roadside in swampy areas. The pliable branches were then bent to form a basket weave across the road. This was covered with sod to produce a natural structure of interwoven roots (Gordon 1836). Even today, road builders in mangrove areas are careful to take advantage of the mangrove root structure (Lay 1985). Unfortunately, the success of these organic methods in particular circumstances led to their far more widespread use than was justified. For instance, repairs were usually made by preferentially filling holes and ruts with rottable organic material rather than with sound stone (Schreiber 1961).

Formal road maintenance was little better. High longitudinal ridges (or *crowns*) were placed on the road surface in the unachievable hope of forcing surface water to run off the road and onto the roadside. The steep profile worsened rather than improved the situation (fig. 3.7). The road degraded under traffic and moisture (as in fig. 3.7b and 3.7c) and then suffered maintenance that consisted of piling the sunken carriageway high with roadside material in a process known as *barreling* because the road looked like a row of sunken barrels (fig. 3.7). This caused all traffic to use the center of the road, leading to extensive rutting (fig. 3.7), which retained water and defeated the original purpose of the crown. The size of the crown in figure 3.7 is not a total exaggeration. In a breath of reason, Robert Phillips in 1737 correctly called these procedures "unnatural processes."

In the eighteenth century some of these roads came to be known as *roof roads* because of their steep cross section (Bird 1969). In a 1778 digest of highway and turnpike laws, John Scott remarked that "the angle in the pantile roof road is often so great as to endanger overturning on the least collision of carriages, and always enough to occasion anxiety to the timorous passenger." A variant of this style was the angular road, which sloped to only one side. Arthur Young's graphic descriptions of English turnpike conditions between 1770 and 1813 (quoted in the next chapter) give further indication of the detrimental consequences of these maintenance practices.

Of course, things were not always this bad. In dry conditions or where local soils were not moisture sensitive, roads could develop by a form of stage construction—with each pass of a vehicle further compacting and strengthening the pavement. Maintenance only required replacing material worn away by the passing traffic. In addition, in wet areas with moisture-sensitive soils, road builders gradually learned to manage conditions a little better. One partially effective form of road construction and maintenance that grew out of the procedures described earlier (fig. 3.7) was the plowman's road. In this form of construction the traveled surface was kept relatively horizontal and raised up to 2 m above the surrounding countryside, in the manner of a Roman road. This was done each spring by using a special plow,

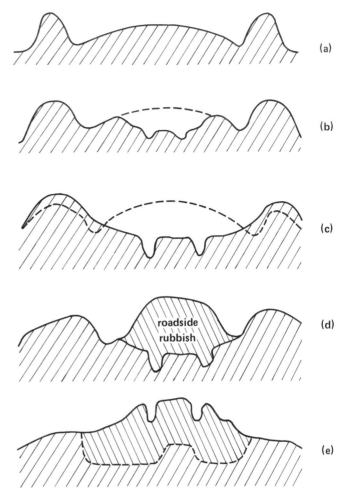

Figure 3.7. Eighteenth-century road degradation and maintenance, based on Law and Clark (1907): *a*, a common convex or barreled road; *b*, the effect of traffic; *c*, an indicted or concave road; *d*, a refurbished indicted road, now a convex road again; *e*, the road in its worst state.

drawn by a team of eight or more horses, to turn over the pavement surface rutted and distorted by winter traffic, throwing furrows toward the center. The furrows were then flattened by harrowing (Bird 1969). The plowing had little technical merit, but it did relevel the road and ensure that the longitudinal drainage ditches were kept cleared and deep (Albert 1972).

Complaints were rife when this form of raised construction and maintenance occurred in the towns. Houses were flooded, ground floors became underground rooms, and dampness was widespread. The alternative of sinking roads to lower their levels by a meter or so became commonplace but brought with it a new set of drainage problems. There were proponents of shaping the surface to produce a concave road (fig. 3.7) to allow fast streams of water to run down the middle of the road and wash away unneeded soil and rubbish (Wilkinson 1934). This was tried without success in Islington in London for a few years after 1717 (Albert 1972) and proposed again by William Jessop in 1808 (Edgeworth 1817). Both travelers and road managers were clearly floundering (Pawson 1977).

There were some early voices of reason and rationality in drainage design.

Thomas Procter's basic philosophy in his road design manual of 1610 was that "water is the [only?] rotting and spoiling of all highways." He emphasized the need to provide deep and effective longitudinal sidedrains, a raised pavement, and a moderate cross fall. The raised pavement was to be made with "much ramming and treading" and covered with gravel or bundles of brush. He advised that carts not be permitted on such roads until the spring. There is little evidence that Procter's sound advice was widely heeded. Procter's book, incidentally, was the first English-language book on roads (Anderson 1932).

By the end of the eighteenth century it had come to be accepted that a convex road surface with only a moderate cross slope was the most efficacious solution for nonurban roads. Urban paving was generally and traditionally sloped toward the center of the street into open longitudinal drains, which were often so wide that they could only be crossed "in the dry" by using small plank bridges. London banned this form of construction in 1662. Up to 1750, Whitechapel road builders made their road surfaces as a series of hills and valleys running across the road in order to encourage transverse drainage. This was called a *waned road*. Another nuisance was that the common repair techniques resulted in a steady raising of street levels relative to house levels, a process that turned rooms into basements and caused major drainage problems for householders (Salusbury 1948).

The craft of street paving remained very labor intensive and reliant on stone masonry procedures, which were precise, time consuming, costly, and not conducive to either extensive paving or broad strategies.

Separate footpaths for pedestrians became a necessity as urban standards rose and many streets became lakes of mud, a condition to which the horse actively contributed. In 1614 London began using sidewalks along the property line. One of the first installations was a high pavement on the Great Road North at Islington, installed to protect churchgoers from being splashed with mud by passing carriages. The carriages were further separated from the pedestrians by the use of a row of vertical posts called *bollards,* a practice first used in 1710. A policy encouraging separate sidewalks was introduced in 1765, although they did not become common until the nineteenth century. Raised sidewalks along the side of the road made it possible to use conventional longitudinal curbside drains. The ride on eighteenth-century London streets would have been rough and noisy and foot passage hazardous and dirty. And yet, the view from the countryside was usually much rosier. As George Colman wrote at the end of the eighteenth century:

> Oh, London is a fine town,
> A very famous city,
> Where all the streets are paved with gold
> And all the maidens pretty.

Harbingers of Change

A marked trend toward centralized development had commenced by the eleventh century, most noticeably as a movement from villages to towns. By the fourteenth century the signs of urban concentration were distinct. Subsequently, for example, the population of London increased from 20,000 in 1400, to 50,000 in

1500, to 220,000 in 1600, and to 1 million in 1800. London did not expand beyond its Roman walls until 1574, but the population doubled in the forty years between 1780 and 1820. Over time this human clustering led to a continuing increase in travel as a consequence of the drawing power of large cities. Indeed, this effect was clearly noticeable from the end of the sixteenth century. Wagon and coach travel increased dramatically between 1550 and 1750 but increasingly came to be limited by the condition of the roads.

Pressures for Change

A large and specific travel demand was created by the need for cities to be supplied with considerable quantities of food from their hinterlands, which in turn provided a strong fifteenth-century incentive for investment in road construction and maintenance (Webb and Webb [1913] 1963). For example, by 1600 London accounted for some 80 percent of English trade, and by 1750 some two thousand cattle and four hundred turkeys a day were being herded down the roads to the London markets. They served well to keep those roads in "a perpetual slough of mud" (Crofts 1967).

Further major travel increases came with the Industrial Revolution in the eighteenth century, military necessity, and a freeing of social restrictions on travel. Certainly by the middle of the eighteenth century the need for good roads and road surfaces was at a crescendo. Fortunately, necessity had remained the mother of invention and a number of talented people were beginning to make significant contributions to road technology, in particular P. M. Jérôme Trésaguet in France and Thomas Telford and John McAdam in Britain.

There were some heralds to their coming: a resurrection of Roman roadmaking practice in France, Eirini d'Eyrinys's discovery of asphalt (chapter 7), George Wade in Scotland, and John Metcalf in England. In 1585 an Italian called Guido Toglietta wrote a perceptive treatise on pavements in which he pointed out how a lighter construction than the Roman type was possible. He recommended a compacted impervious layer of small stones as a surface course on a lime-stabilized base. He also emphasized the importance of drainage. We have already discussed the contributions of Thomas Procter in England in 1610.

However, roadmaking leadership rested with France. In 1622 Nicolas Bergier, a lawyer in Rheims, published a book called *History of the Great Highways of the Roman Empire.* The book gave a detailed technical account of Roman roads and a criticism of existing French practice, which, it said, consisted of "a single layer of stones seated on ordinary sand without other support or foundation than the soil, whatever nature it may be, firm or loose, dry or wet." Unfortunately, Bergier had not heeded Toglietta's wise and perceptive earlier messages. Nevertheless, Bergier's book was widely distributed and its lessons were applied by the new central road administration in France and by various German states, from Swabia to Saxony (Forbes 1958a). It led to a lavish but technically unsuccessful French revival of Roman roadmaking practice (fig. 3.8a).

A third method of pavement construction was introduced by Hubert Gautier in his book *Traité de la construction des chemins,* first published in 1693, the year in which he became the French inspector general of bridges and roads (de Villefosse

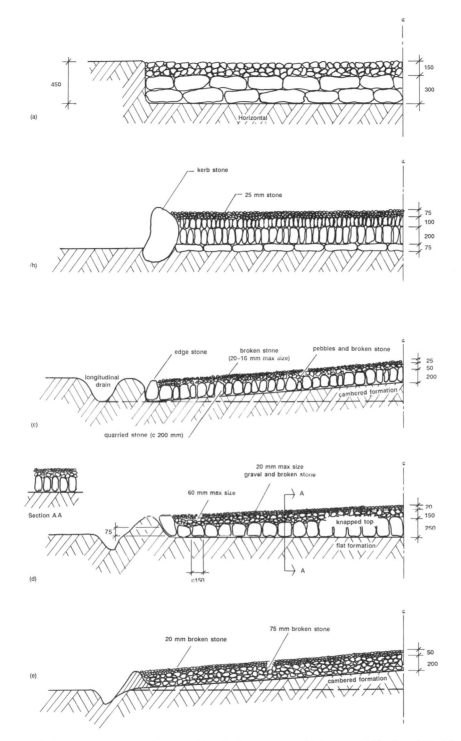

Figure 3.8. Pavement cross sections used by the three masters (Trésaguet, Telford, and McAdam) of the paving revival: *a*, French pavement cross section before Trésaguet and Gautier (note that the base stones are placed on the flat); *b*, Gautier's pavement cross section; *c*, Trésaguet's cross section; *d*, Telford's approach (Telford required the base stones to have their "broadest edge lengthwise across the road." Trésaguet had merely required that these stones be "placed on edge."); *e*, McAdam's method.

1975).[13] His method included placing a substantial layer of large stones to provide a pavement structure and packing this and the surfacing material between parallel, curblike retaining walls (fig. 3.8b).

In 1724 Major General George Wade was appointed commander in chief in Scotland, charged with the task of disarming the rebellious clans and subduing the Highlands (Wilkin 1979). His major technical effort was the building of some 400 km of military road between 1725 and 1737. As there was no shortage of stone in the Highlands and as he had access to ample supplies of gunpowder and unskilled military labor, Wade used the heavy Roman road pavement practices (fig. 3.4). A typical cross section consisted of a base layer of large boulders with broken stone packed into any spaces and an upper layer of at least 600 mm of gravel. Total thicknesses of 2 m were common.

Technically, the results were not good. As Christopher Taylor commented in 1979, the Wade roads "compare very badly with the Roman roads in almost every aspect." In 1802 Telford described them as "in such directions and so inconveniently steep, as to be nearly unfit for the purposes of civil life." They were also roughly surfaced and poorly drained, and Wade is perhaps given more credit than is his due (Wilkin 1979). Nevertheless, the public success of his roads provided a major fillip for a subsequent flurry of English roadmaking (Wheeler 1876).

"Blind Jack" Metcalf—or Metcalfe—would have noted and used some of Wade's roads when he served as a volunteer in Scotland during the 1745 rebellion. John Metcalf was born in 1717 and lost his sight from smallpox at the age of six. He received early motivation to improve England's roads when he walked the 300 km from London to his home near Harrogate in six days. Although this was a long time, it was noticeably shorter than that taken by a coach carrying his friends.

Metcalf subsequently constructed 300 km of roads in northern England from 1753 to 1810. His first major work was 5 km of turnpike built from Harrogate to Boroughbridge in 1765. He emphasized the importance of drainage and developed the concept of surfacing roads with a compacted layer of small, broken stones with sharp edges, rather than with gravel, which is naturally rounded. He also developed with some finesse the old technique of "floating" a road over swampy ground by using a raft made of bundles of brush or heath. Metcalf was clearly a remarkable man; in addition to the above accomplishments, he served in the army in Scotland, built bridges, performed skillfully on the violin, hunted, and played cards with relish.

Along with such individual efforts, there were general signs of an improvement in roadmaking technology throughout the eighteenth century. Widespread common practice had previously been to use branches and other organic matter for pavement construction. Random mixtures of gravel and clay were also common. The records of turnpike trusts show a growing realization of the role of material selection (Albert 1972). For example, screened gravel was being specified in London in the 1730s and "sifted and cleaned" gravel in Kilburn in the 1780s. Screening and sifting gravel produces a more compactible material, and cleaning removes material with poor durability.

Evidence for the development of pavement construction practice also comes from the 1758 edition of the *New American Magazine,* which published observations on road construction gained from "an Irish society interested in the advancement of arts and sciences and based in Dublin." The observations very wisely suggested that roads should comprise a stone foundation covered by a top layer of gravel that "binds into a strong close surface and with an easy fall from the center to the sides into broad ditches" (Labatut and Lane 1950).

In 1768 the famous harbor and lighthouse engineer John Smeaton diverted himself temporarily to roads and built an elevated way across an otherwise impassable valley in Nottinghamshire.

Technological Reactions

Pierre-Marie Jérôme Trésaguet was the first person to bring sound science to roadmaking. A Frenchman, he came from a strong engineering family and worked in Paris on paving matters between 1757 and 1764, directly under the control of two great engineers, Antoine de Chézy and Jean Perronet. For the next eleven years he was chief engineer of Limoges, where he was involved in some 700 km of road construction and had an excellent opportunity to develop a new, better, and cheaper method of road construction.[14] The need was certainly there; contemporary attempts to copy Roman practice were proving costly and ineffective.

Trésaguet was appointed engineer-general of bridges, roads, and municipal works in 1775, and in the same year presented his solution to the roadmaking problem in a memo to the Corps des ponts et chaussées. His memo was subsequently made recommended practice in France and published as a book in 1831. His new pavement drew fairly directly on some of Gautier's ideas but was much thinner and therefore much cheaper.

The Trésaguet pavement employed pieces of quarried stone about 200 mm in size and of a more compact form than in the earlier methods (fig. 3.8c). The pieces were shaped to have at least one flat, narrow side, which was placed on a cambered formation. The pavement structure thus resembled a cobbled road rather than a stack of flat stone slabs. Hammers were then used to wedge smaller pieces of broken stone into the spaces between the larger stones. Next, a further thickness of broken stone was applied to produce a level surface. Finally, the running surface was made with a layer of smaller 25 mm broken stone. The broken-stone running surface performed two functions. First, it permitted a smoother profile to be produced by adjusting the thickness of the layer. Second, it protected the larger stones in the pavement structure from the action of iron wheels and iron-shod hooves. It was far simpler to move a few broken stones than it was to remove and replace a large, rutted stone block. Note the deliberate move away from the naturally rounded gravels to sharp-edged broken stone.

All this sensible and well-engineered structure was placed in a trench to keep the running surface level with the surrounding countryside. This last—very questionable—decision created major drainage problems, which were counteracted by making the surface as impervious as possible, cambering the natural formation, and providing deep side ditches. Trésaguet was certainly aware of the need to keep the natural formation dry, a flaw in earlier modifications of Roman practice.

Trésaguet's important contributions were to recognize the importance of controlling the pressures applied to the underlying natural formation and then to develop ways to assess its ability to resist those pressures. In this he was able to build on Charles Coulomb's recent insights into soil pressure theory. In a paradigmic sense, the change from the existing Roman practice was founded on Trésaguet's realization that a pavement did not have to be a self-supporting structure. In effect, he argued that the natural formation did the supporting and that the pavement had two consequential structural functions: to keep the natural formation dry, and therefore strong, and to protect it from pressures high enough to cause damage.

A common observation of the time was that the use of randomly placed large stones within the pavement structure was inappropriate. They not only provided a bumpy surface but also caused high local contact stresses on the surface of the natural formation, thus resulting in large differential settlements. Trésaguet avoided this by placing large stones over the entire formation, thus ensuring that the formation was subjected to a reasonably uniform and low stress. The system had its problems. The large stones still concentrated the load, mud worked up between them, and they tended to separate laterally. Nevertheless, the method came to be widely used throughout central Europe, Sweden, and Switzerland.

Trésaguet also introduced a system of continuous road maintenance, initiating the roadman method still used today in many parts of the world (Lay 1990). In this technique the roadman was responsible for the maintenance of his length of the road. Writing in 1876, W. H. Wheeler said, "The object of the roadman should be to so manage his road that ruts are never formed."

The science of roadmaking in England at the end of the eighteenth century was not nearly so advanced. Richard Bradley's *Roadmaker's Guide,* published in 1805, stated: "The first principle apparently impressed by the Almighty Power upon all inert matter is a tendency to become proper for the system of vegetation. . . . From this course it becomes the proper system of the roadmaker to select such materials as are most distant from the capacity of supporting vegetables" (Boulnois 1919).

Into this technical vacuum stepped Thomas Telford, a young stonemason from Dumfriesshire in Scotland who was to build his first significant bridge in 1787 when thirty years old and then gain rapid recognition as a leader in British civil engineering. He established an enviable record in constructing bridges, harbors, canals, and buildings and founded the Institution of Civil Engineers in 1820. His ascendance from stonemason to revered engineer reflected the technological optimism and competence that arose from the caldron of the Industrial Revolution.

Telford became involved in roads in 1801 when commissioned by the British government to report on transport measures to halt the depopulation of the Scottish Highlands. His 1802 report led to the establishment of the Commission for Highlands Roads and Bridges, for which he became consultant surveyor and engineer. By 1822 the scheme had produced, under Telford's direct supervision, some 1.5 Mm of road and over a thousand bridges. It continues as the basis of the current road network in northern Scotland. The next chapter describes the major contribution Telford made to the development of the English road network through his direction of the Holyhead Road Commission between 1815 and 1830.

Telford was nicknamed the "Colossus of Roads."[15] In applying his extraordinary skills to roads, he drew significantly from and then extended the work of Trésaguet. He had seen that some of the problems experienced by the French could be avoided by using somewhat more cubical stone blocks, an improvement that came at an added cost. Telford used 300 × 250 × 150 mm partially shaped pitchers, still with a narrow, flat face on the natural formation (fig. 3.8d), but with the other faces closer to vertical than in Trésaguet's method. The longest edge was transverse to the traffic direction, and the joints were broken relative to adjacent transverse courses—

in the manner of conventional brickwork, but with the smallest faces of the pitcher forming the upper and lower surfaces of the course. Broken stone was wedged into the spaces between the slightly tapering near vertical faces to provide the layer with effective lateral restraint.

Telford kept the natural formation level and cambered the upper surface of the blocks, often using masons to knap (shape) the tops of the blocks. Indeed, ex-stonemason Telford's pavement required more masonry work than the Trésaguet system, and he put more emphasis on stone quality than did his predecessors. The natural formation was often called *bottoming,* and the layer of blocks was called *pitching* or *rock bottom.* On top of this rock bottom Telford placed a further 150 mm layer of pieces of stone no bigger than 60 mm in size. Finally, he covered them with a surface course comprised of a mixture of gravel and broken stone.

This arrangement—called Telford pitching—provided a strong and coherent structure able to carry about 10 kg for each millimeter of tire width (Kennerell 1958). In practice and over time, the term Telford pitching came to be used to describe anything between meticulously cubical blocks and pyramidal blocks arranged with points alternatively up or down (see figs. 101 and 102 in the 4th ed. of Coane, Coane, and Coane).

Telford's flat natural formation relied in the first instance on an impervious pavement structure to prevent water from accumulating and thus reducing the strength of the pavement. As further insurance, Telford raised the pavement structure above ground level wherever possible. This was in contradistinction to Trésaguet, who was probably far more constrained by local demands for levels to be maintained. Where the structure could not be raised, Telford drained the area surrounding the roadside. The need to manage water flows and provide good drainage had been widely ignored by previous British road builders. Of course, the Romans had recognized those principles, but their insights had long been lost. Telford's rediscovery of these principles was a major contribution. Conventional practice had been to push material aside and place the road in a trench (fig. 3.8c). Prior to Telford, water was regarded as an unavoidable hazard. For many years after more advanced alternative pavement methods had been developed, the obvious common sense of his method ensured that it continued to be used in areas with poor foundations.

McAdam–Change and Controversy

Despite the contributions of the engineers Trésaguet and Telford, they had not produced the quantum advance necessary to satisfy the growing needs of enhanced and extended travel. That advance was to be supplied by John Loudon McAdam from his distinctly nonengineering background.

The Man and the Times

John Loudon McAdam was born in Ayr in 1756, the youngest of ten children and second son of the impoverished baron of Waterhead (fig. 3.9; Reader 1980). At age fourteen he was sent to the United States to work for his uncle, a prominent merchant, and at eighteen he became the first treasurer of the New York Chamber of Commerce. His role in the United States has been described as that of "an agent for

Figure 3.9. John Loudon McAdam. *An 1825 engraving by Charles Turner, with the permission of the British Museum, Cheylesmore Collection.*

the British Government for the sale of naval prizes, in which position . . . he is said to have made a considerable fortune" (Moore 1876). He naturally supported the losing side in the American Revolution, lost much of his money, and returned to Scotland in 1783.

McAdam became involved with roads when he was made a trustee of the Ayrshire Turnpike in the Scottish Lowlands in 1787. Roadmaking henceforth became his hobby. Over the next seven years, his daily involvement increased and the hobby became a passion.

He signed his name M'Adam. The family name was changed from McGregor for political reasons during James I's reign (Manton 1956). The new name was chosen so no falsehood would be uttered in using it; none could deny a McAdam was indeed the son of Adam (Gregory 1931). Eleven members of the family were to be actively involved in roads: John Loudon McAdam himself; his three sons, James, who was subsequently knighted "for his father's work," William, and John Loudon; his four grandchildren, John, William, Christopher, and James; and close relatives James Shaw, George Pearson, and John Harding (Reader 1980). He died in 1836.

McAdam settled in Bristol in 1801 and became general surveyor for the Bristol Corporation in 1804. He first put his ideas into major practice when he was appointed surveyor of roads for the Bristol Turnpike in 1816. An advocate as well as an engineer, McAdam began actively promulgating his ideas in public parliamentary evidence in 1810, 1819, and 1823. He wrote a booklet called "Remarks (or Observations) on the Present System of Roadmaking," which was published by Longmans and ran through nine editions between 1816 and 1827. His second work, *A Practical Essay on the Scientific Repair and Preservation of Public Roads,* was published in 1819.

The turnpikes of the time had an urgent need for the skills of Telford and McAdam and used those skills well. It was this increasing and unsatisfied need for better road pavements that created the opportunity for McAdam to produce a major paradigmatic shift in road technology. As a consequence of the new methods, for the first time the speed at which Royal Mail coaches could travel was limited, not by the road, but by the ability of the horses. That speed was about 15 km/h.

The Innovation

Let us now look in some detail at McAdam's extraordinary innovation. Concerned by the generally low level of roadmaking knowledge and by the pavement degradation caused by coaches with narrow, iron-tired wheels and relatively high speeds, McAdam had looked for alternatives to the then-current methods of road construction and maintenance. In his empirical observation of many roads, McAdam particularly noted the effectiveness of using small pieces of broken stone, the disruption caused by large pieces of stone, and the efficacy of breaking large stones into smaller pieces. He realized that 250 mm layers of well-compacted, broken, angular pieces of small stone (fig. 3.8e) would provide the same strength and stiffness as, and a better running surface than, a more expensive pavement based on a foundation of carefully made and placed large stone blocks. Further, this course of broken stone would reduce the stresses on the natural formation to an acceptable level, provided the formation was kept relatively dry and drained.

It is possible that McAdam's initial radical departure from the Telfordian use of stone blocks as a base was due to the lack of suitable stone for blockmaking in his part of England, as William Lochhead suggested in a 15 November 1878 letter to *Engineer* (46:358). Smaller pieces of stone would have been used out of necessity, and subsequent observation of their performance would then have shown that the blocks were unnecessary.

Stone size was an important element in the McAdam recipe. For the lower 200 mm thickness of the pavement, the maximum size was commonly 75 mm. However, for the upper 50 mm thick surface course the stone size was limited to 20 mm. Indeed, the stone had to be small enough to fit into the stonebreaker's mouth and was checked by supervisors, who carried in their pockets a set of scales and a stone of the correct mass. A key to the success of the surface was that the 20 mm stone size was much smaller than the 100 mm width of the common iron coach tire.

A McAdam pavement could carry about 18 kg per millimeter of tire width, which was about twice the capacity of a Telford pavement (Kennerell 1958). The strength and stiffness of the course of compacted angular stones came from the structural interlock that developed between individual pieces of stone. The principle is still used in modern highway construction, and since 1820 McAdam's name has been remembered by the term *macadam* used to describe his invention of courses of angular stone held together by natural interlock between the pieces. The angular broken stones were often referred to as "road metal," and pavements made from them have been widely known as macadam or metaled roads. Likewise, the now forgotten term *telford road* was long used to describe Telford's stone-block road shown in figure 3.8d.

McAdam had thus broken away completely from the Trésaguet/Telford pattern. He had observed that a layer of broken angular stones would behave as a coherent mass. No rock bottom layer was needed. By keeping the stones smaller than the tire width, a good running surface could also be made, which could both carry the surface loads and spread them to an acceptably low stress on the natural formation, always assuming that that formation was kept dry. Moreover, the method could often use, as raw material for breaking, the large stones that had beleaguered previous attempts at passage. In this respect, in the 1770s Arthur Young commented of Oxfordshire, "The two great turnpikes which crossed the country . . . were repaired in some places with stones as large as they could be brought from the quarry."

McAdam's technique also avoided many past road-drainage problems because he insisted that the surface of the natural formation be cambered and elevated above the water table. Indeed, good drainage was essential to the success of his method. He was also opposed to the high longitudinal ridges used to ensure surface drainage on most existing roads; he instead preferred the production of a flatter, impervious surface.

The Controversy

Schreiber (1961) has argued that McAdam made little engineering contribution and that his forte was as a tireless publicist and organizer. This view overlooks McAdam's brilliant perception that elaborate layers of carefully placed large stones covered by soil and gravel, as used by successful road builders from the Romans on, could be replaced by a simple, relatively thin homogeneous layer of small broken and angular stones. Even Albert Rose (1952b) in his otherwise excellent summary considered the macadam stone course as only a surface layer and did not acknowledge its role in reducing the stresses applied to the natural formation.

Some argued that others had done it all before, such as the author of a *Westminster Review* article quoted by McAdam's key critic Sir Henry Parnell in his 1838 textbook. Some had, but with little impact or understanding. McAdam had openly based his work on the observation of past successful and unsuccessful practice. Robert Phillips in 1737 presented "A Dissertation Concerning the Present State of the High Roads of England" to the Royal Society, in which many claim he provided the basis for both Telford's and McAdam's methods. However, Phillips still advocated the use of gravel (i.e., rounded stones), and so did not precede McAdam's major concept. McAdam did share with Phillips a reliance on traffic compaction and a recognition of the need to remove loam and other nonmineral matter from the mix.

Metcalf had used a 100 mm surface layer of "small broke and the hardest stone" in the 1750s, and the technique was common in Yorkshire (Pawson 1977). Trésaguet from 1775 on had advocated broken stones as a surface course, and the technique had spread through Europe. Leighton (1972) has suggested that this use of broken stone may have arisen from observations on roads near quarries, where much broken stone would have been available as a byproduct of building-stone production. Similar serendipitous observations later led to the invention of asphalt paving.

A well-known Irish landowner and pedagogue named Richard Edgeworth, who was also famous as a "contriver, inventor, writer, and social reformer" (Clark 1965), published a book in 1817 that some subsequently claimed had documented his successful experience with layers of broken stones (Law and Clark 1907; Jackman 1916).[16] However, Edgeworth actually wrote that "stones, in a common pavement, are usually somewhat oval, from five to seven inches long, and from four to six inches broad. They are laid in parallel rows on the road . . . or alternatively . . . as bricks are laid in a wall." Again, rounded gravel was confused with McAdam's broken stone.

Nevertheless, Edgeworth's modern biographer (Clark 1965) has continued the confusion, wrongly claiming that Edgeworth and not McAdam should have been the household name on the false grounds that "Edgeworth and McAdam followed the same pattern of road construction up to a point, that is, a foundation of large flat stones upon which were laid. . . ." Of course, the very core of McAdam's method was that he did not require any Telfordian layer of large stones.

It has also been said that McAdam drew particularly from the work of John Rennie, William Liester, and James Paterson (Paterson 1825; Moore 1876). According to Paterson, Liester had published a description of a very similar system in about 1790. Paterson, a Scottish road surveyor, documented his work in "A Practical Treatise on the Making and Upholding of Public Roads." Paterson later took some seemingly justified umbrage at the recognition given to McAdam as the inventor rather than as the developer and advocate of the technique (Paterson 1825).

Another writer claimed that the technique had been "used in Sweden and Switzerland long before" (Forbes 1953). In this, Robert Forbes was rather unfairly referring to the adoption of the Trésaguet method in those countries. This method did partially use a layer of broken stones, but only as a supplement and never with the supreme self-reliance of the McAdam approach. Sweden, incidentally, had a reputation in the eighteenth century for the excellence of its roads. Stevenson (1824) noted, "Good rock is generally met with in Sweden; and they spare no pains in breaking it small."

The poet Robert Southey was a good friend of Telford, which presumably led him to write, "Macadamising the streets of London is likely to prove quackadamising." Of course, it proved nothing of the sort. In his 1838 book Henry Parnell takes some pleasure in quoting Benjamin Wingrove—"an eminent and practical road surveyor"—as favoring Telford's method. Wingrove had been dismissed by McAdam from the Leith Turnpike Trust in 1826 for overspending (Albert 1972; Reader 1980).

In summary, there were major differences between the earlier works and that of McAdam, and many commentators did not grasp the difference between rounded gravel and manufactured broken, angular stone. For example, Edward North in 1879 inconsistently wrote, "In Paris the Macadam roads are composed of water worn flint pebbles." Joseph Needham, in his 1971 Chinese epic, incorrectly referred to water-bound macadam produced from compacted rubble and gravel. In 1979 Clay McShane erroneously described macadam as "a technically superior method of using gravel." In many European countries, correctly sized gravels abound naturally in riverbeds, but—like all gravels—they are round-edged, usually smooth, and often of a flat, platelike shape. Gravels were frequently, if unsuccessfully, used on roads and would certainly not have been acceptable to McAdam.

There was much professional debate on the merit of the two methods. North in 1879 wrote: "Macadam seems to have the engineers of France and England mainly on his side. In this country [U.S.], it is believed that Engineers generally prefer a Telford foundation." Certainly the pavements of New York City were then of Telford construction (Moore 1876).

McAdam's refusal to espouse the need for a hard rock-bottom foundation drew much contemporary criticism and scorn (see, e.g., Burgoyne 1861). Others completely misunderstood it. For example, William Lochhead in 1878 and Edward North in 1879 claimed that McAdam copied from the work of Lochhead's father, John, on the Glasgow–Paisley road. Though it is quite possible that McAdam did, in part, the road that William Lochhead describes is of Telford construction.

Like many others, Lochhead failed to understand and too readily dismissed McAdam's dramatic realization that roads did not need to be placed on a heavy base of blocks. McAdam correctly saw the broken stone and dry natural formation as structurally self-sufficient. This view has been consistently misunderstood. Robert Moore, for example, claimed in 1876 that the argument between the two methods had been resolved completely in Telford's favor. Cyril Hartmann wrote in 1927, in a manner that condemns himself far more than McAdam, "The flaw in McAdam's method was that he provided no foundation for his metalled surface. . . . [Telford's] merits suffered neglect in comparison with those of his rival's solely through his misfortune in possessing a less memorable and amusing surname." Such critics would be astonished to find that modern unbound road construction has resolved the issue totally in McAdam's favor (Lay 1990). One of the major economic factors that swung the pendulum was the availability of steam-powered crushing plants, which made broken stone markedly cheaper than placed stone blocks.

The misconceptions have nevertheless continued. Robert Forbes in 1958a wrote that "nowadays [McAdam's] idea of a soft, yielding foundation for the road is not accepted as good practice." This is not true. McAdam's idea is the epitome of good practice, and many successful roads are currently made of broken stone placed on a dry natural formation (Lay 1990). McAdam's "soft, yielding" must be seen as relative to the almost rigid Telfordian foundation. His words did not mean 'soft, yielding' as in rubber or sponge or, more appropriately, as would be attained in a saturated natural formation. No, McAdam insisted that strength and stiffness be achieved by keeping the formation dry rather than by importing material.

How can so many have misinterpreted such a simple and straightforward idea? Telford built Rolls Royce roads suitable for opening up valuable new territory, whereas McAdam built Model T roads suitable for upgrading existing roads (Kennerell 1958). The distinction was not always so clear, and McAdam and Telford each made a few comments on the efficacy of each other's work, particularly on the question of foundations, which Telford preferred strong and McAdam resilient (Parnell 1838; Reader 1980).

The dispute between the McAdam and Telford camps arose from the following events. Telford had supervised the realignment of two lengths of the Holyhead Road at St. Albans near London. The roads were constructed on a natural formation of wet clay and, through a lack of suitable rock, used earth, sand, and gravel as a surfacing. In 1823 Henry Parnell arranged for McAdam's son James to be appointed road surveyor for the trusts involved. James was openly critical of the structure of the road he had inherited. His criticism and his unsuccessful attempts to repair the roads led Parnell in particular to subsequently criticize the McAdams and their method. It does appear that James McAdam's repair method was not good, but neither was the road and there was no basis for then transferring the criticism to the McAdam method, which had not been used.

Construction and Maintenance

Rock breaking had previously been the province of strong men with heavy hammers. In McAdam's method men, women, and children sat on small stools by the wayside using small-handled hammers to break strong stones known as *breakers* to an acceptable size. The broken stones were then placed in piles prior to being spread on the road formation. In 1869 the French introduced an abrasion test to help select suitable stones for macadam pavements (Institution of Civil Engineers 1878).

A key requirement of the process was that no smaller-sized binder material, particularly no soil, was added during this process. This led to macadam being, technically, an open-graded single-sized course. In such a course it is possible to fit smaller pieces of stone into voids between the single-sized pieces of stone (Lay 1990). Presumably, McAdam's strong views against the addition of smaller material arose from observations of their past unsuccessful use. The ninth (1875–1889) edition of the *Encyclopaedia Britannica* despairingly remarked, "The name Macadam often characterizes a road on which all his precepts are disregarded."

As experience with macadam increased, significant interparticle friction under load was observed to abrade the sharp interlocking faces and partly destroy the effectiveness of the open-graded courses. This effect was overcome by introducing high-quality smaller material to fill the voids and produce a well-graded rather than an open-graded course. Indeed, from the turn of the nineteenth century, Edgeworth had been correctly advocating the use of "sharp" sand, which was made up of angular particles. The advent of improved construction equipment made the change increasingly practical.

The material adopted for filling voids in these post-McAdam macadam mixes was commonly called *hogging* or *boggin* and was a mixture of clay and a continuous grading of sand and smaller stone sizes. The clay had crept back in, despite McAdam's advice, and an 1876 report did comment that "the general tendency in Paris is to reduce the use of [filling] material to the lowest quantity possible" (Van Nostrand 1876).

The modified macadam mixes not only lasted longer, but also proved less permeable, more frost-heave resistant, and easier to compact. Their main drawback continued to be that it was easy for horses' hooves to dislodge surface stones. The name macadam stayed with the new mixes and so, ironically, McAdam's name is now often used to describe a product outside his original specification.[1] Indeed, macadam did not become totally effective until bound surface courses were developed (chapter 7).

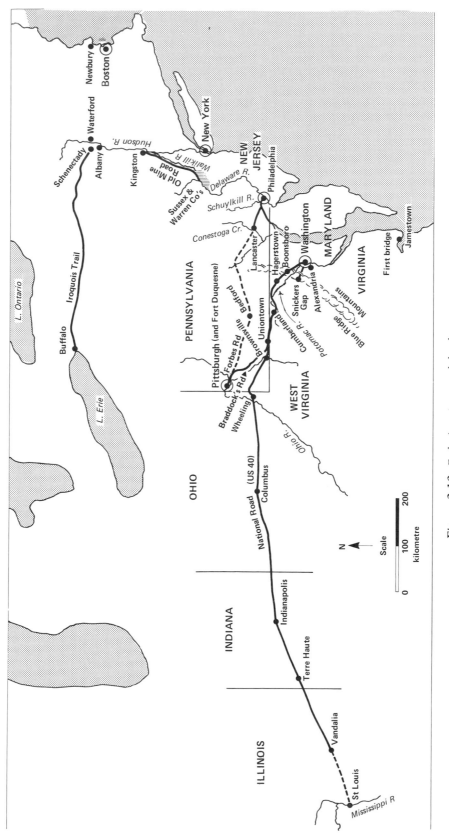

Figure 3.10. Early American road development.

In McAdam's time the broken stones were rammed into place by hand. A hard, smooth traffic surface was produced by extra ramming and by filling the surface voids with sand and stone sweepings from the rock-breaking operation—colloquially referred to as "blinding the open eyes." Cementing tests were developed to measure the effectiveness of various rock types in performing this blinding action (Scientific American 1918). The crushing action of horseshoes and iron tires produced additional fine material to continue to fill any new surface voids. The finer the material the more effective the blinding when wet but the dustier the surface when dry. The resulting material was called traffic-bound or dry-bound macadam. If water was used to aid compaction, then the product was called water-bound macadam. In wet or moist climates, water-bound macadam had an additional advantage in that the surface tension in any retained water acted as an effective glue for small particles.

In the latter half of the nineteenth century a new generation of construction equipment made it easier to obtain, place, and compact large thicknesses of broken stone. This gave macadam distinct cost and construction advantages. Nevertheless, Telford's method was still used for soft-ground construction in the 1950s.

Under heavy traffic, macadam required daily maintenance by roadmen, a process, known initially as "keeping up a road," in which dislodged surface stones were replaced and the surface cleaned (Gillespie 1856). Such treatment would ensure the road lasted about five years before it needed to be *lifted,* a process in which picks were used to merge new stone in with the top 50 mm of the existing surface.

McAdam's methods also depended on a dry natural formation, and Samuel Hughes outlined an effective system of trench drainage that allowed McAdam's objectives to be confidently achieved. The systematic use of subsurface drainage can be dated from Hughes's 1838 book called *Art of the Construction and Repair of Highways.*

Macadam was used on a trial basis in France in 1817 but received little recognition. In 1821 and in 1823 Louis Navier went to England to study suspension bridges but came back as an enthusiastic advocate of McAdam's method of pavement construction.[18] Considerable debate then occurred in France, and the method did not receive a favorable official report until 1830 (Reverdy 1980; Manton 1956). Australia's first piece of macadam was constructed in New South Wales on the road from Prospect to Richmond in 1822, although a letter writer to the *Sydney Gazette* had drawn public attention to the method in 1821 (Lay 1984). Its first use in the United States was in 1823 between Boonsboro and Hagerstown on Maryland's Boonsborough Turnpike, which is now part of U.S. Highway 40 (fig. 3.10; Kirby et al. 1956). Perhaps as a precaution, the stone layer was about 60 percent thicker than McAdam recommended. In 1825 macadam was adopted as the standard for the entire National Road.

Thus, by 1830, McAdam's method had been rapidly transferred to France, Russia, Australia, and North America. In 1834 it was being discussed favorably by H. Bösenburg in the German technical literature. By the 1840s it was in use throughout Europe. In 1870 the United Kingdom had 270 Mm, France 170 Mm, Prussia 90 Mm, and the United States 150 Mm of macadam road.

The late eighteenth and early nineteenth centuries thus saw a period of rapid pavement development, which dragged roads out of a millennium of quagmire. The impact of Telford and McAdam can be seen by the fact that by 1834 English coaches were traveling about 50 percent faster than their French equivalents (Forbes 1958a).

Construction Techniques

Many pavement construction techniques have already been discussed. Early manual roadmaking tools were the pick, hoe, shovel, mattock, wicker basket, handbarrow, long leveling trowel, and large, stumplike hand rammers. A number of these tools can be seen in the frontispiece. The main layout and survey tools were various squares, plane tables, and levels based on either plumb bobs or water-filled tubes (diatropes).

Compaction of the natural formation and of any superimposed pavement courses was mainly by traffic or by hand ramming, a technique in use since at least Roman times. The ramming tool was typically a 1.5 m long, iron-bound 20 kg stump. Hand ramming—the key to a successful road surface—was a trade art. In London in 1846 a foreman named John Pym introduced batteries of men working their ramming tools in unison to time he was beating with a rod (fig. 3.11). The increase in productivity caused a public sensation (Bone 1952). In about A.D. 600 the Chinese had developed a novel and effective compaction technique using large pieces of stone jerked into the air by tightening ropes radially attached to them.

Compaction was sometimes achieved by using the feet of herds of animals, although pigs' trotters had had such a deleterious effect on early Australian roads that pigs were banned from the streets (Lay 1984). Herds of goats were used for dam compaction until at least 1893.

When the situation demanded it, the Romans achieved their desired compaction levels by rolling their roads with large and heavy cylindrical stones drawn by oxen or slaves and hand guided by extended axle shafts. The technique was revived in late medieval times, and a 1725 book by Thiel Leupold called *Theatrum Machi-*

Figure 3.11. Paving blocks in the Strand, on the west side of Temple Bar in London, being ram-compacted to time supplied by a conductor-foreman. *From and with the permission of the* Illustrated London News *Picture Library, 26 April 1851.*

narium contains a drawing of a horse-drawn cast-iron roller. A 3 tonne iron roller was employed on the Boonsborough Turnpike in the United States in 1823 (Rose 1952a, 1952b). There are claims that a Phillip Clay used a 20 tonne horse-drawn roller in England in 1817 (Merdinger 1952), but this use is not noted in Henry Law and Daniel Clark's seemingly comprehensive 1907 review.

John Burgoyne, who was founding president of the Irish Institution of Engineers and chairman of the Irish Board of Works, introduced wooden rollers in Ireland in the 1820s and continued to strongly promote road rolling, particularly in England, into the 1840s (Burgoyne 1861). The 1.5 m diameter cast-iron rollers he favored came via France from Hannover in Prussia. Prussia introduced the water-filled roller in 1853.

Telford and McAdam were among those who regarded compaction by traffic as adequate for their stone courses. Rolling would produce a smoother and less permeable surface, but could also break down the stone surface and hence reduce the life of the pavement. In the 1830s, the French—having accepted macadam—nevertheless also accepted rolling as an integral part of the process. The main problem encountered was that horses could not pull rollers heavy enough to achieve adequate compaction, particular to meet the growing need to compact railway ballast.

Steam traction engines became available in 1842, and the British roller-making firm of Aveling realized that they could be used to replace the underpowered horse. Their first solution was a direct replica of a horse-drawn roller. The logical next step of combining the traction engine and roller into a single unit must have occurred prior to 1850 because the first use of a self-moving roller took place in that year on the avenue de Marigny in Paris (Malo 1866)—perhaps during the cold asphalt trials described in chapter 7.

In 1855 the chief engineer of Calcutta, William Clark, was dismayed by the inappropriateness of the bullocks being used to pull his road rollers. In conjunction with Birmingham engineer William Batho, he specified a new, steam-powered roller (Deacon 1879). Aveling and Batho already had a loose association and subsequently joined together to produce rollers. In 1866 a 20 tonne Aveling roller was successfully used on roadwork in Hyde Park in London. Although a 30 tonne roller (fig. 3.12) was employed in Liverpool in 1867, the weight was found to cause crushing of the stones, thus destroying the macadam interlock (Aveling-Barford Co. 1965). Once a satisfactory rolling weight of about 20 tonne had been established, it was clear to all that some compaction during construction was essential to achieve a satisfactory pavement life under heavy traffic. Thus, the use of steamrollers soon became an inherent part of road construction (Law and Clark 1907; Kennerell 1958).

An Aveling and Porter roller built in 1882 was used in the construction of the M1 motorway in Britain in 1959 before being retired in 1960 after seventy-two years of service (Aveling-Barford Co. 1965). The firm made its last steam-powered roller in 1950. Diesel is now king, but *steamroller*, like *macadam*, has now passed into the language.

The advent of a myriad of motor trucks after the First World War greatly increased the demand for road rolling, for now there was no doubt that pavement courses had to be compacted to maximum density to avoid premature failures. In addition, the new asphalt surfaces needed careful rolling to achieve their desired surface finish. The roller was now a key part of every major road job.

In 1900 John Fitzgerald observed the high compaction levels achieved by a flock of 10,000 sheep running across a scarified, oiled Californian road. This led to

Figure 3.12. The "original" Aveling Steam Roller, built in Liverpool in 1867. In fact, two previous machines had been built, but this one was constructed to an earlier design (Aveling-Barford Co. 1965). *From and with the kind permission of the* Engineer.

the development of the sheepsfoot roller for compacting clayey, cohesive soils (Hoiberg 1965; Hilf 1957). The story needs to be treated with some restraint, for the narrow-tired disc rollers then already in use in the United States would have already moved the technology part of the way toward the sheepsfoot roller (North 1879).

As well as having equipment to compact the newly constructed pavement courses, it is also necessary to have a way of measuring the compaction achieved. The modern compaction test was introduced in 1933 by Ralph Proctor, an employee of the Los Angeles Bureau of Waterworks. His test showed that there was a link between the density of a soil and its moisture content and that, for clayey soils, there was a single (optimum) moisture content at which maximum compaction was achieved. This knowledge allowed road makers to manage the construction process to ensure that the pavements produced were stiff and strong.

Steam was also used to power other construction equipment. There are suggestions of steam devices in use in England prior to 1831, but the first documented application occurred in the United States in 1838 when William Otis produced a steam-powered shovel for railroad work. The next documented use was Eli Blake's jaw-type rock crusher, built in Connecticut in 1858 to produce stone for pavements in New York's Central Park and then for roads in Hartford, Connecticut. The Blake crusher could produce about 5 m^3 of broken rock per hour, which far exceeded the output achieved by breaking by hand. Blake was a lawyer involved in road construction and was related to the inventor of the cotton gin. His innovation was accepted with some reluctance because it was widely believed that hand-broken stones developed better interlock and their production was often a major source of local employment. Hand-broken stone thus remained common until the turn of the century.

Animal power was used for earthworks via horse- and ox-drawn plows, scoops, scrapers, spreaders, leveling drags, and carts (Labatut and Lane 1950). In particular, earthmoving had been by hand-carried baskets (panniers), wheelbarrows, and carts. The wheelbarrow was invented in China in about A.D. 100 and took another millennium to reach Europe (Needham 1971). There was no alternative to loading carts by hand until the development by James Porteous in 1882 of wide, scooplike devices pulled by four horses and able to shift about a cubic meter of material per hour. The capacity of these earthmovers was increased by mechanically lifting the material into carts with special machines pulled by further teams of horses. The modern scraper was first produced by Abijah McCall later in the same decade (Clifford 1977). Trucks capable of depositing their load without stopping were introduced in the early 1900s (Larkin and Wood 1975). Simple blade graders were invented in 1878, and leaning-wheel graders able to counteract the force on the blade, and drawn by teams of mules, horses, or steam traction engines, were first produced by Frederick and J. D. Adams in 1885.

Working in Stockton, California, Benjamin Holt produced a usable steam-powered tractor in 1885, as a prelude to introducing the gasoline-powered crawler tractor with self-laying tracks in 1904. His efforts also led to the founding of the famous Caterpillar Tractor Company. Tracked equipment gained a further impetus at the end of the First World War when tracked military vehicles were converted to civilian use. In 1922 Robert Le Tourneau, also of Stockton, produced the combined scraper and tracked tractor. Self-powered devices arrived in significant numbers in the early 1920s, although tractor-drawn machines were still common into the 1930s. A tractor with a forward blade, the now famous bulldozer, was first used in 1923. Hydraulic controls for the blade were introduced in 1925.

Internal-combustion engines began to play a major role in road construction equipment in 1924. Diesel-powered equipment arrived in 1931 and soon had a major impact on construction practice.

Gunpowders and their kin have been great aids to producing supplies of rock, clearing boulders from the roadway, and making cuttings and tunnels. They appear to have been used by the ancient Greeks and Chinese, but the first recorded mention of their essential ingredient, saltpeter, was in about A.D. 1200. Gunpowder was used in guns early in the fourteenth century, but it does not appear to have been employed for blasting until the seventeenth century. The safety fuse was introduced in 1831 in Cornwall and greatly increased the practicality of using gunpowder for blasting. Thus, the blasting of rock was gunpowder was commonplace when nitroglycerine and nitrocellulose were invented in the 1850s (Gillespie 1856). Within two years, these much superior explosives were in use for quarrying and road clearing. In 1863 Alfred Nobel dramatically increased the effectiveness of nitroglycerine by letting it permeate an absorbent material, and four years later he was conveniently packaging it as dynamite. In 1876 Nobel combined nitroglycerine with gunpowder to produce blasting powder, which subsequently revolutionized hard-rock construction. The steam-powered rock drill was introduced by farmer Simon Ingersoll in 1870 to aid the reconstruction of Fourth Avenue in New York.

Pavement Systems

The difficulty of achieving good grades and good surfaces over long distances was enormous, given the technologies of the time. One particularly appropriate illustration of the problem is that the making of good road surfaces often requires the importation of good stone into an area, but the transport of this stone itself demands a good transport system. Good roads can only be built if good roads already exist.

Telford was the first to tackle this issue in a systematic way, as one of his many contributions to road technology. His pavement achievements have already been discussed, and his magnificent bridges will be described in chapter 8. He was also one of the main instigators of the engineer-contractor relationship and of the use of the engineer as an arbitrator between client and contractor.

In road layout, Telford stressed the importance of minimizing the steepness of a road, rather than devoting all effort to improving its surface. He argued that this was critical to the economics of freight haulage. Whereas McAdam believed that 5 percent was still an easy gradient for trotting horses, Telford conducted trials and then calculated the maximum gradient at which a wagon's rolling resistance was able to counteract gravity. The maximum permissible grade was therefore increased as the roughness of the road increased. This was because the downhill force equals the mass of the vehicle times the slope and is resisted in repose by the friction between wheel and surface. For good roads, the calculations showed that downgrades should not exceed 3 percent (Cron 1976b). However, Telford did not originate the emphasis on 3 percent as a peak gradient; an 1805 Irish act—possibly the first traffic engineering standard—had so restricted the grade of mail coach roads (O'Keeffe 1973).

Downgrades were so emphasized because preventing wagons from running out of control downhill was often more difficult than going uphill. Some inkling of this problem was given in the earlier discussion of Roman vehicles. A number of techniques were adopted to supplement simple braking and avoid catastrophe. One of the more extreme involved tying felled trees to be dragged behind the wagon and placing the payload on top of the tree (Lay 1984).

Apart from the grade, the prime factor determining the payload that could be hauled along a road was the force (T) needed to pull the total mass (M) of wagon and payload. These also depended on the pavement surface friction and were related by a friction factor (T/Mg) that ranged from $\frac{1}{7}$ for loose sand to $\frac{1}{250}$ for smooth stone. Typical values of the time are listed in table 3.1.

On the basis of this understanding, Telford initiated work that led to extensive tables showing the distances over which teams of horse and oxen could pull specified loads along roads of given slope and surface condition. Much of the data came from force (dynamometer) measurements taken on the Holyhead Road by John Macneill, Telford's assistant and fellow Scot. His level-ground data provide much of the basis for table 3.1.[19]

As an illustration of the use of table 3.1, data from it and the previous chapter make it possible to calculate that for dry macadam a well-harnessed nineteenth-century horse could pull loads of (0.05 × 70) = 3 tonne, but only one tonne over badly worn macadam or on a 3 percent slope.

In 1883 Macneill was able to use such data together with haulage capacity and power data to conduct an analysis closely resembling a current-day benefit-cost

TABLE 3.1
Resistance to Iron-tired Wheels of Various Surface Types in Terms of Friction Factors

Surface type	T/Mg
Sand, deep and loose (Coane, Coane, and Coane 1908)	1/7
Dry earth (Aitken 1907), gravel on earth	1/15
Macadam, badly worn or little used (Aitken 1907)	1/20
Broken stone on earth, cobblestones (Baker 1903)	1/35
Solid rubber wheels on reasonable surfaces (Carey 1914)	1/45
Broken stone on paved foundation, sheet asphalt, plank road (Baker 1903)	1/50
Pneumatic-tired vehicles on reasonable surfaces (Carey 1914)	1/60
Well-made pavement, dry macadam (Aitken 1907)	1/70
Brick (Baker 1903)	1/90
Best pavement	1/180
Steel plate or stone trackway (Coane, Coane, and Coane 1908)	1/250
Up gradient of 1/n	1/n

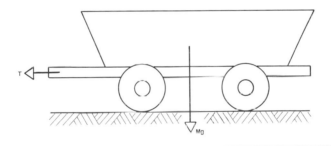

SOURCE: Unless otherwise noted, data are Macneill's.

NOTE: This table relates to the wheel load limits shown in table 2.1. T = tractive force = axle resistance + rolling resistance + gradient; Mg = mass force (weight) of cart and payload; T/Mg = pull needed as a fraction of weight being pulled = friction factor

analysis, in order to assess the economics of Telford's Holyhead Road (Parnell 1838; Cron 1976b). However, it must be said that the principles—if not the procedures—had been established much earlier. For example, U.S. Secretary of the Treasury Albert Gallatin, in his 1808 *Report on Internal Improvements,* clearly enunciated the basis for an economic assessment of roads, including seeing taxes as merely a transfer payment (Newlon et al. 1985). His report was part of a comprehensive plan for a national system of roads and canals, including the National Road (chapter 4). Another major nineteenth-century contribution to transport economics was the development in 1844 in France of pricing based on marginal costs by Arsène Dupuit, chief engineer of bridges and roads (Dupuit 1844; Musgrave 1959). The work was produced for a study of toll bridges.

The wear of roads under traffic became a matter of major debate in the nineteenth century, particularly by the French engineers. The issue was whether the wear was less than linearly, linearly, or more than linearly related to traffic tonnage. Dupuit had, since 1842, argued for a linear relationship, which was confirmed by subsequent work in 1877 (de Saint-Hardouin 1877).

The Americas

In the emphasis to date on European and Asian roads, sight may sometimes be lost of the fifteenth-century achievements of the Incas in South America. The entire Incan road network covered an enormous 23 Mm, some of it through rugged Andean terrain. Sixteenth-century European explorers praised it as superior to that in contemporary Europe.

The Inca Royal Road ran some 5.7 Mm along forbidding heights from Santiago in Chile, north to the capital Cuzco in Peru, and on to Quito in Ecuador, occasionally at an altitude of 3,600 m. It was paralleled in part by a coastal road. The Royal Road was probably the world's longest operating road until well into the nineteenth century. It included suspension bridges, hillside galleries, high embankments, and spectacular series of stone steps. The grades could afford to be steep because the Incas had no wheeled vehicles and some parts were essentially footpaths. Typically, widths were from 4 to 7 m, widening out to 16 m in the northern sector and narrowing to 3 m in difficult terrain (Hyslop 1984).

Roadside lodges were built along the Royal Road at a day's walk apart—typically about 20 km apart, although the terrain sometimes led to a day's journey of only 10 km (Hyslop 1984). There were early reports that portions of the Royal Road were covered with bitumen (Aitken 1907), but subsequent investigations have not substantiated this claim (Spencer, in Hoiberg 1965) and the prospect is not even mentioned by Hyslop (1984). The greatest length of stone paving was about 20 km and, on the rare occasions when it was used, its role was more to control water scour than to improve traveler comfort. The roads were well delineated and sometimes lined by longitudinal sidedrains and rows of planted trees.

The Inca empire could never have been created and sustained without its road system. To appreciate the full significance of these roads it is useful to quote John Hyslop (1984): "To the Incas—the Inca road was a complex administrative, transport and communication system—the roads were an omnipresent symbol of the power and authority of the Inca state." Many Incan roads remain today, and Hyslop describes the network as "prehistoric America's largest contiguous archaeological remain."

Manufactured roads were also found in Yucatan in Central America (Saville 1935). The Spanish came to the Americas in the sixteenth century, and over the next three centuries they created a network of major roads centered on Mexico City, designating each a royal highway, or El Camino Real. Three radiated to the Atlantic Ocean at Veracruz, to the Pacific at Acapulco, and to San Salvador in the south. Three more became the first long-distance roads in the United States, crossing the Rio Grande to reach St. Augustine in Florida via Natchez, Sante Fe in New Mexico via El Paso, and Sonoma in California via San Diego and San Francisco. A later view of this last road is shown in figure 6.2. U.S. 101 now parallels much of its route. Nevertheless, the U.S. portions of these routes were more trails than roads, and the first manufactured road in the United States was probably a 70 km length of road in Florida, running north from St. Augustine to Fort Caroline. It was built by the Spanish in 1565, and the southern end is closely paralleled by Highway 1.

The next North American roads were created by settlers following the extensive Indian trails. A cleared way was built in Nova Scotia by the French explorer Samuel de Champlain. Some minor cobblestone paving using beach stones occurred in Pemaquid, Maine, in 1625 and on Stone Street in New York in 1656. However, roadmaking in the rest of North America was not a matter of high priority for the early colonizers. Not only was it very demanding of resources, but the extensive waterways often provided an adequate transport facility. This situation occurred in many countries, including Canada, Britain, and Sweden. In the United States, significant roadmaking did not begin until about 1720, and no map of the country prior to 1729 shows any roads (Ristow 1961).

The first American highway was probably a 170 km length of road called the Old Mine Road, which was built between 1620 and 1650 to link copper mines in Warren and Sussex counties in New Jersey with a port at Kingston, New York (Wixom 1975). In 1880 it was still said by many to be part of the best route between Philadelphia and Boston and was in use as a road until 1940.

The Dutch linked New York and Philadelphia by the Old Plank Road in 1675, but a regular coach service did not operate over the route until 1756; the journey took three days (Borth 1969). The first postrider traveled between New York and Boston in 1673, in a journey that took two weeks. By 1704 New York and Boston were connected by a bridle path (fig. 3.10; Merdinger 1952). The track became the Boston Post Road and was a major transport route during the Revolutionary War. In fact there were three alternative post roads in operation on the New Haven to Boston link: two passed inland through Hartford and either northeastern Connecticut or Worcester, and a coastal route went via New London and Providence.

The first American road legislation was a 1632 Virginia law describing how roads should be laid out (MacDonald 1928). The oldest extant street is a 600 m section of St. George Street in St. Augustine, Florida, which was built in about 1680 of a concrete made of seashells. Major eighteenth-century American roads are discussed at the beginning of the next chapter.

Another interesting case was the achievement of Peter the Great and his Russian successors between 1722 and 1746 when they built the Prospective or Avenue Road over some 750 km from Moscow to the new town and national capital of St. Petersburg, mainly using corduroy construction (Kennerell 1958). In 1781 Russia began the Siberian Highway from Moscow to Perm, Tobolsk, and Irkutsk (fig. 3.6; Forbes 1958a). During the same period the Russians were developing a form of broken-stone pavement (Merdinger 1952), which in 1797 was first used between Gatchine, just south of St. Petersburg, and the residence of the tsars at Tsarskoe Selo. Twenty years later it was applied to the Moscow–St. Petersburg road. Nevertheless, the roads of nineteenth-century Russia were relatively poor. A review by the U.S. State Department noted that "in some provinces, beyond setting posts every 10 to 15 yards to indicate a highway, little or nothing is done for their maintenance" (Engineering Record 1891b).

Summary

This chapter has traced the development of roads in the period between the invention of the wheel and the beginning of the age of the self-powered vehicle. Throughout this period there was a continual interplay between the need to travel, the means of travel, and the difficulties of supplying an adequate infrastructure for that travel.

The full capabilities of the wheel took some seven millennia to be realized, for those capabilities require relatively smooth and flat surfaces and major power sources. Domesticated animals were largely underpowered for the loads humans wanted to move and the surfaces they were able to construct.

The interplay between road and vehicle can be seen in the activities of the Romans and their less cultured barbarian antagonists. The Romans built a magnificent road infrastructure, but had relatively poor vehicles and riding equipment. The barbarians had no road infrastructure, but good vehicles and riding equipment.

The changes that did occur in vehicle technology in the period covered by this chapter were relatively minor. Once again, system interplay comes into effect, and little change in roadmaking technology can be detected until the eighteenth century, when vehicles began to improve noticeably. Then, by one of those not so coincidental situations that arise throughout the history of technology, a number of major breakthroughs occurred. In this case, of course, the catalyst was the Industrial Revolution. Was this another case of demand and supply, or did the circumstances that gave rise to one also fertilize the other? Later chapters will provide additional opportunities and examples to permit further consideration of this question.

Motives
and
Management

⟶ *Chapter 4*

The Inland Trade of England has been greatly ob-
structed by the exceeding Badness of the Roads
DANIEL DEFOE, *TOURS*, 1726

Some of the major uses of roads were the provision of local access
to food and shelter, as migratory routes to cope with climatic variations, as paths for
religious pilgrimages, and as facilities for interregional trade in such commodities as
flint, tin, and salt. Socioeconomic developments led to the widespread introduction
of wheeled vehicles, which further enhanced travel flows. As society developed,
roads also provided access to employment, education, entertainment, and culture.
In addition, Roman roads had demonstrated that the traveled way provided valuable
administrative, strategic, and control functions.

Military Motives

For most of history, the development of routes from tracks into manufactured
roads has required military motivation. The wheeled military vehicle was developed
in 2500 B.C., and from that time on roads provided efficient transport for attack and
defense and thus a reason for rulers to devote significant resources to their construction and maintenance. In times of peace, such resources were rarely forthcoming,
despite a variety of administrative techniques tried over the millennia, which are
reviewed below.

The Persian Royal Road was the first major example of a road built for military
or administrative reasons, and Roman roads are the most obvious and lasting example of the effectiveness of military intervention and involvement in roadmaking.
The Roman Empire also used roads known as *limes* to define and defend its frontiers,
aided by frequent towers and forts placed along their length. One of the best known
was a portion of the Harran to Ezion-Geber road, which operated against the Persians and Arabs until A.D. 500. The many further examples of the world's military
roads will now be examined.

North America

In more recent times, the first major manufactured road north of the Rio Grande was constructed by the French army between Montreal and Quebec in 1721. Subsequent Canadian military roads, built mainly as a defensive reaction to the consequences of the American Revolution, provided the background to the highway system of modern Ontario and Quebec (Labatut and Lane 1950).

A significant early American manufactured road was also of military origin. From 1743 settlers had used Nemacolin's Indian trail to reach the Ohio River, and in 1753 the route was traversed by a twenty-one-year-old surveyor, the future general and president, George Washington (Merdinger 1952). In 1755 the British army under Major General Edward Braddock used part of this route to build a 3.5 m wide manufactured road over 170 km, from Cumberland in Maryland to Pittsburgh, to aid the capture of Fort Duquesne from the French (see fig. 3.10). Cumberland, located at a river gorge that forms a natural pathway through the Appalachians, was an obvious starting point for Braddock and was subsequently to be near the site of General Washington's military headquarters during the War of Independence. The project used significant lengths of corduroy construction and, although construction of the road was successfully completed, the military mission failed a few kilometers from the fort when a massive ambush destroyed the troop of 1,459 men. Braddock himself was killed (Rose 1953).

Three notable figures in American history were involved in this episode. Washington, by then a lieutenant colonel in the Virginia militia, had advised against building the road, but had accompanied Braddock as a volunteer aide (Rose 1950b). Benjamin Franklin supplied the horses and wagons for the expedition, including the first 150 mass-produced Conestoga wagons. Franklin had surveyed a number of American trails, which subsequently became major roads (Borth 1969); later he was active in recruiting French army engineers to work in the United States. Finally, Daniel Boone, the famous backwoodsman, was a wagoner in the troop. All escaped for further great adventures.

In 1758 General John Forbes established a new Pennsylvanian route from Bedford to Pittsburgh to aid a further attempt to capture the fort. The attempt was successful to the extent that the French burned the fort and left a day before the general arrived. The route was soon known as Forbes Road. Forbes Road is sometimes misleadingly called the Old Glade Road. However the Glade Road, named after the Indian Glade Path, was an alternative route through Mt. Pleasant (Shank 1974). U.S. Highway 30—the Lincoln Highway—now follows much of the route taken by Forbes Road (see fig. 3.10; Rose 1950b). Incidentally, the route from Philadelphia to Pittsburgh used the Nemacolin, Allegheny, and Raystown Indian trading trails for its basic alignment and came to be known as the Pennsylvania Road. The paving of the Philadelphia to Lancaster section of the route was part of the Lancaster Turnpike development in 1793–1795 and will be discussed below.

The federal government mainly used the U.S. Army for its early road construction. Consequently, in many places frontier roads were known locally as military roads, though they were frequently based on existing Indian trails. These roads played a major role in developing the western United States by providing migratory routes and facilitating access to public lands. They were also significant east of the Mississippi; for example, eight territorial roads were built by the army in Michigan, although only one could be justified on military grounds. The army justified the other routes on the basis of speeding mail delivery and encouraging additional sales of public lands. Similar activities occurred in Wisconsin and Milwaukee (Jackson 1982).

Britain

Not surprisingly, military roads also played a major role in Europe. The previous chapter described Wade's technical contribution to Scottish road development between 1726 and 1737, and thus between the two Jacobite rebellions of 1715 and 1745. In early 1724 Major General George Wade was a member of Parliament for Bath and was without a command. He was sent to Scotland to inquire into the situation in the Highlands, particularly the rebellious actions of the Scots, the need to subdue the Highlands, and methods to militarily contain that Jacobite rebelliousness. Wade predicted further rebellion, "partly from the want of roads and bridges." In December 1724 he was appointed commander in chief of Scotland and given a strong charter to build roads to facilitate the movement of troops.

Wade began work in 1725 and built four major roads, three of which were either in, bordering on, or leading to points of military importance in the Highlands (Haldane 1962). After he left in 1737, military roadmaking continued until 1803, when Thomas Telford's commission took over. The term *military road* was officially used in Scotland until 1863, and by then covered some 2 Mm of road. Many remain in use today, such as the A939 from Balmoral to Tomintoul over the Cairngorm Mountains.

The road also played its part in British expansion and colonization. In a lecture to the School of Military Engineering in 1914 Alfred Carey began:

> War, when carried on on the borders of a wild country, almost always leads to permanent improvement in means of communication [Carey then cites the road through the Khyber Pass as an example]. . . . Half the difficulties in the administration of a new country disappear when the road problem is tackled in earnest. After Culloden the policy of cutting throats was superseded by the policy of cutting roads.[1]

The roads did not always assist the British. Work on the U.S. National Road was accelerated by the War of 1812. In 1816 American statesman John Calhoun proclaimed "Let us then bind the republic together with a perfect system of roads." In China, the population was well aware of the strategic role of the centrally controlled road network, and disputatious regional groups cut the roads and destroyed the post houses whenever possible.

France

Just prior to the French Revolution, late eighteenth-century France had some 20 Mm of paved road and 50 Mm of road on which some roadmaking effort had been expended under the leadership of Trésaguet and his predecessors. The stretches of good highway were mainly tools of defense and administration, radiating from

Paris to secondary cities and fortified places (Borth 1969). Strangely, the roads were only lightly trafficked.

The crossing of the European Alps also provides an example of the role of the military in roadmaking. The Romans had used passes at Tenda, Montgenèvre, both St. Bernards, Splugen, Septimer, Brenner, and four to the east (see fig. 3.3).

None were easily mastered and knowledge of their usage often drifted into folklore. The route via Tenda was circuitous, whereas Montgenèvre provided useful links to the Rhône Valley and was used for centuries after the Romans until replaced in the twelfth century by the Mont Cenis pass, which had first opened in the eighth century (Baker 1960). The Great St. Bernard was the most popular pass in the Middle Ages, because it also provided passage into Germany. A narrow road through the Simplon Pass was first mentioned in 1235, but fell into decay, and eventually only foot traffic was possible. Although a packhorse route was revived in the seventeenth century by a Swiss businessman, grades of over 10 percent made the journey difficult and unnerving. An equally perilous packhorse gallery path through the St. Gotthard Pass was opened in 1237, with the completion of the Stiebende chain bridge over the Schollenen Gorge. Its good geographic links soon made it the most popular of the passes. In 1338 the Septimer Pass was opened to very small carts. Given the inadequacy of these tenuous passages, the far longer sea and river route around the Alps was much preferred (Reverdy 1982).

In 1591 the duke of Savoy opened a new packhorse way through the Tenda Pass in order to wage war against the French. The subsequent widening of this way in 1792 provided the first reasonable carriageway through the Alps.

Napoleon saw roads as a key to his military objectives. He adopted many of the Roman attitudes toward roads and gained a reputation as a great road builder. In a memorable project, he required an easier route than either the Tenda Pass or the Great St. Bernard Pass to allow the artillery of his Grande Armée to dominate Italy. Consequently, between 1800 and 1805 he used his troops in the first year and then engineer Nicholas Céard and civilian resources to improve the Roman route in Switzerland by building a new 30 km transalpine road in the Gondo Gorge leading to the Simplon Pass. The design criterion was simply that it had to be possible to take cannons through the pass. Five hundred lives were lost during construction. With twenty-two bridges and seven galleries, it was the first of the major constructed passes through the Alps. The entire route through the pass runs from Brig (altitude 700 m) to Gondo (altitude 855 m), rising in between to an altitude of 2005 m.

Napoleon was responsible for other major roadmaking innovations. He introduced effective provisions for the acquisition by the state of land for roadmaking, and between 1804 and 1812 he spent twice as much on roads as he did on military fortifications (Hindley 1971). He saw the strategic need for roads and brought the French national highway system, *les routes nationales,* to practical fruition. His military efforts promoted a similar awareness in Germany, Austria, and the Low Countries. Indeed, Napoleon had a major impact on European roadmaking.

He favored the ancient Chinese and Roman tripartite road with its three separate carriageways. However, his version consisted of a middle paved with cobblestones for the artillery, one side paved with water-bound gravel for the infantry, and the other side paved with earth for the cavalry. Not surprisingly, few roads of this type were built, although the concept did create a precedent for the later French boulevards (Forbes 1958b). The practice of keeping the pavement for wheeled vehicles and the outer gravel for foot traffic was quite widespread and was embodied, for instance, in a 1780 Danish law relating to main roads into Copenhagen (Kincaid 1986). The military influence on roadmaking was so strong in Denmark, incidentally, that the Danish civil road officials wore military-style uniforms. The multipavement technique was also exported by France to the colonies: for instance, in Java in 1809 Herman Daendels built a great highway along the length of the island with one pavement "kept in prime order" for stagecoaches and a parallel one "for buffaloes and carts" (Bickmore 1865).

A convenient place for new boulevards in many cities was along the site of old fortifications and city walls. Henry IV began this pattern in Paris in the early seventeenth century, and by 1789 the city had some 30 km of ring boulevards, mostly 40 m wide (Stave 1981). Further ring road construction commenced after the First World War. In more recent times, the Paris ring road—the Périphérique freeway—followed the old city walls and, in remembrance of its past, the interchanges are named after the old gates. The word *boulevard* is from a Scandinavian word for *bulwark,* as they were usually built on the line of old city walls or palisades.

The grand boulevards of Paris were constructed between 1850 and 1870 (following the revolution of 1848) and were dictated in purpose and geometry by the potential requirements of a defending army and the need to permit the free movement of troops into areas of Paris that had previously been sources of insurrection. The overall concept can be attributed to Napoleon III, who said it "would slash the belly of this mother of insurrections."[2] It was certainly a change from his prior anti-insurrection tactic of limiting the amount of paving (Law and Clark 1907).

The actual design was by Baron Haussmann, whose work was greatly simplified by the autocratic nature of Napoleon III's administration. In exercising his militaristic charter, Haussmann was nevertheless motivated by nobler ambitions. He saw his great plan for Paris as bringing health to the city, endowing it with "spaces, air, light, verdure and flowers, in a word, with all that dispenses health." The plan had many bitter critics who believed the expense would be ruinous. Although few of its intended benefits were delivered, events have proved the wisdom of the expenditure. Today, avenue Foch is typical of Haussmann's boulevards.

However, construction of the most famous of all the Parisian boulevards, the Champs-Elysées, began in 1667 well prior to Haussmann. The initiative was due to Minister Jean Colbert, who wished to provide a suitable route to the royal palaces on the Seine (Stave 1981).

Haussmann's approach had antecedents almost as old as cities, with numerous examples of streets widened to provide for the army, political strategy, military parades, or religious processions (Mumford 1961). Indeed, in the sixteenth century, the famous architect/engineer Andrea Palladio noted, "The ways will be more convenient if . . . there be no place in them where armies may not easily march." Such avenues were often called military streets.

London had had a similar opportunity for rebuilding its street system after the Great Fire of 1666, but architect Christopher Wren's plans were "frustrated by dilatory surveying and too little capital" (Dyos and Aldcroft 1969). A less charitable view was that Wren's vision was put aside by the "perverse self-interest of the then citizens of London" (Edwards 1898).

Following the Napoleonic precedents, the French continued to construct

grand boulevards. In 1910 a 12 km road was completed between Lyon and Tour-
coing. Employing a truly astonishing concept the boulevard had a total width of
50 m, which included a central 6 m macadam carriageway for heavy traffic, two
3.6 m outer carriageways for light vehicles, and—separating these from the central
way—a 7 m train reservation on one side and paths for horse riders and cyclists on
the other (Engineering Record 1910).

Modern Times

Until the advent of the helicopter and the tracked vehicle, roads exerted a
major influence on the progress of wars, particularly with respect to servicing armies
and permitting rapid troop movements. As late as the First World War, such major
events as the Russian loss at the Masurian lakes in Poland and the far too slow initial
German advance on Paris were both ascribed to the roads being less than adequate
(Everett 1980). In France such pressures led to some 900 km of new road being
built during World War I.

The first freeways had strong military links (Benito Mussolini's pseudomilitary
motives for Italy's autostrade are described in chapter 9). German autobahn construction was accelerated by Adolf Hitler soon after he came to power in 1933,
despite his party's prior objection to the concept. In May of the same year, he announced at Tempelhof that highway building would unite, preserve, and expand the
Reich and then implemented a scheme to build 7 Mm of Reichsautobahnen (Schreiber 1961). Motorization and the associated autobahnen rapidly became major items
in the Nazi PR campaign and a key part of the infrastructure of the German war
machine. In 1936 Nazi Oberstleutnant Nehring vowed "no military motorization
without economic motorization . . . the interests of defence demand the motorization of the economy" (Yago 1984). Idealistically, autobahn construction was seen
as a trial mobilization and as embodying Nazi ideals of national character, spirit,
strength, and beauty. More pragmatically, it was seen as relieving unemployment,
encouraging the car industry, and promoting tourism.

The project was under the control of Fritz Todt, who drew his authority directly from Hitler. The roads were generally divided carriageways with two lanes of
concrete paving in each direction. Construction was aided by the diversion of unemployment funds, the direct appropriation of land, and the use of forced labor. As
a result, about 1 Mm of autobahn were built by 1936 and 4 Mm by 1942, with
another 2.5 Mm under way when construction ceased at about that time. In
1950—as an aftermath of the war—only 2.2 Mm were still in operation. By 1984
the network had expanded again to 8.1 Mm.

The military role in the twentieth-century American road network formally
began in 1922 when the army produced the Pershing Map showing those roads
considered to be of prime military importance. By the mid-1930s the War Department and the Bureau of Public Roads (BPR) had identified some 45 Mm of strategic
highways. From 1935, all U.S. Army equipment was designed to stay within the
American Association of State Highway Officials (AASHO) loading limits for civilian
bridges. In 1941, a defense act was passed that provided specific funds for the construction of roads of defense relevance, including some lengths of freeway. In 1944
a presidential committee recommended that the United States establish the National
System of Interstate and Defense Highways.

In the early years of World War II a number of American urban freeways were justified partly on the grounds that they would facilitate evacuation in the event of air raids, a concept formalized in the 1945 "rural interstate evacuation highway system" (Schlereth 1985). Dwight D. Eisenhower's personal involvement in an evacuation exercise was to influence his later support for the creation of the interstate system.

Consequently, the interstate was initially justified to postwar American Congresses as a national defense system for the movement of military vehicles and the evacuation of civilians. President Eisenhower's Clay Committee told Congress in 1955, a year before the system was approved, that it was "necessary to national defense. . . . In the case of atomic attack on our key cities, the road net must permit quick evacuation of target areas, mobilization of defense forces, and maintenance of every essential economic function. . . . [The system would permit] a considerable amount of evacuation . . . at least several million people." This justification brought with it the requirement that the interstate system's geometry and structures should be able to accommodate and aid the movement of large pieces of military equipment. Indeed, the formal name of the system was the National System of Interstate and Defense Highways. The proposal to build the interstate also received strong support as a ready-made mechanism for postwar reconstruction spending. The development of this freeway system into the world's greatest contemporary road system will be discussed further in chapter 9.

Perhaps the most astonishing and optimistic military use of the car was propounded in a pamphlet produced by the 1969 U.S. National Highway Users' Conference, which suggested that, in the event of an attack on the nation, the individual's car would become "a rolling home. Persons can eat and sleep in it, keep warm and dry, receive vital instructions by radio, drive out of danger areas, and even be afforded some protection against nuclear fallout" (Kelley 1971).

The last of the great military road builders may have been Australia's Len Beadell, who created thousands of kilometers of road in the empty center of Australia in the 1950s and 1960s to accommodate the testing of rockets and nuclear weapons. The roads were noted for their exceeding straightness; in more peaceful times, one that is now in public use is known as the Gunbarrel Highway.

Managing and Financing Roads

Having reviewed the administrative and militaristic antecedents of many roads, it is appropriate to look at nonmilitary alternatives for funding and managing road construction and maintenance. The funding and management solutions tried over a millennium or so can be put into six main categories, which will now be discussed in turn, never forgetting that they usually followed the military methods already examined.[3]

Religious Commitment

One funding and management solution discussed in the previous chapter was to make road and bridge repair an item of religious penance or fervor. In many parts of Europe from the thirteenth century on, the making and repair of roads and bridges were seen by the Church as charitable acts. Keeping a road or bridge in repair was regarded as "a pious and meritorious work before God, of the same sort as visiting the sick and caring for the poor" (Pratt [1912] 1970). Monks routinely collected alms to pay for roadworks, and many parishes offered indulgences—typically, remission of sins for forty days—to those who contributed to such funds (Jackman 1916; Wilkinson 1934). Major causeways in the English towns of Boston and Glastonbury were built by religious orders. Chapter 8 describes a religious order that devoted itself to building bridges on pilgrimage routes. Groups such as the Gild of the Holy Cross in Birmingham dedicated themselves to road and bridge labor.[4]

Reliance on the church to provide roadworks was of limited effectiveness, particularly in England after Henry VIII's dissolution of the monasteries in 1530. Not surprisingly, he failed to pick up the church's road-maintenance charter.

In Scandinavia roadwork also provided family memorials. The Salna Stone records the efforts of three brothers for their father (Simpson 1967):

> While the world lasts,
> Ever will lie this bridge,
> A good man's, broad, firm-based.
> To a father's honor
> Youths fashioned this.
> Nobler sign by the roadside
> There never shall be.

Roman citizens funded roads built as memorials to their own names. Others donated money for roadmaking. In 1474 Maud Heath of Chippenham improved her boggy route to market by using her life savings to build 7 km of elaborate causeway—a way so well built that it remains in service to this day (Hey 1980). There are also occasional records of people leaving money in their wills for road and bridge repair (Salusbury 1948).

The Corvée

A second solution was to use the corvée system, in which each person subjected to it was obliged to provide the state with a certain number of days of physical work each year. In later times it was sometimes known as the statute labor system. *Corvée* in French now means "a chore" or "military fatigue duty." The corvée existed in ancient Egypt and Greece and in many subsequent societies. The Roman version was transferred to Frankish feudal law as a requirement that vassals perform work for the nobleman on whom they depended for their land and housing. In the manor system the serfs and tenants of the manor had three separate work obligations to the lord of that manor. Termed the *trinoda necessitas,* they were commonly military duty, the maintenance of fortifications, and the upkeep of roads and bridges (Rose 1952a).

The method was brought to England by the Anglo-Saxons in the fifth century, with the upkeep of bridges being seen as a military obligation. The precedent was readily adapted by the Normans in the eleventh century as part of their feudal government and manor system (Whitelock 1952). Although inefficient and ineffective, it did not die with the manor house. Instead, common law obligations arose in place of the trinoda requirements and were codified in English law in 1285 in the Statute of Winchester. The feudal relic had become an established general tax.

The approach was progressively extended. The English Highway Act of 1555, while placing the obligation for road maintenance on parishes, provided them with support by requiring parishioners to devote four days a year to corvée-style road-work. The act has been described as a "reasoned statement of tried old principles," but it was also the first act to apply generally to English roads rather than to specific locations (Hartmann 1927).

The effect of the act was that unpaid persons in each parish were put in charge of roads under titles such as surveyor of highways, wayman, and waywarden, a role that probably developed from the position of street cleaner established around 1280. One of the more famous road surveyors was Richard Sadler, who discovered the health spa on which the Sadler's Wells Theatre in London is now based while digging for road gravel. The selection process was that "Constables and Church Wardens of every Parish shall, yearly upon the Tuesday or Wednesday in Easter Week, call together a number of parishioners and shall then elect and choose two honest persons of the parish to be Surveyors and Orderers, for one year, of the works for amending of the highways in their Parish" (Webb and Webb [1913] 1963). In Scotland the appointee was often the local schoolteacher. Paris appointed a master inspector of pavements in 1400.

Although intended as a seven-year trial, the corvée method was made permanent in England in the 1563 Statute for Mending of Highways and was made "perpetual" in 1587. The latter act, incidentally, also increased the work obligation from four to six days. Workers were also often expected to supply their own equipment. The expression Mend Your Ways arose in this context. People fulfilling their corvée obligations were known as statute laborers or king's highwaymen, but their effectiveness can be gauged from the fact that they were colloquially called the king's loiterers (Parkes 1925).

The law remained in force in England for over two hundred years, but the system was doomed to fail. An underlying cause was described in the late eighteenth century by Edward Clarke as "the ridiculous farce of appointing one of the parishioners annually [at no salary] to enforce from his relations, friends, and neighbors a strict performance of a duty which probably he never performed himself; and from which, by showing leniency to his neighbors, he will expect to be excused."[5]

A cotton planter's journal provides a graphic insight into the operation of the system in South Carolina in 1848, showing that the job of the parish road commissioner "required the skills of a peacemaker, slave driver, quartermaster, and general. . . . He directed his neighbors . . . , rounded up horses and wagons . . . , procured lumber and materials, coaxed private men who were rivals . . . with many opportunities for offense. Planters worried that they were asked to give more than their share; that their slaves were worked harder; that their neighbours would benefit more from the project than they would" (Rosengarten 1986). William Gillespie was equally scathing, commenting in 1856 that the corvée, "though nearly universal amongst us is unsound in principle, unjust in its operation, wasteful in its practice and unsatisfactory in its results." The corvée nevertheless remained in use in England—apart from its repeal by Oliver Cromwell from 1654 to 1662 (Albert 1972)—until 1835, and in Alabama until 1913.

In France, the corvée had a long history and took a national role in 1738 when a law was passed allowing it to be extended to the construction of the major road network. In practice, the burden fell most heavily on the peasants and corvée increasingly became a system of privilege and dissent, particularly since much of the work was on roads serving royal residences. Reform of the system was a key factor in the fall of the Turgot ministry in 1776 and was one of the causes of the Revolution of 1789. The corvée came and went a number of times between 1776 and its final abolition in 1791.

The events in France had a close link with P. M. Jérôme Trésaguet and his development of pavement technology, which greatly reduced the need for roadmaking labor. The famous economist Anne-Robert Turgot was an administrator in Limoges between 1761 and 1774, and Trésaguet was chief engineer there between 1764 and 1775. The two were in frequent contact, and together they improved some 700 km of road and abolished the local corvée system. Turgot's political reputation was consequently enhanced by the zeal of Trésaguet's work (Yvon 1985). Turgot had become national finance minister in 1774 and abolished the corvée nationally in his Sixth Edict of 1776. The move had wide political impact, placing new taxes on the upper classes—who forced Turgot to resign from office within months. Trésaguet had been appointed engineer-general of bridges and roads in 1775; however, he too did not receive final rewards. He retired in 1786 and was ignored after the Revolution and died in poverty.

The new American colonies naturally followed English precedents, and from 1607 to 1657 the Anglican parishes in the New World were responsible for roadworks. Virginia then transferred the responsibility to the "Gentlemen Justices of the County Court" (Newlon 1987). Corvée laws were commonplace. Citizens eligible for such statute labor were known as "male laboring titheables," and the process was known as "working out the road tax."

The Romans were one of many communities to use slaves and convicts for roadmaking labor. More recently, in 1800 the government of Upper Canada passed the Stump Act, which required convicted drunks to work clearing stumps from road reservations (Ontario Ministry of Transport and Communications 1984). Stumps were a major issue. In 1804 Ohio made it illegal to leave stumps standing more than 300 mm higher than the road surface. British convicts were used for roadwork in Australia between 1820 and 1870 in what was regarded as a severe form of punishment for recalcitrant prisoners (Lay 1984). In 1829 the state of Georgia repealed its statute labor act and replaced it with a $70,000 allocation to purchase slaves (Better Roads 1980).

Following emancipation, Georgia passed legislation allowing convicts to be used for roadwork, a move that "revolutionized highway construction in Georgia." The use of convicts for roadmaking became fairly widespread throughout the United States (Pennybacker, Fairbank, and Draper 1916). Conditions were most severe in the southern states, where convicts in their distinctive arrowed uniforms were kept permanently chained and spent the daylight hours working and the hours of darkness sleeping in iron cages wheeled to the job site. The chain and attached leg irons were designed to permit a maximum free movement of two meters. For every eight convicts, there was one guard cum construction foreman equipped with a shotgun. The superintending guard also had a disciplinary whip.

The modern equivalent of the corvée is the use of general taxes to pay for specific lengths of road.

Taxes on Abutting Owners

The Greeks and Romans often placed the entire road maintenance requirement on the abutting landowners in a generalization of the corvée. As the Roman Empire decayed, the requirement was increasingly resisted. In the post-Roman medieval era, the requirement became an obligation on adjoining landowners to support the principle of right of way and hence to ensure that roads remained unobstructed. The work could involve smoothing, draining, and strengthening the surface, improving roadside drainage, trimming overhanging vegetation, and removing fallen trees and other debris (Beuscher 1956). Not surprisingly, adjoining landowners usually objected to working on roads intended primarily to serve passersby. On the other hand, the right-of-way precedents permitting travelers to divert into adjoining land if their normal route was impassable provided an incentive for their participation. In 1348 Paris relied on ancient custom to justify requiring adjoining landowners to pay for street paving outside their properties (Parsons 1939).

In towns, adjoining landowners were often required to pave the 10 m of street in front of their property, and a town tax was used to fund the paving of the central strip and the town square (Salusbury 1948). In an English act of 1353, Edward III introduced a general tax to fund road construction in London. The act required the center of the Strand between Temple Bar and Westminster to be kept in repair from money raised by taxes on wool, leather, and wine. The 2 m of pavement on either side of the central strip were maintained by a tax on the adjoining landowner. The measure could not have been particularly effective, for it was followed three years later by a toll levied on freight traveling to Westminster, which was subsequently increased in 1385 and 1446. In 1532 taxes were again levied on adjoining landowners to raise further funds for improving the edges and maintaining the surface of the Strand (Pennant 1813).

Taxes on the Local Community

The fourth funding and management solution extended the previous method by passing the responsibilities for roads from adjoining landowners to the local administrations through which the roads passed. The Greeks sometimes repaired their roads by using slaves publicly owned for the purpose. A more humane version described by Plato (ca. 400 B.C.) required a region to contribute a squad of sixty young men who devoted two years to working on road repair and other municipal projects (Pritchett 1980). The overall method was also used by the Romans.

The technique continued to be applied after the Romans, with local fiefdoms carrying the bulk of the obligation. When growing central control and urbanization caused feudalism to wane from about the twelfth century, the parishes were given the responsibility for their roads and bridges but not for their finance. Not surprisingly, these unfunded local groups had little interest in providing for aristocratic hunting expeditions or for the growing needs of through traffic generated by someone else's centralism and urbanization. Half a millennium later, with the advent of railways and canals, those same parishes often actively oversaw the demise of the arterial road system.

Of course, the burden on the parishes was eased by their direct participation in the corvée system. Another way used to ease their burden was to give them local fund-raising powers. Direct local taxes for roads and bridges were respectively called *pavage* and *pontage*. They were tried spasmodically in England from the thirteenth century, but were difficult in those times to operate and police. Local taxation was given legal status through paving acts for specific towns, beginning with Chester in 1391 when that city was near the peak of its influence (Pawson 1977). The 1531 Statute of Bridges gave local authorities similar power for bridge maintenance.

However, positive taxation powers at the local government level, generally based on land or per capita (poll) taxes, were not widespread until the new generation of local governments arose in the early part of the nineteenth century. Enabling acts were passed as early as 1654 but were never widely implemented (Albert 1972; Parkes 1925). The first effective legislation was introduced in Britain in the Highway Act of 1835. Nevertheless, local taxation did not gain any real substance until public pressure in the latter half of the nineteenth century led the road systems out of their new dark age.

Funding Initiatives

Occasionally, governments used some lateral thinking to find money for roads. For example, from 1556 the French frequently used national salt taxes, beginning with the building of the Queen of Roads between Paris and Orléans. A further funding solution, first employed in the United States, was to devote a percentage (usually about 2 percent) of land sales to finance new road construction. The method was introduced in 1802 with the use of Ohio money to build the National Road (FHA 1976). It is still used today to finance the construction of new roads on many new land developments. However, the system provides no funds for continued road maintenance, and these have been consistently raised from land taxes.

In 1905 New York State funded some 8 Mm of road improvement through the sale of "Good Roads" bonds. The interest on money borrowed to fund road improvements has generally been recouped by either a land tax or by some other tax relating to the increased wealth that the new road has created.

In 1790 the state of Virginia introduced a novel road-funding method when it used a lottery to finance a road from Rockfish Gap to the Fluvanna River (Newlon et al. 1985). Virginia had followed the lead of Philadelphia, which had begun using lotteries to fund street paving in 1761. Philadelphia possibly copied London, which had unsuccessfully tried a lottery to fund Westminster Bridge in 1736 (Hopkins 1970).

The Toll Alternative

The sixth road funding and management solution to be discussed had some basis in economic rationality. It passed road responsibility on to the traveler, via tolls—sometimes called fees, viage, or peage—imposed on those using roads and bridges. Bridge tolls were sometimes called *pontagium* or *brudtholl*. Tolls are first noted in Kautilya's Indian Arthasastra of about 320 B.C.

English Experiences

By the eleventh century, tolls were common throughout Europe; as one commentator remarked, "there were tolls everywhere on everything" (Leighton 1972). It is useful to now concentrate on the well-documented English experience as a case study. Tollgates were specifically recorded in the Domesday Book in 1085. From 1189, murage tolls began to be levied on travelers passing through a city wall and soon became a major source of municipal revenue. Likewise, manorial lords were able and very willing to levy tolls on strangers passing through their lands. They were far less willing to spend any of the money collected on road repair—a factor that caused a major deterioration in the road system. Another method was to impose tolls on freight. The first record of this is of a 1284 tax by the city of Northampton on goods moved.

The toll bridge found wide favor because it was easier to manage, required less frequent maintenance, and was harder to evade than a road barrier. Even at untolled bridges, travelers often encountered hermits begging for alms in return for keeping the bridge in repair. Tolls were introduced on London Bridge in 1286. They were levied over the remaining 545-year life of the bridge and were sufficient not only to maintain it, but also to finance its replacement and to pay for the construction of the Tower, Southwark, and Blackfriars bridges.

In 1274 the Crown granted toll rights on a road leading north out of London, but the precedent had little effect. In 1364 Edward III gave the hermit Philippe Lichfield permission to collect tolls on Highgate Hill on the Great North Road as a reward for his work in graveling the road (Harper 1901).[6] Unsuccessful attempts were made to introduce toll road legislation in England in 1609 and 1622 (Pawson 1977).

The toll road became legal through an act of 1661–1663, which applied to the London end of the North Road. The act was the result of petitions from local parishes and vestries north of London that had been unable to maintain their roads on this prime route (Bird 1969). The tolls permitted by the act were intended to supplement rather than to replace the corvée.

The provisions of the act were immediately effective. In 1663 the first formal tollhouse was opened at Wadesmill 35 km north of London (see fig. 1.1). The main incentive for this particular toll was the use of the road by a large number of malt wagons from outside the parish (Albert 1972). The tolls were collected between 1665 and 1680, when evasion and other factors led them to be removed. To overcome the evasion problem, a 1695 act allowed barriers to be erected in the road to ensure that travelers stopped at the tollhouse to pay their toll. The technique was ineffective, and the scheme soon failed due to continued high evasion levels. The operating trust survived from 1663 to 1733.

The Turnpike Age

In 1706, the first of a long series of acts was introduced, giving toll powers to independent bodies of trustees, rather than to local justices, as had been the case with the Wadesmill toll road. That is, the system moved from public to private enterprise. The acts were motivated in part by the obvious impossibility of making

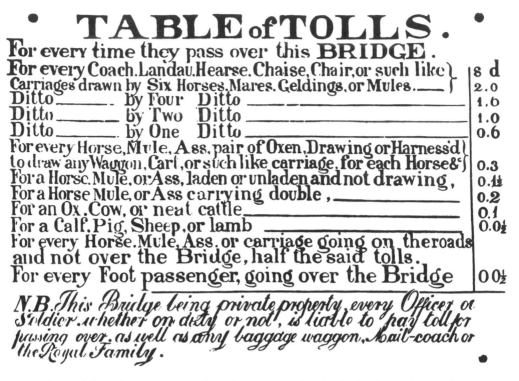

	s	d
TABLE of TOLLS.		
For every time they pass over this **BRIDGE**.		
For every Coach, Landau, Hearse, Chaise, Chair, or such like Carriages drawn by Six Horses, Mares, Geldings, or Mules.___	2.	0
Ditto _____ by Four Ditto _____	1.	6
Ditto _____ by Two Ditto _____	1.	0
Ditto _____ by One Ditto _____	0.	6
For every Horse, Mule, Ass, pair of Oxen, Drawing or Harness'd to draw any Waggon, Cart, or such like carriage, for each Horse &	0.	3
For a Horse, Mule, or Ass, laden or unladen and not drawing,	0.	1½
For a Horse, Mule, or Ass carrying double, _____	0.	2
For an Ox, Cow, or neat cattle _____	0.	1
For a Calf, Pig, Sheep, or lamb _____	0.	0½
For every Horse, Mule, Ass, or carriage going on the roads and not over the Bridge, half the said tolls.		
For every Foot passenger, going over the Bridge	0	0½

N.B. This Bridge being private property, every Officer or Soldier, whether on duty or not, is liable to pay toll for passing over, as well as any baggage waggon, Mail-coach or the Royal Family.

Figure 4.1. Tolls on the Ironbridge. *Photo courtesy of the Ironbridge Gorge Museum Trust.*

any of the other funding methods work. The first two toll roads under this system were both to be parts of the famous Holyhead Road. The legislation was revised again by Parliament in 1711, with provision for tighter supervision.

Such roads were commonly called turnpikes because the usual barrier was a pike (or horizontal bar) attached to a vertical post, in the manner of military barriers of the time. An alternative explanation of the term is that pike referred to the heavy iron spikes placed on top of the horizontal bar (Syme 1952).

The toll-collecting agency was required to maintain the road from the tolls collected. As private enterprises they were also intended to return a profit to their owners. The fees charged were commonly proportional to the number of haulage animals being used.[7] Nineteenth-century tolls used on the Ironbridge are shown in figure 4.1 and the tollgate at nearby Shelton on the Holyhead Road in figure 4.2.

Nearly all the turnpikes were based on the routes of earlier main roads. As these had had to have gentle grades for the low-powered animal-drawn vehicles of the time, they were frequently indirect and tortuously located. Therefore, they rarely provided a significant increment in service. Their strength was that the better turnpike trusts provided more effective road maintenance than did the parishes. Consequently, road haulage costs declined significantly in the first half of the eighteenth century (Albert 1972).

Nevertheless, a satirist was to write of turnpike construction in 1737: "Roads may be considered as made up of three sorts of matter, first stones as big as eggs and less; secondly a yielding substance such as mould; and third, water" (Boulnois 1919).

Figure 4.2. The tollgate from Shelton on the Holyhead Road, recreated at Ironbridge. *Photo courtesy of the Ironbridge Gorge Museum Trust.*

Arthur Young, writing of his travels in 1770, strikingly recorded the following observations:

- To Sudbury. Turnpike. . . . Ponds of liquid dirt, and a scattering of loose flints just sufficient to lame every horse.
- To Wigan. Turnpike. . . . They will meet here with ruts, which I actually measured four feet deep, and floating with mud only from a wet summer; what therefore must it be after a winter?
- To Newcastle. Turnpike. . . . A more dreadful road cannot be imagined.
- Oxford to Witney. . . . Called by a vile prostitution of language a turnpike.
- From Chepstow. Rocky lanes full of hugeous stones as big as one's horse.[8]

Young also had some good things to say about contemporary roads. He found, for instance, that half the turnpikes he traveled over were good and a quarter were middling. The above quotation summarizes the other quarter. He then visited France in 1787 and commented very favorably on the state of the roads on which he traveled (Reverdy 1980). Traveling in England again in 1813, he found substantial improvements and room for some considerable praise (Dyos and Aldcroft 1969).

Wealthy landowners and trustees often regarded the toll they levied on travelers as merely a form of tax, placing no obligation on them to maintain the turnpike surface. The attitude was also common in continental Europe. Turnpike roads were fairly described as "a very galling species of despotism" (Searle 1930?). It was not uncommon for trustees to compulsorily buy land to block existing highways and to erect gates on side roads to force travelers onto the toll road. Writing in 1802, Gray cynically commented that some turnpikemen operated more like highwaymen (Luton Museum 1970). Many citizens were unable to afford the tolls, so evasion levels were high and the turnpikes were often badly received in local communities. Lower-class hostility toward them was high, particularly in Yorkshire, Somerset, Gloucestershire, and Wales (Albert 1972). In the United States, alternative toll-evading routes to the early turnpikes were called "shunpikes."

Riots in 1726 and 1732 and between 1735 and 1749 led George II in 1734 to introduce the death penalty "without clergy" for burning or destroying turnpike gates (Rush 1960). Hangings were common but were not well accepted, and trials and executions were often disrupted by "turnpike levellers." The death penalty had little deterrent effect. For instance, Bristol and Yorkshire mobs destroyed about a dozen gates in 1753. When police took some of the Yorkshire rioters to prison, they were set upon by armed locals and a bloody battle ensued (Smiles 1874). George II, incidentally, had had an unhappy involvement with roads; in a well-publicized event in the 1730s he and his queen were thrown out of their carriage and into the mud of the Fulham to Hammersmith road (Fulham Palace Road) near London.

Despite all this the turnpike system had become increasingly extensive, effective, and significant. As a consequence, road travel increased dramatically during the eighteenth century. From 1751 to 1771, the British Parliament passed some 870 separate turnpike acts in a period of "turnpike mania"; over half the total turnpike length was constructed during this period (Albert 1972; Dyos and Aldcroft 1969; Pawson 1977). Continual technical improvements occurred after about 1760 (Jackman 1916; Wilkinson 1934). Telford and McAdam subsequently had major impacts. Before them, many trusts had been poorly administered and their engineers (surveyors) had possessed little or no expertise. One trust had had a succession whose qualifications were insurance underwriter, old and infirm, carpenter, coal merchant, baker, and politician. Other trustees abused their office, being also associated with suppliers of roadmaking stone, which they placed and replaced at monthly intervals, as in the case of Kensington Road in London in 1765 (British Parliamentary Papers 1765). McAdam, in particular, exposed the organizational and financial folly of many trusts.

As a result, by the 1820s England possessed some 40 Mm of turnpike, covering virtually all main roads and many secondary roads. The largest turnpike was the 300 km Bristol Turnpike Trust, although the average trust managed only about 40 km of road. In 1812 McAdam was given technical control of the Bristol Trust, and by 1823 his family were supervising some 3.3 Mm of turnpike. Some of the turnpikes were extremely profitable. In 1802 the rights for the Hyde Park Corner Turnpike in London were sold for 7,400 pounds sterling.

One immediate and major catalyst for turnpike action in Britain was the rapid development of the post office.[9] During most of the eighteenth century, the postal service used postboys running on foot or riding horses, reaching peak speeds of 10 km/h and using relatively long stages (Vale 1960). In 1784 this system was replaced by a revolutionary operation using new and more efficient coach designs and shorter stages (Copeland 1968). These brought greater speed and punctuality to the service. In 1780 a letter from Bath to London took forty hours by post horse. The change to the coach, inspired by the proprietor of the Bath Theatre, dropped this time to seventeen hours.[10] The post office advertised an average running time for the new service of 13 km/h (Bodey 1971). London to Bath (and Bristol) travel times and the specific coach improvements that occurred over the centuries are described in the next chapter. Meanwhile, the term *postboy* came (ca. 1800) to be used to describe the men who accompanied the post chaisses hired from postmasters by wealthy travelers (Copeland 1968).

The operation of the new mail coach services accelerated the deterioration of the road system and underlined its inadequacies to such an extent that in 1810 the postmaster general brought pressure for the by-then-famous Telford to be given charge of 300 km of turnpikes on the key route from London to the Irish ferry at Holyhead (see fig. 1.1). The route passed through some difficult terrain in north Wales (Belloc 1923). Pressure to upgrade the Holyhead Road also arose from political concern over the implications of a poor link between London and Dublin following the Act of Union in 1800 and from the Board of Agriculture, which saw a need to improve farmers' access to markets (Gregory 1931). The advocacy was led by Sir Henry Parnell, M.P. for Queen's County in Ireland and author of a famous textbook on roads (Parnell 1838). Parnell later became Lord Congleton. Ironically, the Royal Mail did not have to pay turnpike tolls.

A parliamentary committee established in 1810 obtained from Telford "a thorough and melancholy report" of the condition of the road. Consequently, in 1815 Parliament established the Holyhead Road Commission with funding to come from both government grants and from tolls (Trinder 1980). Telford was given charge of the task of planning and building the Holyhead Road and later commented in his autobiography that it had "occupied fifteen years of incessant exertion."

The first Holyhead Road had passed north of Birmingham and through Chester and Denbigh. The new route went from Coventry to Shrewsbury and onto Holyhead, embracing seven separate turnpike trusts. New construction was mainly on this last (Shrewsbury–Holyhead) leg, which was impassable for the mail coaches and barely existed on Anglesey, where it was described as "a succession of circuitous and craggy inequalities." The route chosen is still largely used by the current A5 and includes the famous Menai Straits Bridge. However major upgradings were also undertaken on the London–Shrewsbury segment, with Telford having powers above the seventeen turnpike trusts that operated on that segment. The completed work reduced the London–Holyhead mail time from forty-one to twenty-eight hours and significantly expanded trade with Ireland. One curious side effect was that it increased the importation of Irish pigs to such a degree that low walls had to be built alongside the road to prevent the pigs from eating the hedgerows (Penfold 1980).

The resulting road was often called the Parliamentary Routes and sometimes the Great Irish Road. It was described as "the nearest thing to a national highway ever built" in England and was the last British investment of public funds into road-making until well into the next century (Dyos and Aldcroft 1969).

Telford and McAdam breathed a further forty years of life into the British turnpike system. But the turnpike era effectively ended throughout the world in the 1840s with the advent of workable and highly competitive steam-powered railways and canals (Jackman 1916). For example, the prime pressure for the construction of Telford's Holyhead Road had been from the post office. The road was completed in 1830, but by 1837 the London to Holyhead mail was being carried by rail. Similarly, steam power much reduced the economic effectiveness of Telford's Caledonian Canal (Wilkin 1979). Finally, negotiations with Telford to conduct a project similar to the

Holyhead Road on the Great North Road were dropped in the 1830s as the potential of steam rail came to be realized.

The rise of the canals and railways is traced in chapter 5. The canal system was largely in place by the end of the eighteenth century, which partly explains the emphasis above on the turnpikes for travel and personal message carrying and not for freight movement. By 1840 there were about 7 Mm of canal in operation in England compared with 40 Mm of turnpike.

Thus the turnpikes ended slowly in financial crisis, bankruptcy, and management overreaction. In one attempt to rescue the situation, an 1841 act allowed local taxes to be levied to rescue the turnpike trusts from their financial problems. The move was not welcomed by those required to pay the new tax. When taken together with obvious examples of turnpike maladministration and repression, such as surrounding towns with tollgates "like besieged cities," the ninety-year lull in public dissent against turnpikes came to a tumultuous end.

In 1843 hundreds of tollgates in South Wales were destroyed by rioting mobs and at least one turnpike keeper was killed (Belloc 1923; Copeland 1968). The events were known as the Rebecca Riots because the leader of the rioters usually disguised himself as a woman called Rebecca and was assisted by similarly dressed co-conspirators. The name Rebecca came from a verse in Genesis (24:60)—"and they blessed Rebecca, and said to her, '. . . may your descendants possess the gate of those who hate them!'" The rioters worked at night seeking "good terms for the poor" and when captured were frequently acquitted by sympathetic local juries (Syme 1952). Their actions led to the trusts in Wales being replaced by county roads boards. These boards rapidly began providing initially unheeded evidence of the value of a professional and nonparochial approach to road management.

Only those turnpikes providing feeder and cross services were to remain viable in the age of steam.. This left Britain with a set of scattered, unconnected, and predominantly local roads far different from the centralized system that had developed in France (Webb and Webb 1913). Nevertheless, it has been argued that there was a pattern based on the main London routes and that there was a discernible interrelationship in the expansion of various trusts (Albert 1972).

When London's last turnpike closed in 1871 only 854 of the 1830 peak of 1,116 turnpike trusts were still in some semblance of operation. Interestingly, the last to close on a public road was the Anglesey portion of Telford's Holyhead Road, which collected its last toll in 1895. In 1930 England still had fifty-five toll roads and eighty-eight toll bridges on private land (Gregory 1931), with a few still operating today (1989).

American Application

Prior to the Revolution in 1775, the United States had inherited the English parish, toll, and corvée labor systems of road maintenance. A toll bridge had operated at Newbury, Massachusetts, from 1656. With the end of the Revolutionary War in 1783, growth in traffic and a desire for road links both increased. Individual states were often impoverished, so many turned to the turnpike as the solution to their roadmaking problems (Armstrong 1976).

The first American turnpike, in concept at least, was Virginia's 55 km Little River Turnpike. Although the state legislature had been raising private capital for the venture since 1785, the turnpike did not commence operations until 1802. It eventually ran 110 km from Berryville at Snicker's Gap in the Blue Ridge Mountains via Middleburg and Aldie to Alexandria near Washington. Much of the route is now followed by U.S. 50 (Newlon 1987). The road was probably also the nation's first stone-surfaced road (MacDonald 1928). It was poorly maintained but operated as a turnpike until the mid-1890s.

Between 1793 and 1795 private enterprise turned part of the Pennsylvania and Conestoga roads between Philadelphia and the fertile lands of Lancaster into the 6.3 m wide and 100 km long Lancaster Turnpike (see fig. 3.10). At the time, Lancaster was the largest American city not situated on navigable water. A major incentive for the development of the road was the Whiskey Rebellion in Pennsylvania in 1794, when farmers rioted over a tax on grain used for whiskey-making.

The Lancaster Turnpike has been described as "essentially a local street serving a thickly settled community" (Kirby et al. 1956). A somewhat different perspective is given by another contemporary commentator, who remarked, "Between Philadelphia and Lancaster there was no town but merely a settlement" (Shank 1974). The turnpike had nine tollgates and forty-six vehicle classifications for determining the size of the toll. It was immediately successful, paying between 6 and 15 percent return on investments.

The Lancaster Turnpike was also the nation's first major paved arterial road and its first road built to plans and specifications (Hindley 1971; Rose 1951a). In an example of prescience, the *Encyclopaedia Britannica* states that the road was also the first macadam road in North America. In fact it was made of local limestone with a 450 mm course of gravel on top of "pounded stone" of various shapes and sizes, which was then covered with earth. After this construction method led to early failure, it was changed in 1796 on the advice of a visiting English civil engineer and canal expert, William Weston, to follow then-current English practice. Even this new construction was probably not a macadam road. Indeed, many of the earlier turnpikes followed Trésaguet's technique (Armstrong 1976). Although the turnpike's business declined dramatically, along with that of most other American turnpikes, with the opening of railways and canals and the construction of public roads in the 1830s, the company survived until 1902 (Rose 1952a). Today the route is part of U.S. Highway 30.

Turnpikes became quite extensive in the United States. By the 1830s, Pennsylvania had some 80 companies covering 4 Mm, and New York had 278 covering 7 Mm (Rae 1971). Turnpikes were introduced in Canada in 1805 for a road from Montreal to nearby Lachine and in Australia in 1810 on the Parramatta Road in Sydney. The few turnpikes that were tried in Australia were dismal failures (Lay 1984).

Reviewing this segment of transportation history, it is as if public monies and effort were only spent on one transport mode at a time: up to 1750 on navigable rivers; from 1750 to 1770 on turnpikes; from 1770 to 1830 on canals; from 1830 to 1930 on railways; and from 1930 onward on roads (Pawson 1977). The twentieth-century resurgence of the toll road will be discussed in chapter 9.

Central Control

An obvious seventh solution to the problems of road administration and finance confronting many central governments in the mid-nineteenth century was for central governments to themselves supply the resources and skills. They had, after all, unsuccessfully tried everyone else. Central control had operated effectively in China since 200 B.C. (Needham 1971), in the Roman Empire, and in the Scandinavian countries, beginning with Norway in 1274 ("Proper Road Width," *Engineering News Record* 27, no. 5 [1925]: 107). Some specific national directions will now be reviewed.

The French Initiatives

In France, King Philip II reactivated road construction in 1222 when he resurrected a Roman road. This centralist involvement continued, with various French rulers taking a spasmodic but active interest in roads. This occurred more in France than in England for the following reasons: (1) a reaction to a long period during which central power had been in eclipse, as opposed to the continued strong central sovereignty that had existed in England (Lopez 1956); (2) a heavy reliance on a strong central army, which needed good roads to be effective; and (3) freedom from the major disruptive effects of the Reformation.

As a result, between 1300 and 1900 France probably had the world's best road system.[11] It was unrivaled with respect to national scale, quality, and condition, and its extent was only challenged by the conglomeration of British turnpikes and by the Inca road system. Some 30 Mm of royal roads were built in the form of a star centered on Paris, and the network so formed underlies the current French road system.

Louis XI initiated a national courier system in 1464, which provided a major new incentive for the construction and repair of main roads. It led in 1508 to Louis XII giving jurisdiction over French roads to his treasurer, operating through a decentralized system of road management. Subsequently, Henry II embarked on some significant roadmaking, and in 1552 he also fixed the routes of the main roads and ordered tree planting along them. In 1556 some 14 km of road were constructed between Toury and Artenay in order to link a number of royal residences between Orléans and Tours in the Loire Valley to Paris—for example, the 1496 Amboise château of Charles VIII, the 1519 Chambord château of Francis I, and the 1513 Chenonceaux "bridge" château of Henry II. The new link was part of the Queen of Roads route between Paris and Orléans and was the first of the routes nationales. Despite its royal usage, the work was nationally funded by a salt tax (Reverdy 1982). The 4.5 m wide pavement was constructed of 200 × 165 mm stone blocks on 300 mm of sand.

Faced with a divided France, Henry IV saw a workable road network as a key factor in achieving national unity. He began an active roadmaking campaign in 1594, and five years later replaced the treasurer with a specific chief inspector of highways (grand voyer de France). Provincial voyers were appointed in 1604 and their duties codified in 1607. The first voyer, the duke of Sully, greatly accelerated work on a national network of stone-paved roads. He was subsequently inspired by publication of Nicolas Bergier's book on Roman roads.

Following this initial thrust, the condition of French roads improved steadily until Louis XIII failed to appoint voyers between 1621 and 1645 and the treasurers regained administrative control. This setback was followed by the spasmodic diversion of resources to various wars being waged between 1643 and 1715. A major advance occurred in 1672 when the famous French cabinet minister Jean Colbert established the Corps of Army Engineers.[12] He was prompted in this action by a military defense expert named Sébastien Vauban, who was concerned that his military engineers were unskilled.

The success of the corps and the arrival of peace and technology led to the establishment in 1716 of a central road authority—le Corps des ponts et chaussées—under Jacques-Jules Gabriel. It provided a clear civilian alternative to the Corps of Army Engineers. However, the organization was in considerable early turmoil and its reputation was not enhanced when it proposed a grandiose and impractical system of straight roads radiating out from Paris. Not surprisingly, the corps was reorganized in 1726, 1750, and 1754, partly in an attempt to use the resources recently made available by the extended corvée.

A major related event was the establishment in 1747 of a school for training bridge and road engineers—l'Ecole nationale des ponts et chaussées, a name it took in 1756, some nine years after obtaining the appropriate royal decree. The first director was Jean Perronet, who held the post for almost fifty years until his death. He earned a reputation in France and elsewhere as "the father of engineering education." It is said that his interest in roads was aroused when he discovered the remains of a Roman pavement in the back yard of his house. In 1763 he was also appointed engineer general of bridges and roads. His impressive work in bridge design and construction is discussed in chapter 8. His construction contribution also included being the first to estimate job needs on an hours per task basis. So great was his reputation that he maintained his post throughout the Revolution; indeed, he built his greatest bridge, Concorde, during that period of turmoil.

Perronet was followed as director of the école by Trésaguet, whose major contributions to pavement technology were discussed in the previous chapter. The first great teacher at the école was Bernard Bélidor, and other famous names associated with the school during a remarkably productive period included Augustin Louis Cauchy, Louis Navier, Barré St. Venant, Louis Augustin Vicat, Augustin Jean Fresnel, Arsène Dupuit, and Eugène Freyssinet. Most of these names will reappear in chapter 8, where I discuss their major contributions to bridge design. Cauchy went on to become one of the great mathematicians of the nineteenth century. In 1830 the French began publishing the world's first technical road journal, *Annales des ponts et chaussées*.

The management of the centrally controlled French road network was greatly aided by the continuance of the corvée system. As noted above, from 1776 the corvée came under growing question. When the corvée was finally abolished in 1791, central control was temporarily abandoned and road control and the right to levy tolls were given to the newly created regional administrations. Salt taxes had been used to pay for road construction in France at various times from 1573, and in 1806 Napoleon was forced to reintroduce the tax to pay for regional road construction. These measures all proved so ineffective and unpopular that in 1811 the control of major roads was returned to central government. In an era when rail competition led to serious neglect of roads in many countries, central control kept French roads in relatively good shape. Indeed, an 1836 act ensured that the local road system was upgraded to the high standards of the arterial system.

The centralized French system was copied by a number of other European countries. Austrian rulers, directly influenced by French road developments during the eighteenth century, built a major road from Vienna to Trieste. An initially strong military flavor soon moved toward a system of state civil construction under a central (federal) administration (Forbes 1958a). Jean Marmillod established the basis for the Danish road network from 1764 to 1776. He had trained at the Ecole des ponts et chaussées under Perronet and consciously copied French practice (Hogsbro 1986).

A unique approach developed in Ireland. In the Irish presentment road system initiated in 1765, roads were constructed by interest groups—including contractors—who then presented their finished work to an assessment board for retrospective government funding. The system proved to be a relatively successful example of central road funding (O'Keeffe 1973). In 1822 the Irish Parliament began making direct road construction grants and established a Board of Works in 1831. Following a world trend, the board's road role was abolished in 1851 when the impact of rail became obvious.

Developments in Britain

While a strong central system had arisen in France, little had happened in England except for the following ineffective national appointments (Bird 1969; Hartmann 1927): surveyor of the highways to the king, Thomas Norton, 1609; guide and surveyor of ways, Thomas Norton, 1616; surveyor-general, Thomas Norton, 1625; surveyor of the king's highways, Thomas Hebbs, 1626; surveyor-general of highways, unfilled Cromwell appointment, 1657.

In 1654, Oliver Cromwell's government did pass an ordinance "for better amending and keeping in repair the common highways within this nation." Cromwell also introduced the first compulsory tax for road maintenance, although the ordinance was ineffective and "as much for the ease of military movements and the despatch of intelligence as for the convenience of the public" (Roots 1972). The subsequent slow English move to adopt a central approach was probably due to anticentralist feelings dating from this time and a prevailing feeling that control of the roads was a local right.

By the beginning of the nineteenth century Telford and McAdam had seen the problems of the turnpike system and were aware of the success of the centralized road-management systems in operation in Scandinavia and prerevolutionary France. With others, they argued strongly but without effect for adoption of a centralized system in Britain. In 1808 a proposal for central road control went as far as a committee of the House of Commons; however, another century was to pass before it was adopted in English-speaking countries (see, e.g., Jeffreys 1949; Lay 1984; and Cron 1976c).

Instead, between 1830 and 1860 the rail systems claimed the bulk of long-distance land transport. Road expenditure fell into disfavor, and there was a reversion to the previously discredited second solution: control by local government. The move began with the British Highway Act of 1835, which joined parishes into highway districts and introduced a local land tax for highways. Not surprisingly, road systems under this solution deteriorated rapidly during the latter half of the nineteenth century and in many places became effectively nonexistent. Communities became landlocked—or perhaps mudlocked—for the wet winter months. In 1893 a Massachusetts commission investigating ways to improve its highways noted that roads were "an earlier method of carriage" and pleaded for action to prevent their extinction. As another example of the attitudes, when the old 6 m wide stone-surfaced toll roads were maintained, attention was commonly restricted to only the central 3 m. One commentator of the time noted that "a reason for the excessive cost of labor is the employment, in many parishes, of paupers and old men unfit for work, with a view of economizing the rates" (Wheeler 1876).

County councils were first established in Britain in 1888 to consolidate control and thus help solve the increasing road problem. Nevertheless, some eighteen hundred authorities administered the roads of England and Wales in the 1890s (Jeffreys 1949). In 1908 there were ninety different road authorities within 25 km of Charing Cross, London, and one main road 30 km long passed through ten different road jurisdictions (Boulnois 1919).

Chapters 5 and 6 describe the active lobbying efforts used by the cycle and car clubs to resolve the mess. Under Jeffreys's leadership these clubs united in 1901 to organize propaganda in favor of a central road authority and a national grant for roadmaking. The British government subsequently established a roads board in 1909 and set an important precedent by funding it with a tax on gasoline (Jeffreys 1949). The tax constituted about 36 percent of the price of gas and was halved for doctors and commercial users. The board collapsed in 1918 to 1919, and its role was taken over by the Ministry of Transport.

The United States and Its National Road

A National Road had been conceived in the first days of the Union. In 1802 Presidents Thomas Jefferson and George Washington were among those who advocated its construction as an aid to westward expansion and to assist in holding the large new nation together. The need was heightened by the admission of Ohio to the Union in that same year. Albert Gallatin later added his considerable economic support (Gallatin 1808). After much debate the government decided that the road should be partly financed by a levy on government land sales and built for the federal government by army engineers. Construction began in 1811 and accelerated when the War of 1812 illustrated the need for good communications (Rae 1971).

The proposed new National Road was also variously called the National Pike, the United States Road, and the Cumberland Road—the latter was its original legal name. The road ran from the navigable limit of the Potomac at Cumberland in Maryland, passing south of Braddock's first road en route to St. Louis (see fig. 3.10). By 1818 construction reached Wheeling, West Virginia, on the navigable limit of the Ohio River, resulting in a halving of travel time from the coast and a subsequent doubling of the town's population. The National Road was thus theoretically a portage road (a road used for carrying material between two navigable rivers). Indeed, it first became a portage road when it met the navigable Monongahela River at Brownsville, Pennsylvania.

Travel along the road peaked between 1820 and 1840. In 1820 work began to push construction westward from Wheeling, but by 1824 the traffic load meant that heavy maintenance was needed and much of the surfacing had to be reconstructed. McAdam's method was employed from 1827 on (Wixom 1975; Labatut and Lane 1950). Despite the maintenance, as one commentator remarked, "the stone surface was worn away almost as fast as it was built. The funds available for maintenance were not sufficient. . . . Local inhabitants . . . even stole broken stone from the road bed" (Armstrong 1976).

The road reached Zanesville, Ohio, in 1833, but storm clouds were looming. An army report of 1834 listed eight different ways in which contractors on the job were defrauding the government (Kirby et al. 1956). The financial problems increased and, in an attempt to raise funds, the section east of Wheeling was converted to a toll road in 1835. However, even without the toll, travelers had always preferred the Pennsylvania Road between Philadelphia and Pittsburgh (see fig. 3.10; Rose 1950c). Army engineers stopped working on the road in the same year, and the last federal funding was received in 1838 when the road reached Springfield, Ohio. It was abandoned as a federal project in 1841, having reached Vandalia, Illinois, with its last sections unsurfaced and about 100 km short of the 1.1 Mm needed to reach its St. Louis goal. U.S. Highway 40 now parallels much of the National Road's route; it is steeped in road history, containing parts of Boones Lick Road in Missouri, the Boonsborough Turnpike in Maryland, and the beginning of such overland trails to the West as the Oregon Trail (Rose 1952c; Cron 1976c and Schlereth 1985).

The move to abandon the National Road initiated in constitutional arguments concerning federal involvement in many areas, including banking and roadmaking. It was then sealed by the advent of the far greater expansionary forces created by the transport efficiency of the railroads and the canals (Merdinger 1952). The railhead reached Wheeling in 1842 and the key Baltimore and Ohio line arrived in 1853. By 1856 all federal rights concerned with the National Road had been returned to the states. Nevertheless, despite this seeming negativism, it was by far the best road that had yet been built in North America.

The U.S. Corps of Engineers was created in 1802 and located at West Point. The associated military academy was "prompted by the almost complete absence of military and civil engineers in America." The French aided the establishment of the school, which was largely patterned on the Ecole polytechnique. French was widely taught, and many French textbooks were used (Hill [1957] 1977). It was also quite common for American engineers in the first part of the nineteenth century to study engineering in France.

The professor of engineering at West Point from 1816 to 1823 was Claud Crozet, a graduate of the Ecole nationale des ponts et chaussées in Paris who had served as an engineer in Napoleon's army and been on both the march to Moscow and the 100 Days to Waterloo. He was also very familiar with the work of McAdam. A number of the engineers on the National Road had been students of his and were thus aware of the latest roadmaking practice. Crozet became chief engineer of the Virginia Board of Public Works in 1823 and served in that position for the next eight years and from 1838 to 1843.

The Final Steps

Federal systems of government slightly complicate the development of central control. In the United States and Australia, state road authorities were established as the first step away from local road control. Leading the way were Virginia in 1816 and Tasmania in Australia in 1827. However, by the 1850s these state road authorities were to suffer the same extinction, albeit temporarily, as had the turnpikes (Labatut and Lane 1950; Lay 1984).

A further distortion, demonstrating the same effect, occurred in Germany. German road systems had been chaotic under the individual states. In 1814 the only good road in Prussia was between Berlin and Magdeburg 120 km distant. As national unity strengthened after 1815, the empire took an increasingly dominant role. Indeed, until 1870 it exerted central control over its roads, motivated by military considerations. These roads were excellent, for "broad, smooth roads were necessary for the rapid movement of armies in that time of nearly incessant warfare" (Engineering Record 1891b). Their construction had been an important function of the national government and was administered by a "vast bureau." However, after the 1870–1871 war of liberation, the new imperial government of the German empire considered rail of prime military importance and roads of secondary importance and so gave road control to the largely uninterested states (Gregory 1931). Nevertheless, the need for roads connecting the individual states became a symbol of the struggle against political dispersal.

From 1876 on, the success of the Washington, D.C., asphalt paving program was widely quoted as an example of the benefits of a centralized administration. The details of that program are discussed in chapter 7.

The final move to central road authorities at the turn of the twentieth century was successfully led by a shaky alliance of cycle clubs, car clubs, and farmers' clubs, often under the leadership of Good Roads Associations and with Nationalize the Roads on their banners. The role of these clubs and associations is pursued much further in the next two chapters. Agricultural interests feature strongly; urban roads were relatively good, and the major need for action was in the countryside. The basic objectives of agricultural roadmaking were to improve rural economies, to raise the quality of life, and to prevent rural depopulation.

The continual lobbying had its first formal effect in 1891 when the state of New Jersey passed legislation giving its Board of Agriculture authority to produce farm to market roads. The first American state road authority was established in 1893 when Massachusetts created a highway commission. In the following year New Jersey gave its road powers to the Commission of Public Roads (Wixom 1975). An interesting approach occurred in 1896 when the Canadian province of Ontario appointed a provincial instructor in roadmaking within the Department of Agriculture. All American states had their own highway organizations by 1920, but the importance of central road authorities was not generally accepted until well into the 1920s (Borth 1969). The first Australian state road authority was established in Victoria in 1913 (Lay 1984).

The cyclists, organized as the League of American Wheelmen, successfully lobbied Congress in 1893 for an act to permit the secretary of agriculture to act on road matters. The secretary immediately implemented the act by establishing the federal Office of Road Inquiry to study the technical aspects of road construction and administration. Its first leader was General Roy Stone, a civil engineer who came from the League of American Wheelmen via secretaryship of the National League for Good Roads. Stone left after six years to become president of the National League, which collapsed a year later. The office was renamed the Office of Public Road Inquiries, and a testing laboratory under civil engineer Logan Page was established in Washington in 1900 (FHA 1976). In 1901 the office joined with the railways in providing the Good Roads trains (see chapter 6).

After a period of questionable leadership (Seely 1987), the name of the office was changed to the Office of Public Roads in 1905 and Page was appointed director, then to the Office of Public Roads and Rural Engineering in 1915, and to the Bureau of Public Roads in 1918.[13] When Page died in 1919, civil engineer Thomas MacDonald, who had had a similar role in Iowa since 1904, was appointed; he reigned as federal commissioner for the next thirty-four years (Borth 1969). The office was known as the Public Roads Administration between 1939 and 1949 before reverting to its previous name.

The American Highway Association (AHA) developed in 1912 from the American Association for Highway Improvement, which had been formed in 1910. This latter association was initiated by Page as director of the Office of Public Roads and some thirty officers of a variety of organizations, including the railways. It held its first Road Congress in 1911. The AHA disbanded in 1916 under competition from the American Road Builders Association.

The third congress of the AHA was held in Atlanta in 1914 as a joint meeting with the American Automobile Association (ARBA 1977). The American Association of State Highway Officials (AASHO) was formed at that meeting with the strong support of the Office of Public Roads (HRB n.d.; Paxson 1946).[14] Its goal was to get American traffic "out of the mud." AASHO was subsequently described as "one of the most important [but] least known political groups in the country" (Patton 1986).

The United States led other federal governments toward national road funding with its 1912 Post Office Appropriation Act, which provided financial support for rural roads used for postal delivery. Four years later this precedent led to the Federal Aid Road (or Shackleford or Good Roads) Act, which enabled the secretary of agriculture to provide for both postal and general road improvements (Seely 1987). The pressures behind these moves to federal funding have been described as follows: "The government of the United States put a reluctant shoulder behind the [highway] movement. Driven by the movement, rather than leading it, the government of necessity took charge. . . . It was the government, budget-bound, which required forcing, yielding only when it must" (Paxson 1946). The background to this push is discussed further in the next two chapters.

As a result of the deterioration of the American road system under increased truck traffic following World War I, the U.S. Highway Research Board was established in 1920 as an advisory board of the National Research Council (Cron 1974). Among those who had a hand in its establishment and who were appointed to the inaugural Executive Committee were Thomas MacDonald and Thomas Agg. The first subcommittee formed was one concerning the economic theory of highway improvement with Agg as chairman.[15]

In 1921 the Federal Aid Highway Act allowed each state to receive funding for up to 7 percent of its highway network—3 percent for interstate routes and 4 percent for intrastate routes. The federal government met half the cost of such roads and thus began the establishment of a national highway system. The subsequent development of this act and of federal funding for the interstate freeway system is discussed in chapter 9.

Road financing today falls into three main categories. First, some roads are financed from the public purse using money raised directly from general taxation or public borrowings, such as bond issues. These roads are seen as public goods that will repay the community in some way for investment made in them. Second, some roads are funded by landowners who see the road as increasing the value of their land. These roads may be paid for directly by private enterprise or through taxes levied on specific landowners. Third, some funding for road construction and much road maintenance comes from various charges levied on road users, such as tolls, fuel taxes, taxes on the purchase of vehicles and spare parts, taxes on vehicle ownership, and taxes on possession of a license to drive.

International Moves

PIARC (Permanent International Association of Road Congresses, or Association internationale permanente des congrès de la route, AIPCR), the international road group representing government road authorities, was founded in France in 1908 by the Corps des ponts et chaussées in conjunction with automobile and touring clubs. The first congress was to be held in October of that year in the relatively small Jeu de Paume in Paris.[16] The congress proved larger than expected, and the 1,600 delegates from twenty-seven countries who attended were moved to sessions at the University of the Sorbonne. A permanent secretariat was established in Paris in 1909 and was structured on the same lines as bodies previously established for railways and navigation. The second congress was in Brussels in 1910 and the third in London in 1913; they drew 2,100 and 2,000 delegates respectively. Five congresses were held between the world wars: Seville 1923, Milan 1926, Washington 1930, Munich 1934, and the Hague 1938 (PIARC 1969).

PIARC has a strong European flavor and concentrates on governmental road organizations. The United States joined PIARC at the Milan Conference in 1926, with MacDonald as its official representative. However, its membership was brief, due to some American dissatisfaction with PIARC presidential elections in the 1930s and 1940s. Partly as a reaction to this, the International Road Federation (IRF) was founded in the United States in 1948 and retains its American flavor, with a strong concentration on the industrial aspects of roadmaking. The United States rejoined PIARC in 1991.

Summary

This chapter has examined the various driving forces behind the development of the road infrastructure. Progress has sometimes been stumbling and unimpressive, mainly because the roadway has always been only one part, and a fairly desperate part at that, of the total system with which the community has been concerned. With the noticeable exception of the Romans, the administrative mechanism to permit a systems view rarely existed. And even there, the barbarians were able to show that the Roman system was far from optimal.

The military role of tanks and aircraft in the recent part of the twentieth century has probably blinded our eyes to the dominant role that roads have played in military conquest and defense, and thus to the major influence the military has had on our road systems. Not only did the military create strategic road links but it also led to many of our grandest roads.

A continuing basic problem for the road infrastructure is that it has always been difficult to find a method to fund construction and maintenance. Reasonably, these funds should come from those who benefit from the road, but these people have been difficult to define and even more difficult to tax. The turnpike system flourished for a century or so and overcame some of these problems, but when its monopoly on surface transport was broken by rail and canal, financial uncertainty returned to road funding and remains largely unresolved today. Beneficiaries now commonly pay for their use of roads through fuel and land taxes. The relevant contemporary issues flowing from this are explored further in chapter 9.

By the nineteenth century, dramatic changes were taking place in road vehicles. The following chapter explores these changes and chapter 7 studies their impact on the road system.

A
Surge
of Power

Chapter 5

Fire in the heart of me, moving and chattering,
Youth in each part of me, slender and strong,
Death at the foot of me, rending and shattering,
Light and tremendous, I bear you along.
G. S. BOWLES, "SONG OF THE WHEEL," CA. 1900

*T*he early development of foot and vehicular transport was discussed in chapter 2. We then explored the development of the road system to meet these improved means of transport, the pressing needs they created, and the consequential organizational response to extended road management and investment. This chapter outlines the major changes in transport technology from the Middle Ages to the present day, a period of continued change causing ever-increasing pressures on the road infrastructure. The concentration is largely on methods of personal travel rather than on freight movement, for this was the area of greater change.

Carriages and Coaches

Many post-Roman communities relied for their transport needs on simple two-wheeled carts. These were of low capacity and provided their passengers with a jerky ride as the base of the cart moved up and down in conjunction with the horse's movements. Thus, they were never able to meet the demand for long-distance passenger services, which steadily increased as society became more organized, centralized, and productive.

Early Coaches

In response to market demands, wagons were in common transport service by the thirteenth century. They were large and rode better than the carts, but were also more complex and expensive to operate. They were called long wagons, or wains. *Wain* was the older English usage, and the word *wagon* (or waggon) came from Holland. Toward the end of their time they were sometimes called old coaches or stage wagons. Initially, long wagons were no more than freight vehicles hooded with cloth to become windowless boxes supported directly on unsprung axles and wheels. They carried about twenty passengers and their goods, were typically drawn by four to six horses operating unchanged over the day, and could travel at an average, subwalking speed of around 4 km/h, or about 30 km/day. The absence of brakes on these heavy carriages prevented their use in hilly terrain, where traveling downhill became hazardous. There are eighteenth-century reports of coaches on poor stretches of road being drawn by oxen. Despite intervening advances in technology, long wagons were still at work in the nineteenth century, providing the lower classes with a passenger service to most major English towns at a quarter of the price of coach travel (fig. 5.1).

Once better harnesses, steerable front axles, and light spoked wheels became available in the late thirteenth century, French and Italian coach builders used these technologies to produce special passenger carriages suitable for use on short journeys (Borth 1969). They were basically unsprung wagon trays with various accouterments added above the wagon floor to meet particular purposes. The carriages were often used for ostentation rather than for transport and as a feminine alternative to horseback riding. Men of worth, without exception, still traveled by horseback, and most women of worth preferred the horse litter. Indeed, the carriage did not take over from horse and litter until the seventeenth century (Boyer 1959).

One stream of this development was in the production of ceremonial coaches for official use and social ostentation. In an approach begun in France in about 1245, these vehicles initially followed the above pattern of an elaborate addition to the top of a wagon tray. Ceremonial coaches were subsequently replaced by either elaborate wagons of the type described below or by equally elaborate but more comfortable cabins suspended from a chassis by straps or chains.

Figure 5.1. Passenger wagon service. The reference to a "Flying Wagon" is an ironical reference to flying coaches. *From Smiles (1874).*

Gradually, it became more commonplace to build wagon tray and gorgeous custom-built canopy as a single unit. Queen Mary I introduced ceremonial coaches to England in 1553, riding in a glittering procession of five coaches through the city of London to Westminster on the day before her coronation. Walter Rippon built the first English coach in 1555, and eleven years later he made a considerable advance in cabin structure in order to produce a coach that allowed Queen Elizabeth I to be viewed by her subjects from all four sides. However, her favorite coach was imported from Holland in 1571 for the opening of Parliament (Smiles 1874).[1] Drawn by a pair of horses, it had four curved columns supporting a feathery red canopy (Belloc 1926). Nevertheless, its unsprung construction meant that it was far from comfortable, and on one occasion the queen was unable to sit for several days after using it for a journey. She had had some forewarning—she had once refused to ever ride again in a coach used for an earlier parliamentary opening, such was the discomfort of its ride. Indeed, despite the advances in suspensions, most ceremonial coaches of the time were still wagons with gorgeous canopies.

Returning to real travel needs, by 1374 a number of European wagons were providing an improved ride by using straps or chains to suspend the passenger compartment in the Slavic manner discussed earlier. This led to their colloquial names of jolting, rocking, or trembling carriages. It required considerable skill to make the light passenger compartment perform as a coherent, stable unit. The coaches were also steerable, which meant that their front wheels were much smaller than their rear wheels. Gradually, entry moved from the rear of the passenger compartment to the side.

The word *coach* now used to describe these suspended carriages came either from the name of a fifteenth-century Hungarian vehicle builder, Kotze, who initiated the technology or from the small town of Kocs in which he lived (Rose 1952a; Margetson 1967).[2] The word progressed through Germany (*Kutsche*), Italy (*cocchio*), and France (*coche*) to become the English word *coach*. The Kotze Hungarian coaches were first mentioned in 1417 and later become famous throughout Europe for their lightness and comfort (Boyer 1959). The best known were built late in the fifteenth century for King Matthias I, who used them pragmatically to provide Hungary with a postal service and aristocratically for coach racing, which had become a popular sport in Hungarian society.

Coaches were used in Spain in 1546 to provide the first long-distance wheeled passenger travel of any quality since the days of the Roman officials. Milan led in the development of urban coach travel with sixty coaches in use in 1525. By contrast there were only three coaches in Paris in 1550: one for King Henry II's wife, Catherine de Médicis; one for his mistress, Diane de Poitiers; and one for the Dauphin Jean de Laval de Blois, who was too fat to ride on a horse.[3] By 1600 there were still only four coaches in Paris, although the number skyrocketed to over three hundred in 1610. The French government controlled all coaches between 1575 and 1775.

By 1580 coaches were in general use among the London upper class and had caused an outcry from those who believed that they would destroy horse breeding and lead to effeminacy in Englishmen (Wilkinson 1934). In 1601 the English Parliament attempted to restrict the "immoderate" number of coaches in order to ensure sufficient horses for the army (Bodey 1971). The bill failed on the second reading.

In as late as 1873 coaches were denounced as "among the greatest evils that had happened to the Kingdom, being alike mischievous to the public, destructive to trade, and prejudicial to the landed interest" (Smiles 1874). Similar reactions arose in the German military, and in 1559 Pope Pius IV prohibited cardinals from riding in coaches.

Sixteenth-century coaches were still extremely heavy, demanded large amounts of horsepower, and were more like wagons than our current perception of coaches. Slowly, coach makers learned to produce somewhat lighter vehicles, providing a modicum of comfort. As an example of this change, by 1640 German coaches traveling at walking speed had largely replaced the even slower Spanish coaches and long wagons in catering to the wealthy traveler. By the end of the century they were an established means of urban transport for the wealthy, although they had strong competition from the sedan chair, which for a single passenger provided a more comfortable ride at about the same effective speed.

The Stagecoach

Sixteenth-century coaches could carry up to eight passengers and their luggage on their slow intertown journeys, journeys for which the market demand was strong. To satisfy this demand, the next development in coach technology was the stagecoach, which drew on the old concept of using a series of post houses to provide new teams of horses.

The development of coaches will be explored by examining the changes that occurred in England during the three centuries of coach ascendancy. The first British stagecoach ran from Edinburgh to Leith in 1610, and a regular London–Edinburgh postal service began operating in 1635, followed by a fortnightly passenger service in 1658 (Merdinger 1952). By the end of the century, stagecoach routes were common throughout Britain. They were operated over stages of about 25 km, and a traveler could usually manage two or three stages a day. In 1664 France introduced a further stage system using lighter two-wheeled carts known as chaisses, or shays. In a system known as post travel, travelers supplied their own chaisses and hired horses from the post houses.

The stagecoach presented many new subsidiary business opportunities. For example, the seventeenth century was the age of the highway robber. Those on horseback were called highwaymen and those on foot were called footpads. These "gentlemen of the road" exacted their own tolls from coach passengers. Their prime objective was theft, and physical violence was uncommon. In England, no police force existed to pursue them; the main power of control was founded on the 1285 Statute of Winchester, which required the local shire to reimburse victims for half their losses. The ancient tradition of hanging the highwayman—when he could be caught—at the nearest crossroads was frequently followed. The highwayman had disappeared by 1840 due to a general increase in the speed and volume of traffic on improved roads (Tobias 1967).

As stagecoach operations expanded, the demand for better vehicles increased. Even the old long wagons were improved. A nineteenth-century version called a long stagecoach was more comfortable, thanks to the use of iron springs to carry the passenger compartment. Iron axles and springs were in use in mainland Europe by

1625. Glass windows were introduced in a coach built for Mary, Infanta of Spain, in 1661.

The coach passenger compartments were still suspended by leather straps from four wooden corner posts. One German writer in 1650 described his country's coaches as "like four post bedsteads on four wheels" (Harlow 1928). German industry responded well to the challenge. In the 1660s Philip di Chiesa, an Italian who built his coaches in Berlin, hung leather suspension braces from the top of iron springs shaped like the letter C to produce the first of the famous Berliner coaches. It was the first passenger vehicle to have an effective suspension. Although its style and suspension technology did not find wide use until the middle of the next century, the Berliner provided the model for future coach development and its form can be seen in most modern coaches (Parkes 1925). The improvement in coach-building technology also led to a resurgence of the two-wheeled cart now that they could be made light and comfortable. The first of these was the gig, which was introduced at about the same time as the Berliner coach.

English coach technology consistently lagged well behind that of mainland Europe. Iron springs had been patented in England in 1625, but all the action had taken place with di Chiesa in Germany. To revitalize English development, the newly formed Royal Society undertook a series of experiments between 1665 and 1667 to improve coach design. The society's efforts began with a demonstration by a Colonel Blunt of a vehicle supported on four springs. The multitalented Robert Hooke played an active part in the work; Samuel Pepys somewhat incredulously recorded progress of the work and then bought one of the first spring-suspended vehicles. Pepys's diaries subsequently recorded a number of coaching adventures due to poor road conditions or inadequate road signs.

Other aspects of coach design also gradually improved during this time. The use of iron tires on spoked wheels led to lower rolling resistance, higher speeds, reduced wheel wear, and easier curve negotiation. Iron rim brakes, introduced in 1690, provided further operational advantages. One advantage was that larger teams of horses could be safely employed to power the coaches, with a consequent increase in travel speeds. Iron making was still a craft rather than a production process, so its widespread use in coaches would have been in response to strong market demands.

In 1669, amid changing public perceptions, the new iron-suspended coaches were called flying coaches or, more prosaically, basket coaches. They typically carried four inside and up to twelve outside passengers, but daily travel was still limited to about 75 km. By the end of the eighteenth century, continuing improvement meant that these same vehicles were described as "old heavies." Throughout this period the old long wagons continued to operate, moving freight and carrying poorer passengers. During the eighteenth century, horses and coaches and pavements continued to improve, and the occasional exceptional speeds of the seventeenth century became commonplace (Georgano 1972). Specially bred Cleveland coach horses became widely available. However, over bad roads the traveler would still be lucky to exceed 30 km in a day. In such circumstances walking remained the quickest mode.

By 1750 coaches had taken over from horseback riding as the dominant means of personal intertown travel. In 1754 the following notice appeared in Manchester. "A flying coach, no matter how incredible it may appear, will actually, barring accident, arrive in London four and a half days after leaving Manchester"

(Salkfield 1953). This obviously exceptional event represented about 120 km/day and an average speed of between 10 and 15 km/h. More commonly, stagecoach travel averaged about 7 km/h—"if God permit," as their timetable announcements would read. Although slow by our perceptions, the speed was twice that of earlier wagons.

The dramatic and important decision by the English Royal Mail in 1784 to change from horseback to Berlin-style coach was discussed in the previous chapter. A less than successful mail coach episode is illustrated in figure 5.2. The changeover had a major influence on both coach and road design. In 1792 John Besant introduced his patent mail coach with such novel features as improved turning capacity, band brakes, and a method to prevent the theretofore common event of wheels coming off while the coach was in motion. The new coach was not entirely successful, and an improved design was introduced in 1795 (Vale 1960). The Royal Mail coaches weighed about 2 tonnes when loaded and were close relatives of the well-known American and Australian stagecoaches of the nineteenth century (Georgano 1972; Lay 1984).

The design of coaches continued to advance rapidly, particularly with respect to lightness and suspension. The elliptical spring was invented by Obadiah Elliot in 1804. By allowing the vehicle to be carried on horizontal springs, rather than seated on heavy wood and iron perches, or suspended from wooden posts or large, vertical C springs, the new springs revolutionized coach operations by permitting lighter and lower coaches. Their lightness meant that they could carry twelve rather than four passengers, and their lower center of gravity allowed higher speeds to be traveled without fear of overturning. The most famous of the new coaches was the Concord coach built in Concord, New Hampshire, from 1813 till 1846. The heavier western version carried up to nine passengers in a body suspended from the springs by leather straps called *through braces*.

By 1820 coaches were finally able to provide faster service than riding horseback. Improvements in coach technology can be seen from observing changes in travel times on the route between Britain's two largest cities of the time. On the London–Bristol road—which is closely approximated by today's A4—travel times dropped from seven days using post horses in 1558, to three days or more with the old stage wagons in 1667, to thirty-eight hours with stagecoaches in 1716, to twenty-five hours in 1774, to seventeen hours in 1776, to sixteen hours in 1784 on the first mail coach run, to fourteen hours in 1800, to twelve hours in 1832 (Jackman 1916).[4] The direct distance between the two towns is about 180 km, and in 1905 the new automobiles were to turn the journey into a popular day trip.

Some of this improvement was due to pavement as well as vehicular development, and the former certainly explains the dramatic drop between 1716 and 1774. Indeed, the new surfaces saw travel speeds creeping up to 20 km/h on the roads of Telford and McAdam. Fresh horses were taken every 14 km. The speed record was held by a coach called the Shrewsbury Wonder, which traveled a stage at 27 km/h. Selby in 1888 averaged 23 km/h on the London–Brighton run (Bodey 1971).

Coaches reached their peak in England in 1836 with some 150,000 horses, 30,000 people, and 3,000 coaches providing over seven hundred regular postal runs. On the London–Bristol road, coach departures rose from 25 in 1790, to 70 in 1818, to 150 in 1830. In addition, the 1830 coaches had twice the seating capacity of the

Figure 5.2. The Royal Mail on an unsuccessful run. *From Smiles (1874), after Rowlandson's* The Night Coach.

1790 ones. The death knell for long-distance British coaches was sounded in 1838 when Parliament passed a law authorizing the railways to carry mail. The first use of rail by the post office occurred in 1841.

Some effort was made to maintain urban roads during this period of rail dominance, but rural roads often reverted to their pre-McAdam state. By 1865 road speeds were commonly back to walking pace. It was not surprising to see horses floundering to their bellies in thick mud. Farmers became mudlocked in winter, and it was often impossible to bring produce to market. (The unexpected catalyst causing this situation to change will soon be revealed.) Nevertheless, despite the mud, horse transport continued to play an active and growing role providing feeder and distributing services for the new rail networks. Rail termini came to devote about as much space to stabling as they did to locomotive sheds.

Horse-drawn transport lost its last major transport role with the coming of the car. The last stagecoach in Britain ran in 1908 over some 50 km from Kingussie to Tulloch Station in Scotland. It was replaced by a horse-drawn wagon for six years and by a car in 1914. A horse-drawn mail coach then operated on the route until 1915 (Wilkin 1979). In Australia, the last stagecoach operated in Queensland in 1924–1925 (Lay 1984).

Elegant coaches were much used by the wealthy for personal travel. This use increased dramatically throughout the railway age. Data for Britain show a total of 15,000 such vehicles in 1810, with a leap to 120,000 large vehicles and 320,000 light ones in 1900. Private carriage ownership spread from the upper through the middle class. In 1926 car ownership per capita in Britain finally passed the peak value for carriages, which had been established in 1870 (Thompson 1970).

Urban Public Transport

Within the towns, widespread public transport did not exist until the development of the horse-drawn hackney carriage in late sixteenth-century France. The word came from the French *haquence,* for the ambling horses used to pull ladies' carts, although the vehicle had later links with the London borough of Hackney. The hackney was a relatively minor variant of the burgeoning coach technology and has been described as looking like an "enlarged treasure chest on wheels" (Margetson 1967). Whereas coaches were an upper-class preserve, the hackney served middle-class travelers. In London the fleet of Hackney Hell Carts grew from zero in 1605 to six thousand in 1635, causing such severe traffic congestion and competition to the Thames watermen that Charles I introduced draconian measures to restrict their use. The associated proclamation noted that

> the general and promiscuous use of coaches there, were not only a great disturbance to His Majesty, his dearest Consort the Queen, the Nobility, and others of Place and Degree, in their passage through the streets; but the streets themselves were so pestered, and the pavements so broken up, that the common passage is thereby hindered and more dangerous; and the pieces of hay and provender, and other provisions of stable, thereby made exceedingly clear. (Anderson 1932)

The measures had little effect and, despite a dramatic change in political attitude, Oliver Cromwell was forced to confirm the regal action with an ordinance in 1654. The hackney numbers then dropped to around sixty. By 1662, when they were licensed and regulated, there were four hundred in operation, by 1750 about eight hundred, by 1770 about a thousand, and by 1800 about eleven hundred (Dyos and Aldcroft 1969; Bodey 1971). By this time most hackney coaches were recycled coaches passed down from the wealthy classes. They were becoming increasingly inappropriate and were ripe for overthrow in the transportation revolution soon to occur.

Numbering of the hackneys began in 1814—mainly to aid the recovery of property left in them by forgetful passengers—and from 1838 drivers were required to wear numbered badges. Restrictions on the number of hired vehicles operating on the streets of London, and most other cities, have persisted to the present day. British regulations in 1986 still used the term *hackney carriage* to describe gasoline-powered cabs and still required the cab driver to ensure that his horse's reins were securely held while he was dismounting from his vehicle.

At the beginning of the nineteenth century, street transport was almost entirely for the wealthy; most travel in large cities was on foot. French philosopher Blaise Pascal had operated an urban bus service for a few years in Paris from 1662 using eight-seater coaches running on fixed routes to a fixed schedule and staffed by uniformed drivers and conductors. The technique had been developed by Charles de Givry in 1657, and the service came to be known as *carrosses à cinq sous* (five-cent coaches). Government controls gave Pascal's firm exclusive rights but prevented the coaches from carrying passengers who were not "bourgeois and people of merit" (McKay 1976). The service provided by the buses was slow and expensive, and the operation failed within a year due to lack of patronage. Pascal, who died only five months after the service commenced, had a somewhat ambivalent attitude toward transport. He saw traffic as the source of most misfortunes and commented that, because of traffic, people "didn't know how to have peace in their own rooms."

Urban transport for the masses began with horse-drawn buses, first used in Bordeaux in 1812. The buses traveled at little better than walking speed and generally required a horse for each eight passengers. The word *bus* arose from an operation begun in 1826 to service some hot baths in Nantes, France, and which passed by a hat maker's shop, where the owner, M. Omnès, had a punning sign Omnès Omnibus (Omnès for all). This appealed to the bus operator, Stanislaus Baudry, who took *omnibus* for his own (McKay 1976).

The main distinction between the horse bus and the stagecoach operating in the urban area (the short stagecoach) was that all the bus passengers were accommodated inside the vehicle (Barker and Robbins 1963). Nevertheless, a hybrid short coach cum bus soon developed, based—with linguistic perversity—on the old long-coach technology.

The breakthrough came in 1829 with George Shillibeer's horse bus, which contained major improvements in styling, suspension, and steering. During this period, growing industrialization created major new demands for public transport; consequently, coach-making technology improved rapidly. Typical nineteenth-century horse buses are shown in figures 5.3 and 5.4. Bus fares soon rose, stabilizing at a level that restricted their use to middle-class passengers. Note how the passengers have been returned to the roof, but with increasingly better access and seating. The rear-entry layout of the old long wagons predominated, because it led to much higher passenger capacities than did the side-entry provisions of the upper-class coaches.

A basic problem with the horse bus was that it required relatively large wheels to lower rolling resistance from poor road surfaces. These wheels made the buses awkward and cumbersome. Indeed, service was so slow that London bus operators often provided their passengers with free newspapers and books to read while in transit—it is said that the first free library operated on just such a bus.

The obvious technical solution was to run the wheels on a smooth surface and thus avoid the need to rely on diameter to reduce rolling resistance. (The major consequences that occurred when the rail version of this solution was ultimately adopted will soon be discussed.) Urban road surfaces were steadily improving to such an extent that in London between 1875 and 1900 the horse buses were attracting passengers at a faster rate than the metropolitan railways. Suburbs such as Clapham and Hammersmith expanded under the influence of the horse bus, with land developers subsidizing the initial extension of the services to their estates (Kellett 1969). This not uncommon phenomenon is further explored in chapter 9.

The word *cab* is a shortening of *cabriolet,* which was a two-wheeled passenger cart introduced from London from Paris in 1805. The hansom cab was derived from the cabriolet and in 1834 started to replace the heavier hackney and the short stagecoaches; the last hackney was seen in the streets of London in 1858. The hansom was drawn by one horse and carried two passengers with the driver seated on a raised rear bench (fig. 5.5). A longer alternative introduced in 1838 was the four-wheeled Brougham or Growler, the first four-wheeled vehicle that could be drawn by one horse.

The cab population in London increased from one per 1,000 people in 1840 to one per 350 people in 1900. Hansoms were in operation until the 1920s, but began to feel the pressure from motorized cabs after 1906. London's last horse-cab

Figure 5.3. Wilson's new Favourite bus with deeper internal compartment, launched in May 1846. *Drawing originally from May 1846* Pictorial Times. *Courtesy of the London Transport Museum.*

Figure 5.4. An 1890s horse bus with garden roof seating outside its new direct electric opposition. *Courtesy of the London Transport Museum.*

Figure 5.5. Forder's improved hansom cab, ca. 1890. The design was first introduced in 1873. *Photo from and with the permission of the Science Museum, London.*

license was surrendered in 1947. The alternative word for cab is *taxi,* which comes from a distance measuring device, the taxameter, invented in Berlin in 1894 and used for calculating passenger fares (i.e., a tax on passengers).

Transport in Transition

Once again the developed world's common transport system had reached the limits of its development. The symptoms to be discussed below was threefold: undercapacity, seemingly insurmountable problems, and innovators searching furiously for alternatives.

Problems with the Horse

While providing a useful mobile power source, the horse created many significant problems. The demand for horses was enormous. Coach services effectively required two horses for every 3 km of travel. By the end of the nineteenth century,

Britain had one horse for every ten people, the United States one horse for every four people, and Australia one horse for every two people (Thompson 1970; Lay 1984). Each horse consumed and then discharged about 6 tonnes of sustenance each year and was said to eat as much food as eight men. It required some two hectares of land for its support.

As horse use peaked at the end of the nineteenth century, the problem of horse excreta and horse carcasses in large cities was becoming almost insurmountable, as the following New York data will illustrate. In 1866 the city's famous Broadway was clogged with "dead horses and vehicle entanglements" (FHA 1976). In 1872 a respiratory disease called the Great Epizootic swept through the city and killed eighteen thousand horses. It is quite probable that the manure-laden dust raised from the road surface by the horses' hooves provided a vector for various bacteria. In 1900 horses each day created 1,100 tonnes of manure, 270,000 liters of urine, and 20 carcasses. Carcasses and other transport remnants were commonplace on the streets; the horse had a working life of only about five years, wagons seven years, and harnesses two years.

Similar calculations put the total annual excrement output for English cities at 10,000 tonnes (Thompson 1970). A Dr. Letheby of London analyzed typical road mud and found that the average proportions were abraded stone, 30 percent; abraded iron (tires, horseshoes), 10 percent; and organic matter and manure, 60 percent (Rochester Executive Board 1886). The stench and the attendant flies were often overwhelming. A London doctor writing at that time commented that "the amount of irritation to the nose, throat and eyes in London from dried horse manure was something awful." The effects of horse-drawn traffic on road surfaces will be discussed in chapter 7.

Accidents due to horses were quite significant and caused much public concern. Certainly traffic accidents were not a new problem introduced by the car. Indeed, car riding at about 20 fatalities per terameter traveled appears much safer than horse riding, where the comparable value was 180 fatalities per terameter. In 1540 Francis I of France banned U-turns in city streets to avoid injury to pedestrians by horse-drawn vehicles. London had an active Horse Accident Preventive Society in the late nineteenth century. In one London week in November 1868 seven people were killed on the streets by horses or horse-drawn carriages. In 1867, according to an article in *Engineering,* New York was averaging four pedestrian fatalities and forty injuries a week due to horse traffic ("Pedestrian Bridges," 3:236). Other calculations showed that each 150 sq m of pavement claimed one horse fall per year (Engineering Record 1890). The distance a horse traveled on the streets of London between accidents ranged from 200 km to 550 km, depending on pavement type. This is some orders of magnitude worse than a car driver would now experience. A report in 1900 calculated that horses in the United States were causing some 750,000 serious mishaps a year (Pettifer and Turner 1984).

In addition, residents on main thoroughfares frequently complained of the incessant rattle and rumble of the iron-shod hooves and wheels of horse-drawn traffic, particularly on rough stone pavements. It was common to cover the road with straw in front of houses containing sick people, in order to deaden the noise (Thompson 1970). Cities such as Boston banned horse-drawn traffic from outside of courthouses, so that cases could be heard (FHA 1976). The problem dates from at least Roman times: the poet Juvenal complained in about A.D. 100 that he was unable to sleep because of the noise of cart wheels on stone paving. Many attempts were made to control horse traffic; for instance, London in 1867 prohibited wagons drawn by more than four horses from entering the inner parts of the city. In the same year, New York built a wooden pedestrian bridge over Broadway near Fulton Street ("Pedestrian Bridges," 1867). Many ascribed "nervous diseases" to the noise created by horse traffic.

Horse transport was also quite expensive, certainly to the extent of putting urban public transport beyond the daily reach of most urban workers. The capital cost of a horse in terms of a worker's wages was about three times that of a current car. A coach fare from Paddington to the city of London in 1800 cost one percent of a worker's annual wage. Today, the equivalent bus trip would cost 0.02 percent (Mitchell 1972). The implications of these relative costs are pursued further in chapter 9.

Despite some improvement indicated in the earlier discussion, the key overall point is that, until the nineteenth century, travel times, distances, costs, and capacities had altered little since the invention of the wheeled cart and the harnessing of the horse. Furthermore, there were many signs that the horse was reaching practical capacity and that the cities and the countryside were both reaching the limit of their ability to support any further increases in the horse population. Transport clearly needed a technical breakthrough and it was beginning to appear, but not on the roads. Before seeking this breakthrough, it is first necessary to discuss an essential prior innovation.

Rails in the Street

The haulage capacity of a horse dropped from 3 tonnes over a good surface to less than half a tonne over poor surfaces. On the other hand, a horse could haul about 10 tonnes over a level track (see table 3.1). Such potential capacity improvements provided a ready incentive for innovation directed at measures to lower the rolling resistance of a wagon. The use of special wheel tracks was an obvious solution. Indeed, wooden rails had been introduced in coal mines in twelfth-century Germany. Following improvements in iron technology that occurred during the Industrial Revolution, flat 100 mm wide iron rails were used in 1767 at Coalbrookdale, England, which was by then the leading iron-making center. Plates with raised inner edges to act as wheel guides were first used for freight tramways in 1785. Frequent breakages of the early cast-iron rails greatly restricted their early use. However, improved technology toward the end of the century greatly increased their reliability. The first iron edge-rails were used in 1789 by William Jessop at Loughborough, and in 1799 angle-shaped plateways were used by Benjamin Outram for his Leicestershire tramways.

One obvious way for horse-drawn vehicles to avoid the poor condition of the street surface was to use the newly developed rail technology. The first public goods railway (1801) and the first passenger railway (1807), a horse-drawn service known as the Surrey Iron Railway, operated from Wandsworth to Croydon in south London (Strong 1956). Twenty-five years were to pass before the idea was transferred to the streets with the running of modified horse buses on rails on Fourth Avenue and Broadway in New York. These horse trams provided a welcome service, but the rails protruded above the surface of the street and created considerable public hostility

(Armstrong 1976).[5] Twenty more years passed before a line with its rails flush with the road surface was built in New York in 1852. The delay was mainly caused by the need to await the development of a technology able to produce grooved iron rails.

Given the poor road conditions of the time, rails provided horse trams with a relatively sure and smooth passage through the quagmire of the suburban streets, cutting the horsepower required by two-thirds. Travel speeds rose to about 8 km/h. Horse-drawn trams were thus much superior to horse-drawn buses, and their fares made them more accessible to working-class travelers (McKay 1976). At last public transport had become a reality.

Within eight years of the introduction of the new flush rail, horse trams had taken over from the horse bus; they were to play a significant transport role until the First World War. In turn-of-the-century London they carried 50 percent more passengers than did the local steam rail system (Kellett 1969). The last known horse tram operated in Budapest in 1929 (Robbins 1985). Some typical horse trams are shown in figure 5.6.

Indeed, despite partially successful efforts to power trams with steam, compressed air, and cable, horsepower dominated the world's early tramways. Steam-powered trams were never popular, because they polluted the atmosphere, frightened horses, and deafened citizens. In New York, distressed citizens pulled up the tracks of a number of steam tram routes.

The first real competition for the horse tram came from the electric tram. The catalyst in this instance was the provision of a practical source of electric power, which, nevertheless, was still originally produced by steam. This occurred as a consequence of the invention of the electric generator in 1870, leading to the transmission of useful levels of power by wire. The first almost practical electric tramway was built by Werner von Siemens in 1879 in Berlin to develop methods for supplying power to moving vehicles. Using a ground-based third rail to carry the electrical power, the system proved to be electrically unsafe and was therefore fenced off from the public when it operated in Berlin in 1881. Later in the same year, Siemens demonstrated an overhead supply system in Paris. The next major technical breakthrough was made by Frank Sprague, in Richmond, Virginia, in 1888. His overhead supply system, using a grooved wheel to contact the power cable, was to dominate electric streetcar development throughout the world. Sprague was an ex-naval officer who had worked for Thomas Edison.

From just 20 km of operating track in 1887, eight Sprague-based systems were installed by the end of 1888, and by 1890 more than one hundred American cities were operating or installing electric streetcars running on 2 Mm of track. Development was noticeably slower in Europe, with only 96 km of track installed by this date. By 1902 the United States had some 36 Mm, and by 1903 some 50 Mm. The electric streetcar was adopted in the United States with much greater rapidity than the motorcar, which was invented at the same time, and has been described as "one of the most rapidly accepted innovations in the history of technology" (Hilton 1969).

A major change associated with the electric streetcar was a significant drop in fares, making it truly a travel facility for the masses. Another change was from dispersed private ownership of previous modes to systems under the control of a single large corporation. A third change was an increase in travel speeds to about 20 km/h. The unique combination of low fares and high speeds meant that it was the electric

***Figure* 5.6**. Horse trams in operation in Whitechapel in London in the 1880s. *From and with the permission of the Tower Hamlets Local History Library and Archives.*

streetcar that gave rise to the urban mass transport revolution, causing a huge increase in nonwalking urban travel. Within a decade city diameters had doubled, largely as a consequence of the electric streetcar. This issue will be pursued further in chapter 9; for the moment we stay in the early nineteenth century in order to observe the unleashing of motive power.

Steam Power Carries the World Forward

When in 1270 Bacon predicted, among other things, that "one day we shall endow chariots with incredible speed without the aid of animals," he was jailed for fourteen years by his Franciscan order for being in league with the devil. For five hundred years, the decision must have seemed eminently wise. Bacon's predictions, incidentally, were largely based on stories of self-powered, steerable vehicles operating in China in the Western Zhou dynasty in about 800 B.C.

In 1680 Isaac Newton drew a steam locomotive powered by the reactive force from a jet of steam escaping from a rear-pointing vent. Newton's drawing was based on a common scientific model and had a predecessor in a toy described in about A.D. 50 by Hero of Alexandria in his *Pneumatica* (Fletcher [1891] 1972). Nevertheless, when Newton predicted that people would one day travel at 80 km/h, his views were ridiculed. At the same time as this controversial claim, a Belgian Jesuit priest called Fernando Verbiest, who was then stationed at the imperial court in Beijing, was attempting to impress the Chinese emperor Kang Xi by using steam to blow on vanes attached to a wheel and hence move a carriage (Flower and Jones 1981). There were some obvious alternatives to steam. An Albrecht Dürer woodcut of 1510 shows a human-powered carriage built for Emperor Maximilian I of the Holy Roman Empire. At about the same time, Leonardo da Vinci sketched a cart powered by four men turning crankshafts in his *Codice Atlantico* (folio 296). A wind-powered carriage was built for Prince Maurice of Nassau in about 1600.

The first serious doubts about the Franciscan decision must have occurred just a little later, in 1712, when, after years of experimenting, Thomas Newcomen modified an air pump, developed by Robert Hooke and Denis Papin, to produce a working steam engine. Its purpose was to drive pumps removing water from flooded mines. Newcomen used water to rapidly cool steam, causing it to condense and thus create a vacuum, which provided a useful pressure differential relative to the atmosphere on the other side of the piston head. The engine therefore avoided the difficulties involved in producing effective high-pressure cylinders. It came to be called an "atmospheric" engine. Its application to transport occurred very slowly, with little happening as steam technology slowly improved. The same development stage applies to the entire Industrial Revolution. Did it begin in 1712 with Newcomen or as late as 1738 when rotary steam power first became available?

A major step forward occurred in 1765 when Scottish instrument maker James Watt realized—when asked to repair a Newcomen model—that the losses due to heating and then cooling the cylinder on each cycle could be avoided by condensing the steam outside the cylinder. In 1782 he made a further major advance by moving to a double-acting arrangement in which steam was applied alternately on either side of the piston head. He built some models of steam-powered vehicles but decided not to pursue the idea.

The first motorized vehicle was a 5 tonne steam-powered, three-wheeled tractor commissioned by the French Ministry of War for moving cannon. It was based on a prototype land carriage built in 1771 after some twelve years of development by Nicholas Cugnot, a retired army engineer. The chassis was articulated about a vertical hinge, and steering was achieved by moving the entire forecarriage, which was supported on a single 2.5 m diameter front wheel. Power was supplied by steam pressure on single-acting pistons in a pair of cylinders manufactured by a newly invented device used for machining cannon bores. The pistons caused the front wheel to rotate by moving a ratchet over teeth (pawls) attached to it. This system represented the first solution to the difficult problem of using steam power from a boiler to provide rotary motion to the wheels of a vehicle. The tractor ran for fifteen minutes at a time at speeds of 4 km/h. A famous story concerns the occasion when the tractor's low-geared steering and poor braking led to it demolishing part of a municipal wall, after which the world's first steam vehicle was impounded. As a side result of its continued impoundment, the original Cugnot vehicle survives today in the museum of the Conservatoire nationale des arts et métiers in Paris.

Cornish mining engineer Richard Trevithick provided the first effective way of producing power from steam when he invented an efficient high-pressure, non-condensing steam engine (Kirby et al. 1956). In order to avoid infringing on Watts's patent on steam condensors, Trevithick had to go beyond the atmospheric engine and develop the technologically more difficult high-pressure steam engine. The result arising from this initial frustration more than justified the effort, for in 1801 he was able to build a steam-powered vehicle by placing a steam engine on a simple chassis. This vehicle reached speeds of 15 km/h during its debut at Camborne Beacon Hill, Cornwall, on Christmas Eve of the same year. Unfortunately, it was destroyed by fire four days later when Trevithick left it overnight in a stable with its furnace still burning.

There is room for debate as to whether Trevithick's or Cugnot's vehicle can be described as the first successful self-powered road vehicle. Cugnot's was the first to operate, but was limited to very short (less than a kilometer) trips. Trevithick's vehicle was the first to make a substantial trip and to carry passengers. Both men were skilled engineers, and their work opened the door to the piston engine as the primary source of transport power. Some two hundred years later, that door has yet to be closed.

While all this mechanical engineering had been going on, the roadmaking community had been showing strong signs of complacent self-satisfaction. Telford and McAdam had provided a way to build and maintain the roads, coach technology and horse breeding had provided much improved personal travel capabilities; financiers were lending willingly to the toll road companies; and the Industrial Revolution was creating new wealth and travel demands. Not the least of that revolution's contributions was a new source of industrial power, the stationary steam engine. The rule of the road was clearly assured.

The time was clearly ripe for major change. The perceptive began to realize that society at last had a source of transport power much greater and easier to manage than the horse and the ox. But power on the roads did not come without a price. Steam engines were heavy and large. Although the iron wheel could manage the new loads, few road pavements could carry the iron wheel. To circumvent this problem, inventors first tried bigger wheels. Trevithick's second vehicle, built in 1803, used enormous 3 m diameter rear wheels (Georgano 1972). The large-wheel solution was not effective, and the vehicle created no financial interest. It was dismantled in 1804 and Trevithick died in debt.

Following another Trevithick initiative, the inventors of steam carriages soon turned to placing their iron wheels on the new iron rails. Rolling resistance was greatly lowered and much larger payload-to-power ratios were achieved. However, an initial spate of rail failures demonstrated the need to carefully distribute and spread the heavy loads via the rails, sleepers, and macadamlike ballast. Flanges were placed on the wheels to stop them from coming off the narrow rails. That this method obviated the need for steering was incidental.

Let us explore this progress in some detail. Following the lead of Trevithick, a number of inventors built steam locomotives. In 1825 George Stephenson built his second, the Locomotion, to haul coal over the 40 km journey from Darlington on the Great North Road to Stockton-on-Tees. His efforts were funded by a promoter who wished to sell Yorkshire coal on the London market. Every effort had to be made to minimize the freight haulage costs. Local contractors had previously proposed that the coal be carried by horse-drawn vehicles running on a way made of flat cast-iron plates with raised inner edges. These were called plateways. The Locomotion was therefore viewed with excitement; it was a huge step ahead of both horse haulage and the previous mechanical system, which had relied on wagons pulled by ropes powered by stationary steam engines.

The much greater weight of the Locomotion was carried by iron edge-rails rather than plates, and doubts existed as to whether sufficient driving friction would occur at the iron rail–wheel interface. These were resolved when the train began its operations in 1825, becoming the first public railway to haul both passengers and freight. The operation was not an immediate success, but the promise was clearly there. With the marriage of steam and rail, Stephenson thus achieved a major paradigmatic shift in transport by applying existing technologies in a very new manner.

Thus, as a consequence of the inadequacies of wheel and pavement, steam rail proceeded down its own separate way. In addition, an early railway accident, in which a watching dignitary was killed by a passing train, led to that way being predominantly a private right of way. Thomas Telford, incidentally, could never see the merit of the railway vehicles with their need to run on special, costly, and exclusive ways (Penfold 1980).[6] The duke of Wellington also opposed railways, but for the somewhat different reason that they would "only encourage the common people to move about needlessly." Others regarded steam power and the new travel aspirations of the community as a source of humor and cynicism, as figure 5.7 illustrates.

The feeling was deep enough for Richard Cort to begin the *Anti-Railway* journal in 1835. Apart from the reasons of Telford and the duke, the opposition to the railway companies arose from their effect on the road transport industry and from the unprincipled promotion of many parliamentary acts enabling the companies to compulsorily acquire land (Lee 1971).

Steam railway intercity passenger services began between Liverpool and Manchester in 1830. Passengers were brought to the stations by horse buses operating scheduled feeder services. The carriages were pulled by Stephenson's Rocket and traveled at the previously undreamt of speed of 60 km/h. Speed horizons took a major leap forward, incidentally refuting dire "scientific" predictions that humans would either suffocate or become mentally deranged if forced to travel at over 50 km/h. Such views were only a little more advanced than those of George Washington's physician, who held that any travel at over 25 km/h was a health hazard.

It took some time for society to adapt to these supernatural speeds. People found it difficult to estimate their speed without the benefit of some hard learning experiences, which often led to serious accidents. Typically, people would step off speeding trains as if from the door of a house onto the street. The railways had to react to cover this lack of speed sensitivity. As described by Rolt (1976), "company by-laws forbidding passengers to travel on carriage roofs seem ridiculous today but it was quite otherwise then. Accustomed to the outside seats on stage coaches, passengers invaded the roof tops on the slightest provocation only to miss their footing or to have their brains dashed out against the arches of a bridge."

In the same year as the Rocket's launch, Peter Cooper—a leading inventor and manufacturer—demonstrated the practicality of the steam locomotive to the

Figure 5.7. An 1829 cartoon by Fsop entitled "March of Intellect," or "Lord, how this World improves as we grow older." The six hours taken by the London to Bath steam-powered long wagon was half the coach time. *Drawing with permission of Ironbridge Gorge Museum Trust.*

American public in a race on the Baltimore and Ohio Railroad between his small Tom Thumb locomotive and a horse-drawn railway car. Although a breakdown led to the horse winning the race, the potential of steam rail was clearly demonstrated. Cooper had initiated the race to protect his real estate investments at the rail terminal, which were threatened by the inadequacies of a horse-drawn train operating from Baltimore to the Ohio River some 640 km away, in competition with the Erie Canal.

Steam was first used for urban commuting on a ferry service across the East River between New York and Brooklyn in 1814 (Jackson 1985). However, as with the steam tram, the steam train with its noise, air pollution, and frightening bulk never provided significant intraurban transport. Too big for the task, its strength was in interurban travel.

There was still an urban demand to be satisfied, and by the 1820s a few steam-powered vehicles were operating on the roads (Copeland 1968). Commercial service occurred for some months in 1831 when steam-powered coaches carried fare-paying passengers in London (fig. 5.8). Soon after, another venture survived for four years (Georgano 1972). The services floundered, partly because of the continued difficulty of operating heavy vehicles over poor surfaces. An 1833 newspaper report on the steam coaches noted, "The roads were in a very unfavourable state" (Fletcher [1891] 1972). Writing in 1891, before the age of the car, William Fletcher observed, "These hard roads were a severe test for the wheels and gearing." An additional factor was

Figure 5.8. The Enterprise steam bus built by Hancock for the London and Paddington Steam Carriage Company in 1833. *Drawing by W. Summers, from and with the permission of London Transport Museum.*

that accidents often occurred when the steam coaches struck large stones in the road, one such event killing five people.

A major problem was that the vehicles had difficulty negotiating the loose stones on the unbound surfaces of the turnpikes and met with considerable opposition from the turnpike trusts and the public. The trusts believed that the wheels of steam vehicles did more damage than did horses' hooves—despite contrary evidence from McAdam and Telford. As John McAdam told a select committee, "It is a well known fact that horses' feet do more injury than the wheels of carriages" (Searle [1930]). On a Scottish turnpike between Paisley and Glasgow, steam vehicles were deliberately blockaded by piles of broken stone placed across the road by order of the trustees (Lochhead 1878).

In the latter years (1828–1834) of his life, Telford became interested in a company to run steam carriages on specially constructed granite cartways along the entire length of the Holyhead Road (see fig. 1.1). His colleagues, Henry Parnell and John Macneill, also had an involvement in the scheme. The company conducted one trial trip on the existing road surface, but its vehicle broke down after 90 km (Dalgeish 1980)

In 1836 two steam vehicles built by Walter Hancock of Stratford ran in London for five months and carried thirteen thousand passengers without accident (Bird 1969). However, apart from the opposition from the turnpikes and the public, further problems became evident. First, opposition spread to the competitively threatened stagecoaches and railways. As if that were not enough, the steam road vehicles not only lacked mechanical reliability but also had a range of only 20 km before requiring a fifteen-minute rewatering stop. Indeed, they required about 10 liters (l) of water per kilometer. The much lower power demands of the smooth railways circumvented such concerns. Clearly the time was far from ripe for self-powered road vehicles.

Attention was therefore diverted to other applications of steam power. The first mobile steam engine for agricultural and construction use was produced by Ransomes in 1842. By the 1860s such machines were quite common and came to be known as steam traction engines. Some found ready use as steamrollers. Steam traction engines were easily recognized by their large and elaborate wheel systems designed to reduce contact stresses between the wheel and the ground. The use of large-diameter wheels is not surprising because increasing the diameter of a wheel is about twice as effective in reducing rolling resistance as is increasing its width. Up to the 1930s, steam traction engines were occasionally used for road haulage in Britain (Georgano 1972). They had a maximum speed in good conditions of about 10 km/h.

By the late 1850s a few relatively light steam-powered vehicles were being produced for private, hobby travel. Nevertheless, commercial versions did not arrive until the availability of kerosene and gasoline meant that it was possible to replace coal as the heat-producing fuel. These "steamers" are discussed later in the chapter.

Steam-powered vehicles continued to meet opposition in Britain. An 1836 act placed heavy taxes on steam coaches, and an 1861 act limited rural speeds to 17 km/h and urban ones to 8 km/h (Rose 1952a). The year 1865 saw the introduction of the notorious Red Flag Act—properly the Locomotive Act—which required that road locomotive speeds not exceed 7 km/h in the country and 3 km/h in the city; that the vehicle be preceded by someone walking with a red flag by day or a lantern by night; that each vehicle have a minimum of three operators; and that the vehicles "consume their own smoke," an impossible condition that was not imposed on railway locomotives. The Red Flag Act was not significantly relaxed until 1896, when the effective speed limit in Britain became 20 km/h.

No other European country had anything like the Red Flag Act, and it was a major restraint on British motorcar innovation during that time (Syme 1952). However, the state of Tennessee had a rival turn-of-the-century law that required a week's notice of any impending car trip (Pettifer and Turner 1984). Nevertheless, Britain exported steam-powered road vehicles from 1865 on, and British enthusiasts such as Colonel Rookes Crompton ran their steam vehicles in India (Bird 1969). Crompton's machine reached speeds of 33 km/h, and he went on, in 1869, to use rubber-tired steam engines to operate an Indian transport system based on trucks pulling strings of wagons. This road-train technology was transferred to many other countries, with Daimler-Renard producing an internal-combustion version in 1903.

Nevertheless, the demand for steam was clearly on the railways. The enormous improvements in interurban travel times and freight capacities that the railways initiated are sometimes forgotten. The changes were quite dramatic. From the 1840s rural roads became mere feeders for railway stations, and arterial roads fell into rank disrepair and disrepute. The early rail protagonists were well aware of their primary competitive target; for example, the world's first comprehensive road traffic survey was carried out under John Burgoyne in Ireland in 1837 as part of the planning for the Irish rail network. The survey team recorded in detail data on carts, produce carried, and people traveling on the principal roads. The resulting report introduced the variable-width traffic-flow charts still employed today. It also correctly predicted the dramatic decline in road use which was to occur between 1840 and 1900.

Bicycles and the Democratization of Travel

Improved transport arose not only through new power sources, but also through improved mechanical efficiency arising from the application of the skills created by the Industrial Revolution. We often forget the great social and technological impacts resulting from the invention of a usable bicycle in the closing third of the nineteenth century.

The first known two-wheeler was built by Count de Sivrac in France in 1791. His unsteerable wooden machines were known as hobby horses and were powered by the rider's feet running along the ground.[7] A steerable bicycle with a front wheel fork was invented in Germany in 1815 by Baron Karl von Drais de Sauerbronn of Mannheim, a gentleman of the court of the grand duke of Baden. A famous inventor, he was popularly known as the Professor of Mechanics. The vehicles attained speeds of 16 km/h and soon became popular in Berlin, Paris, and London where they were known respectively as Drais Laufmashin, draisiennes, and dandy horses. By 1818 Milan city officials had been forced to ban their nighttime use.

According to an article in the *Engineer*, in 1830 a French post office official, Dreuze, supplied postmen with bicycles powered by turning the axles rather than by the feet running on the ground ("Velocipedes," 19 March 1869 27:197). Kirkpatrick Macmillan in Dumfriesshire in Scotland introduced the first pedal-powered bicycle with a mechanical advantage in 1839, using cranks and rods to transmit the pedal power to the rear wheels. It was the first bicycle on which the rider could stay permanently out of contact with the ground. Macmillan clearly understood that a single-track vehicle could become stable above a certain speed, whereas it would fall over when traveling slowly. His bike was also the first to have a brake, but this did not prevent the Glasgow police court from fining him for "furious driving" when his bicycle struck a child in 1842. Macmillan's major new ideas were not developed further for another twenty years. Despite the potential of the various bicycle variants, investors and inventors were drawn away from them by the then-current railway mania (Whitt and Wilson 1982).

The first commercial bicycle was built in Paris in 1861 by carriagemakers Pierre and Ernest Michaux, who were oblivious of Macmillan's work and used foot pedals and a crank-based front-wheel drive. Bicycles of this type were known as vélocipedes and "bone shakers." With wooden tires, iron wheels, and large frames they weighed over 50 kg and gave a truly bone-shaking ride. The first cycle race was held in France in 1868; the prize was a gold medal donated by Napoleon III. In the following year, an international bicycle race was staged over 140 km between Paris and Rouen. Major advances in bicycle technology occurred with the invention of tension-spoked wheels between 1870 and 1876 and the improvements in ball bearing technology in England due to Joseph Hughes in 1869 and William Brown in 1878.

A desire to improve the impractical gearing associated with directly pedaling at a conventional wheel axle then led to the penny farthing bicycle. The wheel dimensions in the penny farthing were an attempt to obtain reasonable distance traveled per pedal rotation with direct hub pedaling. However, it raised the center of gravity of the rider far above that of the machine and made sudden stops quite perilous.

The safety bicycle with pedal chain drive supplying power to the rear axle was introduced in 1877 by Henry Lawson in London, reaching where Macmillan had been thirty-eight years earlier. Lawson was motivated by the concerns of his clergyman father for the safety of cyclists. The chain drive allowed the gearing necessary to avoid the large wheels of the penny farthing, although Lawson's bike still had a rear wheel 20 percent smaller than the front wheel. The "safety" in safety bicycle referred to the fact that riders were far less likely to be thrown forward over the handlebars than they were on a penny farthing. Lawson's ideas were only slowly accepted by the cycle manufacturers, and the safety bicycle did not become popular until the advent in 1885 of the first recognizable modern bicycle, the Rover designed by John Starley of Coventry. Starley's uncle James had invented the tension-spoked wheel and the safety bicycle.

Cycle manufacture flourished in Coventry while the rest of the developed world—not fettered by Britain's restrictive regulations on powered vehicles—concentrated on the new motorcars. However, by 1890 the United States alone was producing over a million safety bicycles per year. The practical development of bicycle technology effectively stopped in 1900, when a committee of cycle racers meeting in Paris determined rules for cycling that included restricting the form of racing bicycles to that currently in use (Dawson 1986). Once again regulations had struck a blow against progress.

The first motorcycles were steam-driven bicycles produced in 1859, by Pierre and Ernest Michaux in France and by Sylvester Roper in the United States (Rose 1952a). They were driven by single-cylinder steam engines powering the rear wheels via a belt and pulley. Gottlieb Daimler produced the first internal-combustion motorcycle in 1885.

It is strange now to learn of the alarm with which many greeted the bicycle. Cycling was said to be bad for the health. It was pointedly noted that the cyclist, although using the road, "was usually not even a ratepayer." Cyclists were damned as "cads on castors" and were said to have scared horses and pedestrians, raised dust, scattered mud, and traveled at excessive speeds of up to 20 km/h (Webb and Webb [1913] 1963). They were frequently, and not accidentally, struck by coach drivers' whips. An 1888 British regulation required all cyclists to carry a bell that would tinkle continuously while the cycle was in motion (Bird 1969). Cyclists were not legally permitted on the streets of New York until 1887. The associated New York act was used as a model by other states. The duke of Teck wrote to the British home secretary complaining of cyclists as "Maniacs, Persons in a state of madness." The invective flowed in both directions. Writing in the 1880s, the early American cycling activist Karl Kron commented on the "Great American Hog . . . in whose mind the mere act of purchasing a horse creates the curious hallucination that he simultaneously purchases an exclusive right to the public highways" (Hill 1980). Clearly, road hogs predated the motorcar.

Whereas steam was providing the world, or at least the prosperous world, with long-distance travel, the bicycle was offering efficient and effective short-distance travel to all, in a way that the horse and the steam train never could. The bicycle was defined as the "poor man's carriage" and initiated humanity to that new twentieth-century right, the freedom of all citizens to travel when and where they please.

Bicycle publishing began with *Le Vélocipede illustré,* which was produced in France between 1869 and 1872 (Duncan 1928). Cycle clubs were forming at about the same time and soon became active forces in the community. The first was London's Pickwick Bicycle Club, created in 1870. In the United States, various local cycle clubs joined together in 1880 to form the League of American Wheelmen, holding their first state road convention in Iowa City in 1883. In the 1880s the cycle clubs were particularly active. They produced the first modern road maps, were the first group tourists, were the founders of many of today's automobile clubs, and were the first organized protagonists for better roads through the Good Roads Associations that arose in most developed countries (Reader 1980; Lay 1984).

The first record of a dedicated cycle path being built is in Brooklyn in 1895. The 2 km path proved extremely popular (Armstrong 1976).

The bicycle also contributed a great deal to the new automotive technology. For example, it gave practical birth to steel-tube framing, ball bearings, chain drives, differential gearing, pneumatic tires, and tension-spoked wheels. It was the basis for such car companies as Peugeot, Opel, Morris, Rover, Winton and Willys. Some of the greats of American car manufacture, such as Charles and Frank Duryea and William Knudsen of General Motors (GM), began as bicycle mechanics (Flink 1985). The pattern was even stronger in less industrially developed countries such as Australia (Lay 1984).

The bicycle was to have another impact. In 1688 a device called the *vinaigrette,* or *brouette,* was introduced into Europe (Belloc 1926). This was a passenger cart hauled, not by horses, but by humans. The Asian equivalents were large passenger wheelbarrows (fig. 5.10). However, these carts and barrows were generally too heavy for their human haulers and thus provided an ineffective service. In 1870 the

Figure 5.9. The first two-day bicycle tour in the United States, the Wheel Around the Hub, took place 11 and 12 September 1879. The tour was organized by the Boston Bicycle Club (founded 11 February 1878), the first American bicycle club. *Photo courtesy of the Schwinn History Center, Chicago.*

Figure 5.10. A typical Chinese passenger wheelbarrow. *Photo with permission of Paul Popper Ltd.*

Japanese took the new bicycle technology and invented the jinrikisha, rikisha, or rickshaw—a two-wheeled vehicle for one or two passengers, pulled by a person on foot or on a bicycle. In some ways, it was a miniaturization of the hansom cab. The key to its success was its light body and low rolling resistance. The jinrikisha proved extremely popular, with some two hundred thousand in operation in Japan by the turn of the century (i.e., one for every two hundred Japanese). In addition, about ten thousand per year were exported to other parts of Asia. Jinrikisha production was thus a forerunner to Japan's modern industrialization and export success (Rimmer 1986).

Tires and the Humanization of Travel

The earlier development of solid leather, wooden, and metal tires provided a hard and unforgiving running surface.[8] Safe, fast, and comfortable travel required a far more elastic material. Rubber had that potential but could not be readily worked. The breakthrough came when Charles Goodyear discovered the practical vulcanization of rubber in 1839.

The next major step could have occurred when a twenty-two-year-old Scottish engineer named Robert Thompson invented and patented the pneumatic tire in 1845 in London. He fitted the tires to a brougham, in which he traveled some 1.7 Mm without exceptional problems. However, the idea did not take root and the patent lapsed. Lest it might be thought that this was because Thompson had been employed by the Stephensons, of railway fame, it should be added that his solid india rubber tires were used by the railways for the wheels of station handcarts (Fletcher [1891] 1972). The main problem was that the potential manufacturer, Charles Macintosh (of waterproof macintosh fame), was unable to make an airtight tube. In 1847 Macintosh's partner, Thomas Hancock, also used solid india rubber tires on carts, and in 1865 these tires were applied to the new bone shaker bicycles (Forbes 1958a). Others seeking a more comfortable bicycle ride less successfully attempted the alternative route of developing spring wheels. Nevertheless, by 1867 the solid rubber tire had developed to the extent that it became feasible to run heavy steam locomotives on highways, with contemporary writers claiming that the solid rubber tires were better than any spring system.

The solution was far from perfect, however, and it is important to realize that the bicycle became popular and widespread largely through a reinvention that changed the face of the transport world. This was the Scottish veterinarian John Dunlop's development in Northern Ireland in 1888 of a workable pneumatic tire, an invention made specifically for the bicycle. Indeed, the first pneumatic tire was built by Dunlop for his young son's bicycle, using lengths of 1 mm thick sheet rubber formed into a 50 mm tube with rubber solution, tubing from a baby's bottle, and a football pump. Dunlop's aim was to improve the speed rather than the comfort of his son's bicycle (Duncan 1928). His objective was met: when the tires were first used in a bicycle race in Belfast in 1889, the pneumatic-tired bicycle won all four races. The main difference between Thompson's tires and Dunlop's was that, in the interim, the bicycle had created a demand for the invention.

The pneumatic tire made the bicycle a usable and useful tool. This in itself was important, but the key long-term effect of the pneumatic tire was that it overcame the millennia-old narrow wheel/high-contact-pressure problem. The pneumatic tire allowed high loads to be applied to wheels in the knowledge that the tire would spread the load out over an area such that the contact pressure would approximate the tire inflation pressure.

A small calculation will demonstrate this. Wheels with solid tires could carry loads of up to 2 tonnes. They significantly damaged pavements, and the use of iron and then steel tires had exacerbated the problem. For a 2 tonne load and a typical steel tire width of 100 mm, the contact pressure between tire and pavement would be about 2 MPa (megapascal). On the other hand, the modern truck wheel can carry double the load with contact pressures of only 0.7 MPa, and with far less impact than the solid wheel, thus significantly reducing the actual stresses caused in both pavement and vehicle.

In practice, the favorable load-distributing effect of the pneumatic tire was far more dramatic. Pavement engineering uses the concept of equivalent standard axles, or ESA (Lay 1990), to compare the damaging effects of various vehicles. The ESA value of a particular wheel configuration is the number of passes of the standard axle that would do equally as much pavement damage. Table 5.1 gives some damage equivalents for the turn of the century, when two dramatically different transport technologies were overlapping. The advantage of the rubber pneumatic tire is very obvious.

TABLE 5.1
Damage Equivalents for Various Wheel Types

Vehicle type	Equivalent damage per trip[a]
Unharnessed animal, or single animal harnessed to a vehicle with pneumatic tires	0.2
Single animal harnessed to an unloaded vehicle with steel tires	0.5
Single animal harnessed to a vehicle with steel tires or a car unable to travel at over 30 km/h[b]	1
Two (or *n*) animals harnessed to a loaded vehicle with steel tires	2 (or *n*)
Car able to travel over 30 km/h	3
Truck with steel tires	25
Steam traction engine with steel tires	36

Source: Based in part on Cron 1974.

[a] To give the unit value, one (1) is regarded as the "normal" case.

[b] This related to the damaging effect of speeding vehicles when traveling over poor surfaces.

The damage equivalents must be assessed with respect to their frequency of occurrence. At the turn of the century, the pavement damage problem was most frequently one of surface wear and deterioration or of overloading by a single vehicle, rather than of long-term structural response. Today, failures are usually structural and due to the accumulation of deformation or fatigue cracking, and hence are more related to the frequency of traffic.

The pneumatic tire also offered major advantages to the early road user because its lower contact pressures meant that it deformed the pavement less and therefore was not continually climbing out of a hill. The pneumatic tire therefore had a much lower rolling resistance when traveling over poor ground. Of course, the energy being lost in rolling resistance was causing pavement damage, and so the two sets of advantages are closely related.

The first application of pneumatic tires to a powered vehicle was to a de Dion-Bouton steam-powered bicycle in 1887. Seven years later the Michelin brothers, André and Edouard, used pneumatics on their homemade Blitz entry in the Paris–Rouen road race. Although they had to patch their tires twenty-two times in the course of the 130 km race, the incentive was there. In the next year, high-pressure demountable tires were successfully used in the Paris–Bordeaux race and also installed on a Peugeot, the first conventional four-wheeled car to use pneumatics.

However, solid tires were strongly preferred by early motorists, with many predictions that pneumatics would never be widely used on cars (Harding 1980). The prophesies were doomed because, without the pneumatics, speeds in excess of 30 km/h were not possible. In addition, solid tires were far from perfect and often broke, stretched, or disintegrated and rarely lasted more than 2 Mm.

Practical pneumatic tires did not become widely available for cars until about 1900. A major improvement in the 1920s was the introduction of low-pressure balloon tires. The early trucks were too heavy for the first generation of rubber tires, which could only carry about half a tonne. They therefore ran predominantly on solid steel tires until reliable solid rubber tires became widely available in 1910. Of course, the solid tires were only marginally less damaging to the roads. Michelins produced the first pneumatic truck tire in 1912, but a number of difficulties were encountered. Technological success came about in 1916 with tires using cord rather than canvas as reinforcing, but these did not make a significant impact until the late 1920s.

Although the pneumatic tire brought the advantage of applying lower vertical contact pressures and better stress distributions to the road surface, the superior surface friction of both solid and pneumatic tires compared with iron or wood tires was of critical importance because—coupled with the greater available torque—they allowed markedly better acceleration and deceleration. The resulting speed changes applied much greater horizontal loads to the pavement surfaces than had previously been encountered.

Worse was to come for, as braking capabilities increased, so did travel speeds. This led to drivers traveling around curves at high speeds, producing yet another set of aggressive new forces. In addition, the tires of the speeding vehicles began to create significant surface suction, causing the uplift of material from the pavement surfaces, thus creating a new form of surface damage. Finally, the mode of pavement failure changed. Whereas the old iron-tired wheels had produced longitudinal ruts, the new speeding pneumatics introduced the world to transverse corrugations in the pavement surface. Thus, the speeding, braking, accelerating, skidding, and cornering new vehicles began to seriously abrade, degrade, and deform the old road surfaces, a threatening theme which is continued in the next chapter.

The Incessant Power of Internal Combustion

Before exploring the effect of the tire further, it is necessary to consider one far from forgotten invention, the internal-combustion (IC) engine. In the steam engine, combustion occurs outside the cylinder containing the piston and is used to heat water to produce steam to drive the piston; whereas in the IC engine, combustion occurs within the cylinder and works directly upon the piston.[9] A better but more difficult solution, IC also depended greatly on advances in other technologies.

The First Attempts

Before discussing the process of invention and development, it is necessary to establish definitions. The first IC engines fell into one of two categories:

1. Atmospheric, relying on atmospheric pressure to create the power stroke of the piston by opposing a vacuum created in the cylinder by the combustion explosion. The engine therefore avoided the difficult design problems created by high pressures.
2. Noncompression, in which the combustible mixture of air and fuel was not compressed prior to ignition. Such engines were modeled on James Watt's popular double-acting steam engine, but used pressure from the explosion to drive the piston.

Later, Otto's four-stroke with its precompression of the combustible mixture prior to ignition was to offer a third category.

Internal combustion as a source of power was invented in 1673 by the famous Dutch scholar Christiaan Huygens, who was working in France at the initiative of Minister Jean Colbert. Faced with the need to make a machine to pump water for the palace at Versailles, Huygens synthesized a number of existing concepts. He built on an idea first proposed by Leonardo da Vinci, on Jean de Hautefeuille's gunpowder-based water suction machine developed in the 1670s, on a piston and cylinder air pump invented by the German physicist Otto von Guericke in 1650, on the power of atmospheric pressure discovered in 1643 by Evangelista Torricelli, and on a device used by the military to calibrate gunpowder mixes. The resulting atmospheric engine used gunpowder as the explosive device. The device was not practical, and Huygens's assistant, the French mathematician and physicist Denis Papin, turned to using steam to create the vacuum via condensation. He had worked with Robert Hooke in England for a number of years and drew in part on Hooke's development of the air pump. Papin's work, published in 1695 also contributed to Thomas Newcomen's development of steam power.

Coal gas, or illuminating gas, is a combustible mixture of predominantly methane and hydrogen made by heating coal in a retort. It had been toyed with for various purposes throughout the eighteenth century. Alessandro Volta's use of methane to fire pistols in 1777 led thirty years later to Isaac de Rivas—a Swiss officer in Napoleon's army—patenting an atmospheric IC engine that ran on coal gas. He built the engine in 1813 and subsequently used it to power a vehicle, which he drove on Swiss mountain roads (fig. 5.11; Cummins 1976). It was thus the first IC car. However, both the valves and the ignition device were hand operated, so coordination of the engine timing sequences proved a major problem. The power transmission system was also relatively primitive.

In 1825 Samuel Brown in England made and sold a few two-cylinder atmospheric engines producing 3 kW and running on either coal gas, hydrogen, or hydrogen sulfide.[10] One was fitted to a chassis in 1826 and successfully climbed Shooter's Hill in the London suburb of Blackheath. However, its gas consumption was far too high for it to be practical (Bird 1960). Mainline interest was then diverted for half a century by advances with the new steam engines.

Liquid fuel was first used to fire an IC engine in 1841 in a naphtha-based machine developed by Luigi de Cristoforis in Italy. In 1859 a Belgian enameler called Etienne Lenoir produced in Paris the first useful IC engine, using coal gas from the town gas supply to power a double-acting engine.[11] The key to his invention was to draw a mixture of gas and air into the cylinder and then to cause expansion at half a piston stroke by igniting the mixture with an electrical spark. It was effectively a one-stroke engine because the double-acting arrangement meant that work was done on each piston stroke. A basic inefficiency was that the working portion occupied only the second half of a stroke. Lenoir's 250 W, 1.7 Hz engines created much interest and were quite widely sold. In Paris alone 143 were in use by 1865. Nevertheless, their size, heavy fuel consumption (about 12 cu m per hour of gas per watt of power produced), and frequent cylinder overheating meant that they did not prove a major market success.

Lenoir also developed engines powered by the coal tar byproduct benzene, which had been commercially available since 1849. He probably built the first practical IC-powered motorcar when, in 1862, he attached one of his benzene engines to a carriage chassis and drove the 10 km from Paris to Joinville-le-Pont at about 4

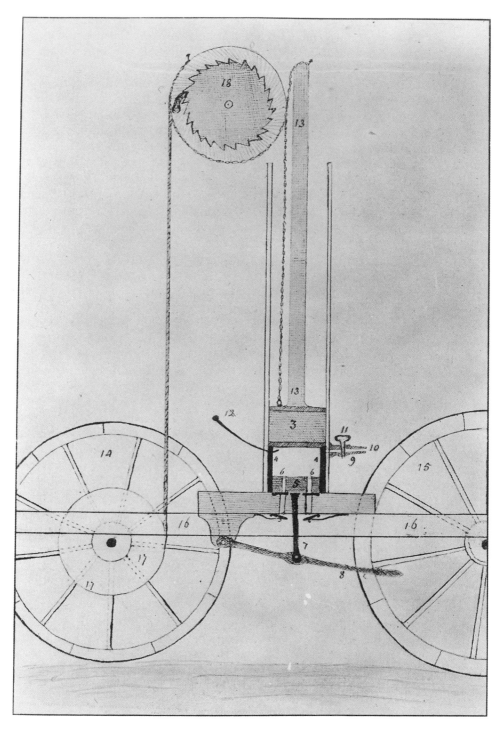

Figure 5.11. The first internal-combustion car, the de Rivas atmospheric engine and power transmission system. *Drawing from and with the permission of the Science Museum, London.*

km/h. Lenoir sold only one of his cumbersome vehicles—to Alexander II, tsar of Russia—in 1864. The other vehicle was destroyed in the Franco-Prussian War in 1870–1871, and Lenoir lost any further interest in the development of powered transport. He died in relative poverty in 1900.

The fine engineering concepts in the Lenoir engine were counterbalanced by the inadequacies of the available fuels. This situation was soon to change. Petroleum products were widely used in the Mesopotamian empires from at least 3000 B.C. A number of early religious groups made regular visits to fire temples based on petroleum and located by the Caspian Sea and in Pakistan, India, and Burma (Frazer 1914). The Chinese began drilling for oil and gas, using bamboo rods, in around 200 B.C.

Petroleum was not in high demand in the West until the first technically efficient oil lamp was produced by Aimé Argand in 1784. Its quantum jump in output initiated a new industry. Initially, whale and coal oil were used as fuel for the lamps, but they were subsequently replaced by kerosene, which did not flicker while burning.

Sensing a growing market, commercial oil refining began in 1857 using seepage material available near the San Buena Ventura mission in California. Two years later Edwin Drake drilled the first successful oil hole beside Oil Creek in Titusville, Pennsylvania, striking oil some 21 m below the ground. Indeed, the drillers had a number of clues as to where to drill; a 1775 map of Pennsylvania shows the word *petroleum* close to Titusville, which was to be the world's first oil town (Welty and Taylor 1958). The first high-volume refineries began in 1865.

Kerosene for lighting and heating was the major product of these first refineries. Consequently, many early IC engines were built to take advantage of kerosene's availability. Gasoline was also produced during the process and was initially burnt as a waste byproduct. In 1870 Julius Hock in Vienna became the first to use gasoline for power, burning it in a Lenoir-type engine.

A key advantage of the new IC engines was that they easily used these new petroleum distillates, which were most attractive power sources, providing liquid energy that was readily dispensed, transported, and consumed and which possessed a very high energy-to-volume ratio. Their main drawback was that they had previously been considered too flammable and volatile to be useful. Thus, their adoption was slow and cautious.

Given the pace of subsequent developments, kerosene remained the dominant commercial refining output until 1911, when finally supplanted by gasoline. The first gasoline pump was produced by Sylvanus Bowser at Fort Wayne, Indiana, in 1885, and the first dedicated facility selling gasoline for cars—a filling or service station—was opened by A. Borol in Bordeaux, France, in 1896.

At the turn of the century, vegetable alcohols were also promoted as an IC fuel, particularly by the French, who declared 1902 an Alcohol Year. The relevant ministry officially announced that "gasoline had had its day." However, alcohol was never able to match the power range supplied by gasoline.

German Intervention

The practical development of the IC engine is very much a German story. It begins with Nikolaus Otto, a grocer and traveling salesman who was impressed with the potential of Lenoir's invention. In 1860 he used a family inheritance to attempt to convert a Lenoir engine to liquid fuel. This project proved a long-term one, for it was to require the development of a practical carburetor and many other new concepts.

In the interim Otto and Eugen Langen, a mechanical engineer, became partners in a firm in Deutz. They developed an atmospheric engine powered by coal gas drawn in at the beginning of the piston stroke and then ignited by a flame. The explosion drove a free piston on a rack and pinion to the end of its stroke. The violent recoil required heavy foundations. Power was produced on the return stroke. Whereas Lenoir's engine had an expansion ratio of two, the Otto and Langen achieved close to ten. It gained this advantage by eliminating the inactive first half-stroke of the Lenoir engine. Thus, it had half the fuel consumption of Lenoir's, while running at twice the speed and producing three times (800 W) the power. Its negative features were that it weighed over one tonne and was extremely noisy, due mainly to its rack and pinion drive (Cummins 1976). These did not outweigh its many benefits, and in 1867 the Otto and Langen engine won a gold medal at a Paris exhibition and went on to become the clear market leader.

In 1872 the partners were joined by production engineer Gottlieb Daimler and design engineer Wilhelm Maybach. Daimler had trained as a gunsmith before taking a course in industrial engineering (Simsa 1986). He had also been excited by Lenoir's gas engine work and had visited him in Paris in 1860. In 1861 he worked in England with Crossleys, who were to be licensees for the Otto and Langen engine, and with Whitworths. Daimler became managing director of the Otto and Langen firm.

Otto discovered the importance of precompression and selective mixing of the charge within a cylinder by observing the decreasing density of smoke as it left a chimney and then making an erroneous deduction from his observation (Simsa 1986). This good fortune allowed the team to introduce the modern precompression IC engine in 1876, the famous Otto Silent, a name that distinguished it from the Otto and Langen engine. The Otto Silent had a horizontal cylinder and noticeably retained some of the features of both the steam engine and the Lenoir engine. Its major change was that it used four strokes for each combustion cycle rather than the one or two of its predecessors. In the first stroke fuel was drawn into the cylinder, in the second it was compressed, in the third it was ignited so that the pressure increased and power was produced, and in the fourth the spent gases were exhausted. Otto patented the four-stroke invention, an act that unfortunately stifled development of the technology.

The need for three nonpowered strokes was countered by the greater efficiency of the engine. Although the power output remained at about 800 W, the engine was much smaller and quieter than its predecessors. The dream of self-powered vehicles came close to reality as power-to-weight ratios improved by a factor of three. The dream was not without its sceptics. Many engineers believed that engines producing over a kilowatt of power could not be placed on a vehicle with a suspension because engine vibration would shake the vehicle to pieces. Indeed, the early cars did vibrate a great deal, although reports of the time indicate that bystanders often believed that the associated shaking of the passengers was due to fear rather than to any resonance with the engine. To overcome this problem, improvements had to come not only in power, but also in balance (Paterson 1927).

When Otto insisted on his name being attached to a four-cylinder engine developed by Daimler, Daimler and Maybach left Otto's group and established themselves in Cannstatt in 1882, initially in a workshop in the garden of Daimler's house. Daimler was then forty-eight years old, so the move was not that of impetuous youth. Indeed, he maintained a large financial stake in the Otto and Langen firm. During the early days of the new firm, its neighbors suspected that it was producing counterfeit currency and had the premises searched.

The two businesses were not without competition. Karl Benz had once worked for the same Karlsruhe firm as Daimler, and in 1861 he had also seen a Lenoir gas engine at work. Driven by the appeal of the concept, he established an engineering works in Mannheim in 1871, using his wife's wedding dowry and her continued moral support. Benz began producing spark-ignited 656 ml two-stroke gas engines in 1879 and sold his first engine in the following year.

Between 1882 and 1886 Benz and Daimler independently developed the first gasoline-powered internal-combustion engines suitable for transport. Daimler probably won this particular race in 1885 with an engine that produced just over 1 kW of power at 10 Hz and weighed over 50 kilograms (kg). The equivalent Benz engine had a slower engine speed of about 4 Hz, although even this was still much faster than any of its predecessors. Both therefore represented major improvements in power-to-weight ratios (Cummins 1976).

Daimler produced his first vehicle in 1885, a wooden motorcycle with outriggers. It was powered by a 260 ml single-cylinder, air-cooled IC engine with the fuel ignited by a tube kept hot by an external spirit burner. The engine produced 400 W at 10 Hz, made possible in part by the first modern carburetor. Power was transmitted by a pinion acting on an internal gear rim mounted on the rear wheel. The vehicle could travel at 12 km/h. Maybach drove it for 3 km from Cannstatt to Unterturkheim on its inaugural trip in November 1885 (Simsa 1986).

Benz used his new engine to power a tricycle, employing tiller steering of the single front wheel in order to avoid the unsafe and unyielding pivoted-axle steering carried over from the horse carriage. The vehicle is now in the Munich Museum. Benz conducted some private trials of the vehicle in 1885, making four circuits of a cinder track beside his factory and subsequently venturing onto the road in a surreptitious night timetrip through some Mannheim streets (Nixon 1936). Emboldened by this success, he attempted a more public voyage later in the year but drove the vehicle into a brick wall adjacent to his house.

Otto was unable to successfully maintain the patent on his invention of the four-stroke because Alphonse de Rochas, a French civil engineer and friend of Lenoir, had published a lengthy and rambling leaflet in 1862 giving the theory behind the four-stroke, including the benefits of precompression. Although he was aware of Lenoir's engine, there is no evidence that Otto knew of de Rochas's theory when developing his invention. Nevertheless, German officials in 1886 declared Otto's patent for his four-stroke invalid, and the invention became available for all to use. The officials were not very perceptive. The librarian of the Imperial Patent Office commented that the "internal combustion engine has as little future as steam for motivating road vehicles" (Nicholson 1970). Otto died in 1891 in much distress over his continued failure to achieve a patent.

Patents were also to have a major impact on the early American auto industry. A lawyer called George Selden had patented a two-stroke car in 1876, after seeing a stationary two-stroke IC engine at an exhibition in Philadelphia. Although the patent was totally based on prior devices, it affected American manufacture until 1911 (Flower and Jones 1981).

Both Benz and Daimler had eagerly awaited the decision on Otto's patent. In January 1886 Benz received a German patent for his first public vehicle, a two-seater tricycle powered by a horizontal single-cylinder 984 ml four-stroke IC motor, developing about 500 W at 7 Hz (fig. 5.12). The engine ran on gasoline, had an electric ignition, and was water-cooled—the last technique was made possible by Benz's invention of the radiator. Power was transmitted to the rear wheels by belt and chain.

In June 1886 the vehicle made its first public voyage, traveling at about 13 km/h over a short distance. On 3 July 1886 the Mannheim local paper commented with some foresight that the new vehicle should have a good future, for it would appeal to business travelers "and possibly to tourists as well." In 1888 his redoubtable wife, Bertha, was the first to drive the Benz motorcar over any significant distance when she took it for a half-day, 100 km drive from Mannheim to Pforzheim, without her husband's knowledge, purchasing the fuel for her journey from wayside pharmacies. Given that this was the first significant motorized trip, Bertha Benz was arguably the world's first motorist.

The saddle of the motorcycle subsequently caught fire while Daimler's son was riding it. Daimler henceforth concentrated on cars and in 1886 used a vertical single-cylinder, water-cooled four-stroke IC engine to power a four-wheeled American horse carriage. Blessed with four forward speeds, it made its first journey on the road in September 1886. It has been argued that this Daimler engine was "undeniably the direct ancestor of the present day car engine" (Bird 1960). It was also a significant advance over all its predecessors.

Figure 5.12. The first Benz car, called the Benz Patent Motorwagen, produced in 1886. *From and with the kind permission of Daimler-Benz AG Archiv.*

In 1886 Daimler built the famous Steel Wheel Car, or Motorkutsche, depicted in figure 5.13. Clearly, it was not an engine on a carriage but a new vehicle designed from the ground up. Its 470 ml horizontal single-cylinder, water-cooled engine was also a major development, producing some 2.5 kW at 12 Hz, permitting a vehicle speed of 16 km/h. The prototype did not go into production, and Daimler, who was under financial pressure and had little interest in cars, then temporarily put aside car making to concentrate on developing his successful 1.5 kW engines (Nicholson 1970). To fill the vacuum, in 1890 Peugeot, a French cyclemaking firm, began building cars powered by the 1.5 kW Daimler engines.

In reviewing this retrospective "race" between the two men, it can be said that in 1885 Daimler built the first gasoline-powered engine and Benz built the first gasoline-powered vehicle. Their emphases certainly differed; Benz focused on the whole vehicle, designing his vehicle from scratch to suit the engine, and Daimler concentrated on the engine and then looked around for a chassis to which to attach it. Although Benz and Daimler never met, the firms established by the two men finally merged during hard times in 1926. Daimler died in 1900 and Benz in 1929.

For many years it was thought that a German-born Austrian engineer named Siegfried Marcus had operated an IC-powered wagon at the 1873 Vienna Exhibition, after a previous limited trial in 1864 (Field 1958). However, it has since been shown

Figure 5.13. The first Daimler car, called the Motorkutsche, with Gottlieb Daimler sitting in the rear and his son Adolf driving. It was produced in 1886. *From and with the kind permission of Daimler-Benz AG Archiv.*

that the vehicle was not built until 1888, with the 1873 story started in 1900. It used a single horizontal cylinder producing 500 W. The vehicle covered some 12 km to a nearby town, traveling at about 7 km/h. It was then banned by the authorities for the noise it made. The same fate awaited Australia's first self-powered vehicle (Lay 1984). The Marcus vehicle is preserved in the Technisches Museum in Vienna.

It is easy, but unfair, to gloss over the enormous contributions of Otto, Benz, Daimler, and Maybach. They not only had to develop a workable engine with good power-to-weight ratio and low fuel consumption, but they also had to invent or produce effective electrical ignition, water-cooling, carburetor, differential transmission, gear box, and steering.[12] Nevertheless, it must be remembered that such essential features as gearing, differentials, universal joints, and independent suspensions drew heavily on the technology developed for steam vehicles half a century earlier. In addition, the first cars still used wooden brakes and chain drives and vibrated enormously.

The dramatic IC developments did not occur in a commercial vacuum; the market potential was widely recognized. As the author of an article in the November 1886 issue of *Scientific American* perceptively commented:

> To produce a small engine that can be operated by the combustion of petroleum is a problem that has received the attention of quite a number of our best inventors. That there is a great demand for small power machines there is no question; and the almost unlimited supply of petroleum, and the low price it is sold for, induce the seeker for a cheap, small power source to turn his thoughts in the direction of petroleum for a fuel.

The French Takeover

Faced with a somewhat sceptical local German market, opposition from a government that favored railways, and a lack of interest from the wealthy classes, Benz sold his third motorized tricycle, together with sales and manufacturing rights, to Emile Roger in Paris in 1887. It was the world's first car sale. Curiously, the first American car sale occurred in 1893 when Ransom Olds sold a steam vehicle to a British patent medical firm for use in Bombay (Flink 1974), and the first German to buy a car was later declared insane (Beasley 1988). The first independent car dealer appears to have been Willam Metzger, a bicycle dealer who set up shop in Detroit in 1896. Sears began selling cars by mail order catalog in 1903.

The Benz business expanded, and by 1888 he was employing fifty people and exporting most of his relatively cheap IC tricycles. In the following year an improved model was successfully exhibited at the Paris Exhibition. Benz produced his first four-wheeled vehicle, the Velo (fig. 5.14), between 1891 and 1893. Technically the Velo was little different from the tricycles, although, perhaps inspired by his earlier steering mishap, Benz introduced the stub-axle forward steering system still used in today's cars, although it was operated by a tiller and not a steering wheel. Though they were rated at 1 kW, the evidence is that the single-cylinder 1,045 ml engines running at 7.5 Hz produced closer to 2 kW. The Velo was variously called the Doctor's Car, the Benz dogcart, or—in England—the ha'penny Benz in reference to its running cost per mile. The Velo was probably the first standard-production motor vehicle. One-third of the output was sold in Germany, one-third in France, and one-

Figure 5.14. The Benz Velo, the first inexpensive passenger car and the first standard production motor vehicle. The driver is Benz's daughter, Clara. *From and with the kind permission of Daimler-Benz AG Archiv.*

third in other countries. The Benz Victoria, first marketed in 1893, was much larger than the Velo and used a 4 kW engine.

Benz became resistant to improvements and did not change his Velo model until 1901. In 1898 he lost his position as the world's largest car manufacturer—a role he had held for the first ten years of the car's history—and by that time he had been technically passed by many other manufacturers of small cars, led by de Dion-Bouton and Renault in France.

After some pressure and after seeing Benz and Daimler vehicles at the 1889 Paris Exhibition, Emile Levassor, an engineer with the French powersaw manufacturers Perrin and Panhard, acquired Daimler's patents in 1891. The famous Panhard-Levassor was the first designed-for-the-purpose car. Earlier four-wheeled cars had been IC engines carried on a horse carriage or straddled between a pair of bicycles. However, in the Panhard-Levassor, the 2.5 kW Daimler engine was moved forward to provide more weight on the steering wheels and hence increase steering adhesion. Power was transmitted to the rear axles by a chain drive. This arrangement permitted much larger engines to be used and came to be known as the Système Panhard, or Levassor. It also overcame the visual void created by the absence of the horse, a void that some earlier designers had filled by attaching a replica of a horse's head to the front of their carriage.

Panhards were big, heavy cars reflecting Levassor's two notorious sayings: "Make it heavy and you'll make it strong," and "It's brutal, but it works." The first Panhard was produced in 1891. Levassor is remembered by the 1907 monument in his honor at Porte Maillot in Paris, which has been described as "one of the most grandiose tributes ever paid to an automobilist" (Silk 1984).

In 1893 Maybach developed the modern jet-type float-feed carburetor. This permitted smooth changes in engine power and speed. In the same year, Rudolf Diesel patented his explosion-type engine, although it did not appear in passenger cars until the late 1930s. A year later, Panhard produced a new car, in which power was transmitted to the rear axle by a solid shaft driven by a forward-mounted, two-cylinder, four-stroke engine running at 13 Hz. It is said to have been the first modern-looking car, although to the modern eye it seems much closer to the horse-drawn carriage than to the motorcar.

In 1895 Panhards began using the Daimler-Phoenix engine, which produced 3 kW of power at 12 Hz and provided top speeds of 30 km/h. Meanwhile, Daimler-Maybach was still making its heavy rear-engined cars. The 1894 Panhard and 1895 Daimler, both Daimler-powered, stand side by side in London's Science Museum, and the dramatic external design improvements introduced by the Parisians are very apparent. In yet another technical innovation, Panhard introduced the steering wheel on a raked column in 1898. The Panhard firm went into slow decline after World War I and was eventually acquired by Citroën in 1955.

In 1895 de Dion-Bouton, converted from steam engines, produced the first compact IC engine and a year later patented the synchromesh gear box. Up to 1894 the power output of cars ranged between 0.5 and 5 kW, depending on vehicle size. In 1896 Frederick Simms and Robert Bosch introduced magnetic ignition, an invention that had a major efficiency impact. Bosch, who made a number of contributions to automotive electrics, had trained under Thomas Edison. In the following years car design improved dramatically, with an increase in power-to-weight ratio of twenty-five to one between 1895 and 1903, mainly due to the exceptionally rapid development of Otto's four-stroke engine (Aitken 1907; Flink 1985). The advantages of the IC engine were soon widely recognized. They were feverishly placed on wagons, in carriages, and between bicycles to produce a whole range of self-powered vehicles, increasingly running on the new pneumatic tires.

The first modern car in all essentials is said to have been the 1901 Mercedes (fig. 5.15), designed by Maybach in 1900 and produced by the Daimler firm. It is not coincidental that it appeared a year after the death of the conservative Daimler and was very much Maybach's creation. The Mercedes was named after the daughter of the Austrian-born Parisian banker and car dealer Emile Jellinek, who was also Austrian consul in Nice. Jellinek had already used his daughter's name as a nom de plume when he indulged in car racing. He had urged the construction of the new vehicle, supported by a sales order for the first thirty-six (Flink 1985). His main requirement was that the mass of the new vehicle be closer to one tonne than to the

Figure 5.15. The first Mercedes, built in 1901 in Germany. It had a water-cooled front engine. *From and with the kind permission of Daimler-Benz AG Archiv.*

2 tonnes of the preceding Daimlers. The new name was chosen because Jellinek did not think a car with a German name would sell well in France.

The Mercedes was based on the Système Panhard, but the one feature that distinguished it from earlier cars was its relatively low ground clearance, reflecting the greater availability of good road surfaces. Its four-cylinder 5.6 l Phoenix motor could produce twenty kW and permitted the vehicle to cruise at over 70 km/h. The engine power was soon raised to 26 kW. The Mercedes also had the first effective (H gate) gear change. The model was widely copied in both Europe and the United States (Hughill 1981).

The IC invention was initially exploited commercially by the French rather than by the Germans. At the time, the French road network and petroleum distribution system offered many more motoring opportunities. The French were to stay the world's major car makers until 1905. The United States was more like Germany, where U. S. car manufacturer Albert Pope commented in 1903, "The American who buys an automobile finds himself with this great difficulty: he has nowhere to use it" (Flink 1974).

By 1905 the car had developed into a truly independent transport technology. The dangerous short wheelbase and high center of gravity associated with horse-drawn carriages had given way to longer wheelbases and lower centers of gravity. Convenient side-entry doors and enclosed bodies had emerged. Water-cooled, forward-placed engines operating at about 16 Hz, friction clutches, and self-centering steering operated by a wheel on a raked column were common (Nicholson 1971). Drivers no longer had to be physically strong. The bodies of many early cars were custom-made for their wealthy owners by *carrossiers,* or coach builders, who readily catered to their wealthy clients' common desires for individuality and flamboyance. In 1908 Henry Ford began his mission to change the exclusivity of the car.

American and British Beginnings

The development of operational IC cars was due in the United States to the two Duryea brothers, who produced a one-cylinder car in 1893, in Britain to John Knight in 1895, and in Australia to Harley Tarrant in 1899. Charles and Frank Duryea, who were bicycle mechanics, were inspired by reading of the Benz vehicle in an 1889 issue of *Scientific American* and by then seeing one exhibited at the 1893 Chicago Exhibition. On the other hand, in 1896 Henry Ford—who by then had built two IC cars of his own—traveled from Michigan to New York to see a Benz on display in Macy's store in New York in 1896. He commented, "It had no features that seemed worthwhile" (Ford 1924). The first Duryea operated in Springfield, Massachusetts. Serious production began in 1896 when thirteen were made and sold. Demand soon began to increase dramatically. In 1900, the United States produced 4,192 cars: 1,575 were electric, 1,681 were steamers, and 936 were IC.

Self-taught mechanical tinkerer Henry Ford built his first, primitive, IC-powered quadricycle at his home in 1893 while employed by Detroit Edison. The vehicle produced about 2 kW and could travel at 40 km/h. Six years later Ford left Edison, taking his patents with him, to join a small racing-car company. He did not achieve prominence until 1902, when his ability as a race-car driver was established and his 999 car had won several major races. The firm went on to build Cadillacs, but Ford parted company in 1902 to begin a new venture. The Ford Motor Company was founded in 1903, a week after a major victory by the 999. For five years the company made and sold relatively up-market cars, beginning with the two-cylinder Model A. Seventeen hundred Model As were sold in 1903. Ford introduced Models B and C in 1904, Models F and K in 1905, the four-cylinder Model N in 1906, Model R in 1907, and Model S in 1908. By this time the annual sales had risen to ten thousand.

The development in Europe of high-strength vanadium steel permitted Ford to redesign his earlier heavy cars down to the lighter and smaller Model T, a move that was already evident in the Model N. These new Fords followed the high ground clearance of the pre-Mercedes design approach, giving them an advantage on rough rural roads. Ford began producing the 15 kW, 2.9 l, four-cylinder Model T Ford (or Spider or Tin Lizzie) in 1908. It had a top speed of about 70 km/h.

Up to this stage—and some twenty years after the first car—manufacture was still very much on a car-by-car basis. There were two reasons for this. First, cars remained a luxury device for the wealthy. Second, the methods of mass production did not exist; indeed, they were to be introduced to the world by car manufacturers. Fifteen million Model Ts were made over the next eighteen years. While the competitors' cars were selling at $2,000, the price of the Model T dropped from $825 in 1908 to $260 in 1925. Sales peaked in 1923. A famous saying of the time was One day, One Dollar; One Year, One Ford. For more than twenty years Ford built the same mass-produced car in "any color so long as it is black" (Ford 1924). Black was chosen, incidentally, because it was the only paint color of the time that did not fade.

Ford's competition was predominantly Alfred Sloan's decentralized General Motors, which had been founded in 1908 and was now firmly decentralized. GM stole the market in the 1920s and 1930s by producing a range of cars very different from the unchanging, ever-cheaper Model T. The new General Motors cars changed annually and offered a range of models to satisfy the pretensions of upwardly mobile citizens. Car owners could progress from Chevrolet to Pontiac to Oldsmobile and, finally, to Cadillac. Even if they stayed within one class, their new car would be very distinguishable from the neighbor's one-year-old car. It is also worth noting the remark attributed to Charles Kettering of General Motors when defending his firm's relatively slow introduction of improvements in technology and safety: "It isn't that we are such lousy car builders, but rather that they are such lousy customers." Nevertheless, in any year during the 1920s American firms never made less than 90 percent of the world's cars.

In exploring the relatively slow involvement of the otherwise inventive British engineers, it must be recalled that Lenoir, Daimler, and Benz had all previously worked on the manufacture of gas engines, rather than on the steam engines that had tantalized British inventors. Of course, the Red Flag Act also had a major stultifying effect. Another factor inhibiting British engine development was that British law up to 1900 prohibited the sale of low flashpoint fuels. This meant that gasoline was relatively impractical and that British engines had to run on kerosene (Cummins 1976). The kerosene market had been contracting since Edison's 1879 invention of the incandescent light, so Benz and Daimler's invention came at a very appropriate time for the petroleum industry.

The English were readily able to rationalize their situation. The new IC cars were popularly referred to as "foreigners' playthings" and "nasty-smelling horseless carriages," and car racing was "a new French sporting craze."

Racing and Records

Cars entered racing in 1891 when a Peugeot successfully followed the field in the Paris–Brest cycle race. The first real car race was sponsored by *Le Petit Journal* and run in 1894 over the 130 km from Paris to Rouen. Vehicles could only enter if they were "easily controlled, safe . . . and not too expensive." The 102 initial entries had first to show that they could travel at least 13 km/h. This reduced the actual field to 21. The winner was decided by a jury using speed, ease of handling, and running cost as criteria, rather than by first across the line (Bird 1960). Equal first place was awarded to Daimler-powered vehicles built by Panhard and Peugeot. All thirteen of the IC-powered vehicles finished the race, whereas only four of the eight steam-powered vehicles were able to do so. Nevertheless, the fastest vehicle and first across the line was a de Dion-Bouton steamer at an average speed of 20 km/h. It had lost the points needed for victory because it operated as a tractor trailing its passenger compartment.

The first conventional race was held in 1895 over the 1.2 Mm from Paris to Bordeaux and back. The field of forty-six included twenty-three IC, thirteen steam, and two electric cars. Only nine finished the race, and the first eight places were filled by IC cars. The initial winner was Emile Levassor, driving his Panhard-Levassor with its Daimler-Phoenix engine at an average speed of 30 km/h. However, he was disqualified on a technicality, and the race was awarded to the driver of a Peugeot whose mean speed was almost 20 percent slower than Levassor's. A year later Levassor died as a result of an accident during a Paris–Marseilles–Paris car race. The Paris–Bordeaux race received worldwide publicity and convinced many of the viability of the IC vehicle as a means of fast long-distance transport.

Soon after this first car race, the organizers, led by Albert de Dion and others of great wealth and power, formed themselves into the Automobile Club of France and became the world's first automobile club. In 1904 the club took a new lead and organized the first international meeting of automobile clubs.

The demise of the Red Flag Act in Britain in 1896 was celebrated by the first of the famous London to Brighton car rallies, said to have "started in confusion and ended in chaos." The event was possibly won by the American Frank Duryea driving his own gasoline-powered vehicle, however the dominant car was Levassor's winning Panhard from the Paris–Bordeaux race (Bagwell 1974).

Despite the evidence of the race results, the IC cars faced serious competition and the 1898 and 1899 world speed records of 65 and then 106 km/h were held by electric cars. In 1902 the record was lost to a steam car traveling at 125 km/h. IC engines held the record between 1903 and 1905, and steam returned in 1906 when a Stanley Steamer traveling at 196 km/h rewrote the record book. The official railway speed record in 1904 was 165 km/h, held by a GWR (Great Western Railway) train traveling between Plymouth and London. IC motors did not begin to dominate the speed record list until the arrival of the Blitzen Benz in 1909, which pushed the record to 235 km/h over the next two years.

Arguably, the first sports car was the 1903 Mercedes 60 with its 9.25 l engine running at 17 Hz. The first car produced specifically for racing was designed by Ettore Bugatti and built in 1905 by the old Alsation metalworking firm of De Deitrich. Its engine produced 45 kW, representing a further dramatic increase in available power.

In 1900 passenger car trips were being undertaken over the 1.7 Mm from London to Edinburgh and back. In 1903 a 15 kW Winton car driven by Vermont physician Nelson Jackson and his chauffeur Sewell Crocker spent sixty-five days in total, and forty-four actually traveling and hurdling millions of obstacles, to become the first vehicle to motor across the United States, traveling from San Francisco to New York (Harding 1980). A formal road across the country was not available until the Lincoln Highway (or Route 30) was opened in 1923. Australia was crossed from west to east in 1907 (Lay 1984). That same year saw the epic 14 Mm Peking to Paris rally. Four cars completed the journey, taking between sixty and eighty-one days. By 1914 all inhabited continents had been crossed by car.

Competition for Internal Combustion

The availability of kerosene and gasoline had also made steam-powered vehicles far more effective than their 1830s predecessors. This new technology was first evident in such European steamers as the Gardener Serpollet, which was produced in the final years of the nineteenth century.

The most famous of all was the Stanley Steamer, which hit the road in 1897. Modeled on European steam vehicles and made in Boston by the twin Stanley brothers, Francis and Freelan, from Maine, it burned gasoline for fuel and produced 4.5 kW. The Stanley Steamer also introduced light tubular-chassis construction to the car industry. It proved reliable and more effective than the electric cars, which then had about half the market. The brothers sold their patent rights to Amzi Barber of asphalt fame and magazine owner John Walker in 1899, but within a year had resumed steam-vehicle manufacture. Barber's Locomobile Company made some five thousand steamers, but changed to IC engines in 1902. Stanley Steamers were produced for about a quarter of a century; their best year was 1912, when some 650 were sold.

Steam-powered vehicles had some lingering advantages over the IC engine. They were quieter, did not need a gear box, and once the engine was in operation, a single lever would stop, start, and reverse motion. However, the advantages did not outweigh the disadvantages of weight, requiring ten minutes to warm up, and needing frequent and significant amounts of water. The early steam boilers were also prone to explode. The difference was not as evident as it might now seem, because the early IC vehicles were inefficient, noisy, and evil smelling, and did not start instantly, but took some minutes, even in good conditions.

The first experiments using electric batteries to power vehicles began in the 1830s, but it proved impossible to provide an adequate power output. A Vermont blacksmith called Thomas Davenport built the first electric-powered car in 1834. A series of desultory and unsuccessful experiments followed, including work by Charles Page funded by the U.S. Congress during the 1850s.

The invention of the generator overcame the battery problem for fixed-track electric vehicles. Battery-powered electric cars did flutter onto the transport stage in 1896 and held the world speed record in 1898 and 1899, but, in retrospect, their challenge was never a serious one. For the ordinary driver, their major selling point was that they were far easier to start than either the steam car, which required its water to be boiled, or the IC car with its frustrating and often unsuccessful and wrist-breaking manual crank starter. However, the batteries were heavy, the fumes offensive, and the power insufficient to give good performance over reasonable distances. Electric cabs were introduced into Paris in about 1895, but were never particularly successful, and were soon replaced by two-cylinder IC Renaults.

The simplicity of the electric car appealed to women far more than to men. However, the introduction in 1911 of electric starters in IC cars, initially called a "ladies' aid," and of enclosed car bodies after 1919, stole even that market from the electric vehicle. Sales of electric vehicles peaked in 1912. In 1913 the United States had thirty-five companies making electric cars and thirty-seven thousand such cars on the road.

The Linguistics of Power

The word *petroleum* is a combination of *petra* (rock) and *oleum* (oil) and has been in use since at least 1526 (*OED*). The French word for petroleum is *petrole,* and the word *petrol* was probably coined by Eilhardt Mitscherlich in Germany in 1823 (Simsa 1986). The American term *gasoline* arose from the fact that many of the early IC engines were gas engines and did indeed run on gas.

The word *car* is a very old one and derives from the same Celtic and/or Indian root that gave rise to carry, chariot, carriage, and cart. Before the sixteenth century the word was generally applied to wheeled vehicles, but then came to be used for poetic and grandiose purposes. For example, John Milton wrote of the "gilded car of the day." The word was thus picked up by the ambitious new railways and from 1837 was applied as *rail car* to their more prestigious passenger vehicles. Subsequently, the terms *streetcar* and *tram car* came into use. Indeed, the first motorcar was a motorized rail car and the term was not used for a road vehicle until 1895.

The word *automobile* was coined by the French Academy in 1875 to describe steam buses. It was adopted in the United States as a substitute for the term *horseless carriage* at about the same time as the words *motor wagon* and *motorcar* came into use. The word *automobile* was attacked in a *New York Times* editorial on 3 January 1899 for "being half Greek and half Latin. [It] is so near indecent that we print it with hesitation." Although the *Times* was quite correct, relative to the horse-drawn carriage the new IC car must truly have seemed to offer automatic mobility. A more alliterative early American term was "benzene buggy," a reference to the common use of benzene (once spelled benzine) from coal distilling as a fuel in early IC engines. The word *benzene* came from the material's original source in the natural resin benzoin—its link to the Benz name is a coincidence. A number of the early IC vehicles were merely horse carriages with an engine mounted on them (see, e.g., fig. 5.12), and so it is not surprising that the names horseless carriage, horseless buggy, and benzene buggy came into early vogue. The English, for a while, favored the word *autocar.*

Daimler produced the first IC motorbike in 1885. However, commercially produced motorcycles did not become available until 1895. Since many of the early motorcars were heavily based on cycle technology, the word *motorcycle* was often used in the early days to describe four-wheeled as well as two-wheeled vehicles.

Summary

The self-powered vehicle was very much a creature of the nineteenth century and of the Industrial Revolution. Just when the horse and the ox had reached the practical limits of their transport capabilities, a new set of motive power sources appeared on the scene. Was this just another technological coincidence? The initial steam engines associated with these new power sources were far heavier than existing road vehicles, and the ways of the time soon proved inadequate to carry them. However, motive power was so attractive that innovators soon borrowed rail technology from the mining industry and the railways were up and running on their own fixed rails. The advantages of powered travel on the railways were so attractive that the road systems fell into ruinous decay.

But steam as a power source had its problems. A heat engine needs heat, and supplying it by burning coal and steaming water has inherent drawbacks. In addition, other processes of the time meant that alternative and more convenient heat sources such as coal gas and the relatively useless light "end" of the newly mined petroleum were becoming available.

Inventors did not fail to realize that the Industrial Revolution and the widespread desire to travel had created large potential markets for devices of low mass and high power. This set of circumstances threw up history's normal set of on-the-spot exceptional human talent, ready and eager to meet the need. When Lenoir fired the starter's pistol with his gas engine, the race was on.

In particular, the internal-combustion engine provided the world with a dramatic leap forward in transport capability—a leap for which it was largely unprepared. The story is a fascinating one of competition, cross-fertilization, market adaptation, and great inputs of human ingenuity. And what would have been the influence of men like Daimler and Benz had they been born at some other time in our mechanical history? How is it that history consistently provides the right people at the right time? Does the world always have a reservoir of untapped talent or does each circumstance produce the talent needed?

There can be no short summary of the astonishing impact of the IC engine on world transportation. Instead, the next chapter covers short-range responses and chapter 9 addresses long-range effects. Nevertheless, it is useful at this stage to ponder two questions. First, where might the world be today if petroleum had never existed? A century after its first application as a power source, no useful substitute fuel has yet been found,and so how would all those market pressures have been met without widespread degradation and disruption? While it is proper to decry the often untrammeled pollution and division caused by the IC-powered car, there is little doubt that without it we would in net terms be a poorer civilization. Some further insights on this issue are given in chapter 9. Second, what a pleasant coincidence it was to find that the production of fuel for the new vehicles produced a byproduct so admirably suited to meeting the new demands that the IC vehicle was placing on the roads. This factor occupies much of chapter 7.

Power in the Road

Chapter 6

*If we cannot control the speed of motor traffic in the
infancy of the motor era, how can we expect to cope
with it in the future?*   LOGAN PAGE, 1910

*T*he previous chapter discussed the many problems created by the
growing nineteenth-century use of the horse for transport. The problems had often
seemed insurmountable, and the new motorcar was seen by many as their solution.
The following resolution from Britain's 1898 Annual Municipal Engineering Conference illustrates the faith of many in the car solution:

> That this Conference of Municipal Engineers assembled in connection with
> the congress of the Sanitary Institute . . . is of the opinion that the intro-
> duction and use of efficient motor vehicles should be encouraged by country,
> municipal, urban and other authorities, in view of the fact that the extended
> use of such vehicles would contribute to the general improvement of the sani-
> tary condition of our streets and towns. (Bagwell 1974)

Municipal engineers might have been partly jumping from frying pan to fire,
but overall they were right. Commentators have concluded that, at the time, the
developed world had reached horse capacity and any appreciably greater numbers
of horses would have been quite literally insupportable, and that the motor vehicle
had raised the quality of urban life by driving the smell and squelch of the horse
from the streets (Thompson 1970).

The Arrival of the Car and the Truck

The development of the IC car began relatively slowly as an additional hand-
crafted plaything and convenience for the wealthy, used for touring and pleasure
rather than for commerce (Flink 1974). Gradually, it came to be usefully employed

by the professional classes, ensuring that it remained a sign of affluence and authority. The floodgates opened when manufacturers realized the potential size of the middle-class market.

Mass Production

The presence of the car could no longer be missed. From two motorcars in 1886, the fleet had grown to eleven thousand by the turn of the century. In 1906 the world was producing fifty thousand vehicles a year, 60 percent of them made in Europe. Two major events increased the use of the car in the United States. First, during the 1906 San Francisco earthquake it provided emergency services, giving outstanding performance and receiving great publicity (Flink 1974). Second, the car and the truck then made a major contribution during the First World War. Consequently, the rate of car usage changed dramatically after the war. By 1928 the world had twenty-six million cars and three million trucks, most of them mass-produced. In 1909, the car makers ranked twenty-first among all American industry in terms of the value of their production. By 1925 they were on top of the list.

Large-scale car manufacturing began in Detroit in 1901. By 1904 Detroit was producing 20 percent of the American car output, and within eight years the city was claiming the title Motor City as the heartland of American motor vehicle manufacturing. Mass production was begun by Ransom Olds of Oldsmobile fame in 1902 and was furthered by Henry Leland of Cadillac, who made a car with completely interchangeable parts. Mass production began in earnest in 1911, when Henry Ford built on the ideas of Olds, Leland, and others. The moving assembly line was introduced by Ford in 1913, and three hundred thousand cars were sold in 1914.

Some important dates and events in the development of the mass-produced car in the United States are given in table 6.1. The introduction of the enclosed, all-steel body did not occur until relatively late, because it was first necessary for IC engines to become sufficiently powerful to be able to carry the associated increase in body weight. The most common car type in Europe in the 1920s had a light, soft top with only rudimentary protection from wind, rain, and snow. Cold weather required occupants to wear thick protective clothing. Only 10 percent of American cars in 1920 were enclosed from the weather. By 1927 this had risen to 83 percent. Lighting was also slow to change; many trucks in the late 1920s still used kerosene or acetylene lamps.

Trucks—Internal Combustion at Work

The original trucks were lowly developments of the wheeled sled. The word *truck* came to be applied as a verb to trading, particularly in market produce. In the United States the word was then turned into a noun and applied to the vehicle. The British prefer the word *lorry*, a north country perversion of the name—Laurie—of the 1838 inventor of the flatcar.

Following the precedent of the great, lumbering, steam-powered traction engines, the first practical trucks were also steam-powered. They were in use on the roads until about 1930, although Britain with its abundance of coal did not stop building steam trucks until 1950 (Gibbins and Evans 1958). Daimler built the first IC truck in 1891 by slinging an engine under the tray of a wagon (fig. 6.1). Daimler trucks came on the market in 1894 and still had the bulk of the market in 1900. They were not universally well received. According to one contemporary commentator, the IC truck "barked like a dog and stank like a cat" (Bagwell 1974).

By 1910 it was generally recognized that the truck was superior to the horse-drawn wagon, where adequate pavement surfaces were available (Page 1910). In economic terms the tonne per kilometer costs in 1914 for hauling freight over short distances were, relative to the IC truck, 30 percent higher for rail and 100 percent higher for the horse. One further advantage of the truck was that it was a far more efficient user of road space, carrying the same load but using only 20 to 40 percent of the road space. It thus had a major and immediate favorable impact on urban congestion.

As in many other parts of this story, the needs of the military played a leading role in the development of the truck. The French began using IC vehicles in military maneuvers in 1897 and the Germans in 1899. The first use in battle was in the Boer War between 1899 and 1902. The end of the horse era was surely the 1911 announcement by the British War Office that the motor truck was to replace the horse on a large scale in the British army (Georgano 1972). The message did not filter down rapidly to the regiments. Osbert Sitwell has recalled that in 1912 officers of his hussar regiment were still giving lectures to recruits on such subjects as "The place of the horse in the twentieth century. & How the horse will replace mechanical means of transport" (Sitwell 1948).

The future military usefulness of the truck was recognized in many countries. The governments of Britain, France, and Germany paid citizens an annual subsidy of about 20 percent to purchase and maintain trucks suitable for military use. The French led the way with measures commenced in 1907. By the time war broke out in 1914, the French army possessed some six thousand IC vehicles. In 1911 the U.S. Army began testing trucks powered on all wheels. In the following year the offerings of eleven truck makers were put through a 1.5 Mm cross-country trial. Only one vehicle passed, and it was subsequently deemed fit for wartime ambulance and escort duty (Labatut and Lane 1950).

The outbreak of war saw many vehicles requisitioned for military use and provided further incentives for truck development. For instance, in 1917 the United States began making a standard military vehicle called the Liberty Truck. The one- and 3-tonne versions became the most widely used trucks in the war. In an effort to get the first Liberty Trucks to the battlefront as quickly as possible, they were driven from Detroit to Baltimore in midwinter at a peak speed of 23 km/h. Twenty of the thirty starters successfully made the trip.

The truck was not the first motor vehicle used in World War I. That distinction belongs to the cars and cabs requisitioned and hired by French general Joseph Galliéni to take his troops to the first battle of the Marne in 1914 (Labatut and Lane 1950). Galliéni at the time was military governor of Paris. The tactic allowed the general to launch a successful surprise attack on Alexander von Kluck's German troops at Meaux. In the same year, London motorbuses were used in France to take troops to the battlefront. In all, Britain took four hundred motor vehicles to France in 1914 and had used four hundred thousand by the end of the war, finishing with a working fleet of eighty thousand trucks.

These leftover trucks provided a major vehicle stockpile after the Armistice. Postwar government sales of cheap trucks and the return to civilian life of many

TABLE 6.1
The Development of the American Car

Year	Event	Number of cars registered (in thousands)	% of vehicles with closed body	Motor vehicle operating-cost index	Motor vehicle purchase-price index	Adult Americans per car	U. S. households per car
1900	Forward engine, steering wheel	8					
1901	First all-steel coachwork						
1905	Shock absorbers, acetylene headlights	77	—	—	—	5,000	—
1906	Built-in baggage compartments						
1908	Model T Ford introduced, steering wheel on left	194	—	—	—	—	—
1909	Workable fabric tops	306	—	—	—	—	—
1910	Headlights, horns, and brakes made compulsory in some states; ignition timing linked to speed	458	—	—	—	113	44
1911	Electric starters	619	—	—	—	—	—

Year						
1912	Windshields common, electric lighting	944	—	100	100	—	—
1913	Installment financing	—	—	—	—	—	—
1915	First all-steel body	2,332	—	—	—	25	10
1916	Hand-operated windshield wipers	—	—	—	—	—	—
1919	—	—	10	—	—	—	—
1920	Shock absorbers	8,226	17	55	101	7.5	3.0
1922	Balloon tires	10,448 (1921)	30	—	—	—	—
1923	All-steel body common. Power-operated windshield wipers	—	34	—	—	—	—
1924	"Duco" in use—cars painted different colors	15,000					
1925	Bumpers standard equipment	17,431	57	46	70	3.8	1.6
1926	Car heating, safety glass	—	72	—	—	3.6	1.5
1927	—	—	—	—	—	3.4	1.4
1928	Synchromesh gearboxes	—	—	—	—	3.3	1.4
1929	—	—	—	—	—	3.1	1.3
1930	—	22,800	—	37	63	2.9	—

SOURCES: Based on Flink 1974 and Jones 1985.

Figure 6.1. The first truck, built on a Daimler Riemenwagen chassis in Canstatt in 1891. *From and with the kind permission of Daimler-Benz AG Archiv.*

ex-farm boys whose only usable skill was their ability to drive and maintain the new trucks provided a major impetus toward the complete motorization of postwar communities. In Britain, the "light subsidy type lorry" stemming from the prewar measures played a major postwar role. In the United States General John Pershing returned from the war convinced of the future military importance of trucks and determined to further demonstrate their effectiveness (Wixom 1975). To this end, in 1919 he organized a convoy of seventy-nine military trucks and almost three hundred soldiers, including a Lt. Col. Dwight D. Eisenhower. The convoy traveled from Washington to San Francisco, initially following the route of the National Road, and took fifty-six days and averaged 80 km/day.

Disposal of the surplus war trucks was a major problem. The United States government sold many at quarter price to European governments, the post office used 5,700 as mail carriers, and 24,500 were given to state highway authorities for road construction (Borth 1969). The trucks thus played a major role in developing the American road network.

This was just as well, for the rapid increase in truck traffic after 1918 had had a major impact on road systems. In the United States, "hundreds of miles of roads failed under the heavy motor truck traffic within a comparatively few weeks or months. . . . These failures were not only sudden but complete, and almost overnight an excellent surface might become impassable . . . characterized by an almost simultaneous destruction of the entire road structure." The failures were officially described as "massive," and there were angry demands for load limits and higher truck taxes (FHA 1976). The truck industry responded with the slogan Build the Roads to Carry the Loads. One constructive consequence was the establishment of the Highway Research Board. The truck makers, in response to public pressures, agreed to voluntarily limit truck capacities to 7.5 tonnes to protect future roads (MacDonald 1928). So much for voluntary limits.

Nevertheless, as trucks moved from solid rubber to pneumatic balloon tires in the 1920s, road makers came to realize that it was indeed cheaper to build roads for the new IC trucks than for the old horse-drawn carts and wagons. Thus, many jurisdictions began to positively encourage the new trucks and discourage the old. This move incidentally accelerated the swing of freight away from the railways and onto the road.

A regular traffic census on the London–Folkestone road showed that the percentage of freight moved by truck rose from 41 percent in 1911 to 95 percent in 1922. The major impact that trucks were having on the railroads is demonstrated by a 1923 report of a New York railroad company, which stated with some prescience that "a large part of the high classification shipments such as thread, machinery and brass parts, has been transferred to motor trucks. The railroad has been left with low grade commodities such as coal, trap rock etc" (Hatt 1923, 59).

Whereas the train had displaced the stagecoach in a relatively explicit manner, given the large investment in even the simplest rail track, the displacement of train by truck was neither explicit, easy, nor rapid. The debate and the search for equilibrium between train and truck continues to this day.

Bus developments were closely linked with truck technology. An electric (battery-powered) bus ran unsuccessfully in London in 1897–1898. IC buses were first produced by Benz and operated in 1895 out of Siegen in Germany, and Daimler-powered buses with steel wheels began operating in London in 1899. However, IC buses were not in effective use until about 1905, largely due to the lack of reliable solid rubber tires (Bird 1969). By 1910 they had taken the market from the horse buses, and by 1912 only 11 percent of London's registered buses were horse-drawn—the last such bus stopped running in 1914. Double-decker IC buses were introduced in 1905 in Berlin.

The Impact of the Car

The car seemingly not only offered the world a way around the problems created by the horse, but also presented a major jump in individual travel and freight capacity. Given the moves to private political freedoms occurring throughout the world, its timing could not have been better. Was this just another coincidence?

Initial Reactions

By 1910 there were many who were quite perceptively predicting the impact that the car would have on society. The accuracy of these forecasts is not surprising, given not only the rapid changes that were visibly occurring with the car, but also that the world had already seen similar progress with the bicycle and the streetcar, technologies that had both advanced and pervaded.

Despite the problems with the horse and the undoubted potential of its replacement, the turn-of-the-century car was not treated with universal acclaim. Indeed, quite the opposite often occurred. Writing in 1913 Sidney and Beatrice Webb recounted how

> the first outcome of this new invasion of the roads was a storm of opposition,
> a persistent wail of complaint, from all who did not happen to use the new
> vehicles. . . . Horses were frightened, foot passengers were startled and some-
> times terrified, the King's highway ceased to be a place in which people could

saunter, or children play, with a degree of safety. . . . In dry weather the flying cars raised clouds of dust, and in wet weather they scattered mud . . . cottages [were] rendered almost inhabitable. . . . Public feeling against the new users of the roads was strong and bitter . . . all cursed the swarms of motorcars that infested the great arteries of traffic.

Rudyard Kipling described the car as a "petrol-piddling monster." Queen Victoria called it a "very shaky and disagreeable conveyance altogether." When opening the first international motor show in Paris in 1898 President Félix Faure of France commented that he thought cars were "quite ugly and foul smelling." Incidentally, the growing extent of the industry can be gleaned from the seventy-seven brands of cars on display at the motor show. As had the cycle before it, the car had a tendency to frighten horses and thus cause great antagonism in other travelers and bystanders. A frequent, joking explanation of the horses' reactions was, "How would you act if you saw a pair of pants coming down the street with no one in them?"

An important event occurred in 1909 with the ratifying of a "Convention with respect to the international circulation of motor vehicles" by most of the major countries of Europe. This convention contained qualitative restrictions on noise and smoke emissions, including the requirement that the noise should not cause ridden or led animals to be terrified.

In addition to the appearance and the smell and the noise and the fright, the most obviously annoying characteristics of the car at the turn of the century were that it was dangerous and it raised dust, with the latter looming larger in the eyes of the public. Once car speeds exceeded 30 km/h, dust covered everything within 20 m of a highway. Adjacent property values dropped by about 30 percent (Bagwell 1974). The dust, which was usually the sand and gravel blinding material relied upon to seal macadam surfaces from water entry, also caused structural damage to the roads. With the blinding gone, water entered the pavement structure and weakened the natural formation, leading to early failure of the road (Page 1910). The 1910 PIARC meeting was primarily concerned with the control of dust from vehicular traffic. It was the first technical recognition of the car as an issue in pavement design (McShane 1979).

The French delegate to the first PIARC meeting in 1908 disputed the general view that the motorcar would ruin macadam pavements, stating that "there must be a traffic of at least forty cars per day to have any effect on the pavement and it only becomes serious if the traffic exceeds one hundred cars per day. Out of the 36 Mm of the national highway system less than 2 Mm have this magnitude of traffic."

The need to alleviate the dust menace led to road trials aimed at producing dust-free surfaces and to the development of a variety of novel dust-suppressing devices attached to the undersides of cars (Plowden 1971). The next chapter shows how the incentive to produce dust-free surfaces led to major advances in pavement technology. The vehicle-based solutions were far less successful.

The End of Horse-drawn Travel

In 1895 popular publications began carrying articles showing that the car was cheaper to operate than the horse. Perhaps the car truly entered the arena in 1896 when it began displacing the horse as the top attraction at major American circuses.

By 1905 the United States was building as many cars as carriages. In 1909, in a clear harbinger of the future, the car replaced the horse as the official method of transport for American presidents. Massachusetts traffic census data give a graphic illustration of the general takeover, showing that the percentage of horse-drawn vehicles on the road dropped from 58 percent in 1909, to 35 percent in 1912, to 16 percent in 1915, to just 6 percent in 1918 (Cron 1974). Despite this relative decrease, the absolute number of horses in the United States continued to increase dramatically, passing four million in 1910 and ten million in 1920. Nevertheless, motor vehicle numbers increased even more rapidly, and by that time the United States had as many motor vehicles as it had horse-drawn carriages and wagons. By 1924 the shift in general transportation from horse to truck was effectively complete.

Two factors adding to the decline in horse-drawn traffic were that many of the new asphalt roads being built to accommodate the car were, first, too smooth to provide adequate traction for the horse and, second, subject to permanent hot-weather damage by the hooves and iron-tired wheels of horse-drawn vehicles (Page 1910).

Road Accidents

The first road accidents involving cars were reported in London and New York in 1896, the same year in which Emile Levassor was killed in a French car race. The London accident occurred on 17 August at Crystal Palace when a woman pedestrian named Bridget Driscoll was killed by a car driven by one Arthur Edsall. Edsall claimed to have been traveling at only 7 km/h, to have shouted, to no avail, "Stand back!" and to have rung his bell before striking the unfortunate Driscoll. At the inquest, the coroner expressed the wish that such an event would never be repeated. In the American accident, a car collided with a cyclist in New York City, breaking the cyclist's leg. The driver spent a night in jail. The first American fatality due to a car occurred on 14 January 1899, also in New York City, when an electric cab hit and killed real estate salesman Henry H. Bliss as he was assisting a lady passenger off a streetcar. Perhaps not coincidentally, the first car insurance was also underwritten in 1896, by the English Law Accident Insurance Society.

Motor vehicle road accidents soon became a factor of growing significance. The number of London motor traffic fatalities increased from 161 in 1904, to 344 in 1908, to 561 in 1912 (Carey 1914). The American figure for 1909 was one death per annum for every 250 motor vehicles. In terms of fatalities per distance traveled, table 6.2 shows that a steady improvement in effective road safety has occurred since 1920 (Lay 1990).

By comparison with the earlier horse accident figures, on an accident per distance traveled basis the car soon became safer than the horse. Indeed, in 1902 an American car manufacturer was to comment: "The introduction of the mechanical carriage has been relatively quiet. To be sure, there have been some accidents. . . . Yet, if we look back to the early days of the trolley car and the bicycle, it will be fairly evident that the introduction of the automobile has been very free from unpleasant incidents" (Flink 1974).

The concern for road safety steadily increased in the early 1920s, and in 1924 Herbert Hoover, then American secretary of commerce, convened the world's first road safety conference, the First National Conference on Street and Highway Safety.

Although sometimes demanding, conventional driving now requires little

TABLE 6.2
Kilometers Traveled in the United States Per Traffic Fatality

Year	Gm (millions of km)
horse data	6
1909	8
1920	6
1925	9
1935	10
1945	14
1950	23
1955	25
1965	29
1975	46
1985	62

physical skill, and most people are capable of driving. Nevertheless, drivers operate their cars with no external constraints forcibly placed on them and in a manner that is essentially self-paced. Thus, they privately and personally decide on their traffic operating conditions, and this requires them to exercise considerable self-control. It is the driver's foot pressing on the accelerator that directly determines the speed of the car. Drivers lose this control, and some necessary driving skills, when under the influence of alcohol, drugs, and/or youthful exuberance. These driver self-control factors—commonly labeled human error—are currently, at 80 percent, by far the major single contributor to road accidents.

In some circumstances, road accidents are a probabilistic consequence of the impact forces created between two objects moving at relative speeds of over 10 km/h. Although the human body is obviously not designed to withstand such impact levels, measures to soften and control both the vehicle and the roadside occurred slowly.

Successful countermeasures have either reduced the propensity for error—e.g., through better visibility, better signs, divided roads, better intersection design, and the diminution of alcohol-affected driving—or reduced the consequences of error, measures such as the provision of seat belts and softer car interiors and the removal of roadside hazards. Seat belts were first introduced in France in 1903. Their use was first made mandatory in Victoria, Australia, in 1970.

Consequently, in both absolute and relative terms, road safety has steadily begun to improve over the last decade. Indeed, in relative terms road safety in the developed world has been improving for all this century (Lay 1990).

Traffic Congestion

Traffic congestion did not begin with the car; the mix of nonpedestrian traffic and city life has never been a compatible one. There were attempts to manage traffic in the towns of early India. Rome first tried one-way streets to control its traffic, but increasing congestion in 45 B.C. led Julius Caesar to declare the center of Rome off limits between 6:00 A.M. and 4:00 P.M. to all vehicles except—of course—those of officials, priests, high-ranking citizens, and visitors. Claudius I in A.D. 50 extended the ban to all Italian towns. Things continued to decline, and in about A.D. 125 Hadrian was forced to further limit the number of vehicles entering Rome. In about A.D. 180 Emperor Marcus Aurelius broadened the bans to all towns of the Roman Empire.

Increases in the numbers of hackney coaches were a common cause of traffic congestion in London. Market days were also regular sources of congestion as both stock and customers thronged the streets. The problem was commonly tackled by limiting the hours during which particular markets could operate, a measure that has its modern counterparts (Salusbury 1948).

Following the Industrial Revolution, as workplaces became more distant from homes, even pedestrian traffic became tightly jammed in peak hours. It would seem that there was some realism in Dore's nineteenth-century view of a London street (see fig. 9.1). Similarly, congestion was reported on Broadway in New York City in 1850.

To help alleviate the problem, London in 1867 banned the driving of livestock through its streets. Nevertheless, the slack was soon taken up; a twelve-hour traffic count in Cheapside, London, in the 1870s recorded a remarkable 1,030 vehicles per hour (News 1894). Between 1875 and 1905 there was a 25 percent drop in the average speed of London traffic, with only a very small percentage of motorized vehicles present in the traffic. This led to the establishment in 1906 of the Royal Commission on London Traffic (Thompson 1970). In 1911 the British Post Office noted, "Whether it will ever be possible to obtain an average reliable speed of over eight miles per hour [14 km/h] during ordinary business hours in Central London even with motor vans is a matter of extreme doubt" (Mogridge 1986).

The increasing number of cars inevitably led to greater traffic congestion. In 1920 Fifth Avenue in New York was carrying six lanes of traffic, and yet "in the busy part of the afternoon, progress is so slow that walking is often quicker" (Lay 1920). In 1924 urban traffic congestion was under serious study in Chicago (Cron 1975a). Outside the cities, significant traffic congestion was being experienced on some American intercity routes by 1920 (Cron 1975c).

Traffic counting began in Ireland in 1837. The technique was formally resurrected for the car era in Maryland in 1904 in a scheme that grew to a network of some 180 counting stations in 1923 (Armstrong 1976). In that year and with good data behind it, Maine began serious predictions of future traffic growth, based on cars per head of population. The work was quite perceptive. In a review fifty-six years later, Heightchew noted that "very little has changed in such studies [conducted] today" (1979). In 1923 the Bureau of Public Roads was promoting the use of pneumatic-tube traffic counters and punched cards for data recording as a means of easing the traffic-counting tasks (Hogentogler 1923).

Social Reactions

Chapter 5 told how the numbering of hackney cabs and their drivers began in 1814 and 1838, respectively. The subsequent recording of individual drivers and vehicles in a register was usually introduced as a revenue-raising measure. In 1867 the French began licensing all Parisian drivers of horse-drawn carriages, and Chicago initiated the licensing of car drivers in 1899. Chicago also had the novel idea of prohibiting licenses to people who wore spectacles, lest the spectacles become dislodged during driving (Flower and Jones 1981). Ontario, Canada, began licensing chauffeurs in 1909, but for the next eighteen years did not require an owner-driver to be licensed because he would be more "inclined to exercise greater care in driving his car" (Ontario Ministry of Transport and Communications 1984). The word *chauffeur,* incidentally, is French for "fireman" and comes from the earlier role of operating boilers on steam vehicles.

In 1901 New York State required vehicles to be registered on payment of a fee, thus introducing specific motor vehicle taxation. The law also required owners to display their initials in a conspicuous place. Both the American Road Makers (later the American Road Builders Association) and the American Automobile Association were formed in 1902, the latter to combat the new taxes and other legislation that its members saw as restrictive and anticar. Their objections were not merely to the registration tax, but also to the requirement that each vehicle should carry an identifying number.

A distinguished French commentator on American ways remarked, "Historians of the twentieth century should see that Ford's revolution is far more important than Lenin's" (Bruckberger 1959). In a similar vein, a noted politician and author argued that "Ford and Edison shaped human experience more broadly and enduringly than Lenin and Hitler" (Jones 1985). In a more restrained review, Flink (1985) commented, "The automotive idea, combining a light, sprung, wheeled vehicle, a compact, efficient power unit and land-surface roads, has been far more revolutionary in its impact than any [other] technological innovation emerging from the 'Second Industrial Revolution.'" And, on another tack, George Bernard Shaw asked in 1898, "What Englishman can give his mind to politics as long as he can afford to keep a car?" General Motors copywriters took up the theme in 1924 when they asked, "How can Bolshevism flourish in a motorized country?" (Patton 1986). From the earliest days there was widespread popular interest in the motorcar, although only an elite few were early car owners.

The growing interest in cars is seen in the media of the time. The first motoring journal was *La Locomotion Automobile,* which began publication in 1894.[1] In the following year the editor of the much older *Cyclist* journal began publishing the *Autocar* in an England that had only seventy cars. Later in the same year, the journals *Motorcycle* and *Horseless Age* commenced publication in the United States and in England David Solomon formed the Self-Propelled Traffic Association and held the first motor show at his home (Bird 1969). The association merged with the newer Automobile Club in 1897 and became the Royal Automobile Club in 1907. Chapter 5 showed that the Automobile Club of France formed in the same year, so both can lay claim to being the first car club. The first American automobile show was held at Madison Square Garden in New York City in 1900.

Motorists' motivations in forming into clubs were rarely altruistic. The first motorists were usually rich and powerful people who formed their associations to support—often quite blatantly—the view that driving should be free of regulation and restriction. In England they had been goaded into this attitude by excessively restrictive laws such as the Red Flag Act. Whatever the reason, the ostentation of many British car owners was said by the United Kingdom government's chief industrial advisor to be one of the reasons for British industrial unrest between 1900 and 1920 (Bagwell 1974). In the United States, the Automobile Club of America was founded in New York in 1899 and was described one year later as "an ultra-fashionable coterie of millionaires, who have taken up the new and expensive fad of auto-locomotion and banded themselves together for its pursuit and the incidental notoriety attributed to all the functions of upper swelldom" (Flink 1974). In 1905 its membership list included such families of public wealth as the Astors, Vanderbilts, Rockefellers, and du Ponts.

For most early purchasers, the car was clearly an item of conspicuous consumption, to use the term Thorstein Veblen coined in 1899. Woodrow Wilson, when president of Princeton in 1906, noted that nothing would spread socialism more than public resentment of the ostentatious displays of wealth by motorists (Patton 1986). Citizens, he said, saw cars as "a picture of the arrogance of wealth, with all its independence and carelessness" (Pettifer and Turner 1984). An article in a 1911 Swiss newspaper described motoring as a showoff's sport (*protzensport*).

This clearly perceived market segmentation was widespread. In 1901 Daimlers forecast that the world market for cars would never exceed one million, because only a tiny portion of the working class could be educated as chauffeurs. The early attitudes are epitomized by Harry Graham's poem of the time:

> Once as old Lord Gorbals motored
> Round his moors near John o'Groats,
> He collided with a goatherd
> And a herd of forty goats.
> By the time that he got through
> They were all defunct but two.
>
> Roughly he addressed the goatherd:
> "Dash my whiskers and my corns!
> Can't you teach your goats, you dotard,
> That they ought to sound their horns?
> Look, my AA badge is bent!
> I've a mind to raise your rent!"[2]

Lt. Col. Moore-Brabazon, speaking in a British House of Commons debate in 1934, said: "No doubt many of the old Members of the House will recollect the numbers of chickens we killed in the old days. We used to come back with the radiator stuffed with feathers. It was the same with dogs." King Farouk of Egypt is said to have had a car horn that imitated the squeals and howls of dogs being run over (Keats 1958). In the early 1930s, Prince Nicholas of Romania had a special horn fitted to his car, "which, when activated, obliges all other vehicles to give way." Perhaps it was such behavior that caused the *London Times* to comment in 1902 that "the number of drivers of motorcars who are not gentlemen would seem to be unduly large"?

The arrogant, threatening attitude of many early motorists was further epitomized by Kenneth Grahame's car-driving Toad in the *Wind in the Willows,* published in 1908. Toad drives off in a car "as if in a dream, all sense of right and wrong, all fear of obvious consequences, seemed temporarily suspended." He is described as "Toad the terror, the traffic-queller, the Lord of the lone trail, before whom all must give way or be smitten into nothingness and everlasting night. He chanted as he flew."

The repeal of the Red Flag Act itself tells an interesting story. When Lord

Harris introduced the new Locomotives on Highways Act in the House of Lords in 1896, he said that the drafters had prepared it in the dark, "owing to the fact that they had no experience of these light locomotives." When it was introduced into the House of Commons, the speaker hypothesized that "it was even possible that these motorcars might become a rival to light railways (laughter)." One opposition speaker predicted that the bill would be the ruin of the horse-breeding industry (Plowden 1971).

Far more prophetic remarks were being made in the United States and France. The director of the U.S. Office of Public Roads, Logan Page, warned in 1910 that "legislation now against a rich man's toy may later on be against a poor man's economy." In France, Anatole France perceptively commented in 1908 that the car would

> place itself at the disposal of the whole people and behave like a docile, industrious monster. It is indeed true that to prevent it from continuing to be harmful and to make it beneficial, roads must be built which meet its specification, highways which it can no longer tear up with its wild pneumatic tires which then fling poisonous dust into the breasts of mankind. Slow carts as well as all animals would have to be banned from the new roads. Garages and viaducts will be constructed so that order and harmony are maintained on the roads of the future.

Early cars were indeed predominantly a plaything of the wealthy, used for touring and pleasure rather than for commerce (Flink 1974). That a common early practical application was to provide mobility for doctors on their rounds merely added to the car's aura of exclusivity. This trend continued even though commentators since 1895 had been demonstrating that a car was actually cheaper to run than a horse and carriage. As the economics of their operation slowly became more obvious and as Ford from 1913 on began to dramatically drop prices, cars expanded to become an essential transport mode for the rural community. This change was most dramatic in countries, such as the United States, Canada, and Australia, with sparse, rural areas where the alternative to the car was the horse-drawn carriage. It was least dramatic in the densely populated areas of Europe, already relatively well served by rail transport (Schaeffer and Sclar 1975). Indeed, in Europe the car remained a tool and toy of the wealthy and powerful long after it had been democratized in the New World. From its rural New World base, the car then moved into the city.[3]

Working with Cycle and Rail

Bicyclists were the first organized groups to take a national interest in the condition of roads (Jeffreys 1949). In Britain, the two national cycle clubs joined forces to create the Roads Improvement Association in 1886. The association prepared and circulated pamphlets on how to improve and maintain roads, mainly emphasizing McAdam's principles of stone breaking and surface compaction. The local, and largely untutored, road managers (road surveyors) resented this intrusion into their field. The more ignorant they were of roadmaking, the greater their resentment. The association remained active until just after the Second World War.

In the United States, the League of American Wheelmen combined with the farmers' group called National Grange of the Patrons of Husbandry (composed of state granges), to found the National League for Good Roads in 1892. The farmers were particularly interested in the league as an avenue for gaining support for improved farm to market access and for upgraded rural mail deliveries. The league held its first national conference in 1894, and President Theodore Roosevelt attended its 1903 convention. The league operated in 1905, a time when the cyclists' role began to decline relative to that of the motorist.

From the beginning the National League had published the *Good Roads Magazine* and, within a couple of years, a New York civil engineer and lawyer, Isaac Potter, began to turn it into an influential journal furthering the prime cause of the League of American Wheelmen. *Good Roads* described itself as "the first publication in the world devoted strictly to road improvement." After forty years it changed its title to *Roads and Streets*.

The League of American Wheelmen was also instrumental in 1902 in starting the American Road Makers, an organization that changed its name to the American Road Builders Association in 1910 and which has been active to the present day (Wixom 1975). The league itself became inactive in 1907.

A Good Roads Association had formed in Australia in 1890, again on cyclists' initiatives, and merged with the automobile clubs in the 1920s. The Canadian Good Roads Association survived as an active technical support and political lobby group for over half a century. Typically, these associations actively campaigned for central control of road matters. For example, in Australia cycle-club activists became the leaders of the new motor clubs and persuasive lobbyists for central road authorities (Lay 1984). It can fairly be said that the bicycle led the road systems out of the stagnation that had begun in about 1840.

The major dissension that arose within the Good Roads movements was not between owners of different vehicle types, but between urban and rural interests. The farmers naturally resented the intrusion and disruption by the city dwellers in their noisy cars and the damage that the cars caused to roads largely paid for by the rural community. The dissension did not diminish until car ownership became more widespread in the country and more central road funding occurred.

The growth in car usage meant that the expanding push for good roads and central control, which the cyclists had begun, became even more necessary as those roads that did exist began to deteriorate under the new traffic. The legacy of sixty or so years of inattention during the steam era had left no effective infrastructure for the dazzling new invention. Roads had become no more than feeder roads to railway stations, and even many of these were already in such bad condition that they had begun affecting railway business (fig. 6.2). Freight charges by road were about fifty times higher than by rail, and many farmers could not afford to get their produce to the railway siding.

Initially, the motorized vehicles were enthusiastically welcomed by the cyclists as a means of extending the effectiveness and pleasure of their mode of travel. Little did they know what a tiger's tail they had grasped. The railroads also did not recognize the tiger's tail and often joined forces with the cyclists' and motorists' clubs to advocate increased roadmaking. Their advocacy was based on building roads to serve railway stations. There were, of course, other roadmaking advocates at work, such as those who correctly saw roads as improving land values and others, with more altruism, who saw them as improving local health and virtue (Rose 1979).

In 1901 the American and Canadian railroads began operating Good Roads

Figure 6.2. El Camino Real (U.S. 101) in Ventura County, California, in 1912. The vehicle is an early Cadillac. *Photo by kind permission of Caltrans Library History Center.*

trains in cooperation with manufacturers of roadmaking equipment, the federal Office of Public Road Inquiries, and Good Roads Associations (Labatut and Lane 1950; FHA 1976). A typical train had two carloads of officials and experts and ten carloads of roadmaking machinery (Wixom 1975). They traveled the American countryside promoting roadmaking as a virtue and a skill. The last Good Roads train toured Iowa in 1916. Good Roads Sundays were also popular, with clergy encouraged to draw attention to the link between good roads, good living, and Christian progress.

The railways played another role. In the United States the Populist movement, led by silver-tongued William Jennings Bryan—the first major politician to campaign from a car—had widely attacked the "public be damned" attitude of the large railroads, which were seen at the turn of the century as arrogant, impersonal, unsavory, land-grabbing, and monopolistic (Wixom 1975). David Belasco commented that "to the more optimistic generation of the 1870s and 1880s the train had represented the path to a glorious future, by the early 1900s it came to symbolize the all too alienating present" (Daniels and Rose 1982). The newly offered car, on the other hand, promised to return travel to a simpler, more individualistic, and preindustrial lifestyle in which success in travel was a consequence of personal struggle and effort. Another private attraction of the car of the time was that it did not require collective political or community action for its service to materialize (Foster 1981).

One writer predicted in 1903 that "the stagecoach will be avenged upon the railways by the motor." Others likened cars to gypsy vans and peddlers' wagons. Motorists described themselves as motor hobos and motor vagabonds. In 1903 the journal *Outing* explored the possibility of the car leading to a rediscovery of the

United States. The *Brooklyn Eagle* in 1910 saw the car as "the last call of the wild." In addition the car promised to restore the human scale to travel and to machinery. The 1915 Lincoln Highway guidebook warned that traveling on the highway was "still something of a sporting trip, and one must expect and put up cheerfully with some unpleasantness" (Daniels and Rose 1982). Autocamping became a popular fad. One famous weekend trip in a Model T included Ford, Edison, Firestone, and President Harding (Patton 1986). It was also widely expected that the car would improve the quality of rural life and stimulate a back-to-the-land movement (Jackson 1985).[4]

Riding in cars was additionally seen as a direct source of good health. An American commentator noted in 1903 that the car was "the greatest health giving invention of a thousand years. The cubic feet of fresh air that are literally forced into one whilst automobiling rehabilitate worn-out nerves and drive out worry, insomnia, and indigestion. It will renew the life and youth of the overworked man or woman and will make the thin fat and the fat—but I forebear" (Flink 1974). An Australian politician of the same time, Thomas Bent, justified spending public money on his administration's first government-owned car with the argument that, when he was not using it, the car could be used to treat tuberculosis patients by driving them in the open air.

It is now a truism to record that the car did indeed cause major social change. The IC engine changed transport and, when the car moved from being a scarce commodity to an abundant one, transport then changed society. As noted by E. L. Armstrong in 1976: "Until the twentieth century, travel for the average person was a luxury; now it is a necessity and a major part of life." And it soon became a pervasive part of everyday life. John Steinbeck remarked in *Cannery Row* in 1945 that "two generations of Americans know more about the Ford coil than the clitoris. . . . Most of the babies of the period were conceived in Model T Fords."

Managing Travel

Strong creative tensions have always existed between the infrastructure needs of the vehicles of the day and the ability of civilization to produce the requisite level of infrastructure to satisfy those needs. Usually the vehicle was adapted to the infrastructure, although numerous long-term infrastructure adjustments did occur. The coming of the steam train in the 1830s delivered a quantum leap in usable power. It therefore became much more important to begin to adapt the infrastructure to meet the capabilities of the powerful new vehicles. A whole new discipline of railway engineering evolved, concentrating particularly on track alignment, cross section, and geometry and on signaling systems between the vehicles and the network managers. With the invention of the IC car half a century later, similar skills were needed in the road industry. Here the discipline that arose was called traffic engineering and, not surprisingly, it borrowed heavily on the developments that had occurred in the railways. Its history will now be recounted.

Giving and Taking Priority

Despite the above introduction, traffic controls well preceded the invention of the steam train. Some were encountered in the earlier discussion of traffic congestion. In about 700 B.C. the Assyrian king Sennacherib forbade illegal parking on the

Royal Road in Nineveh. Roadside poles carried the instruction "Royal Road. Let no man lessen it." The offense might now seem minor, but the penalty was death by impalement on a stake. Parking restrictions were also applied in Pompeii, and Julius Caesar introduced requirements for off-street parking. By 1920 cities such as Philadelphia had reintroduced parking controls and one-way streets, but this time to manage the car (Lay 1920).

The strongest and most persistent and successful advocate of the need for logical methods of traffic control and regulation—and particularly of one-way streets—during the early years of the car was the American William Eno, who was made a chevalier of the French Legion of Honor for his work in solving the traffic problems of Paris (Montgomery 1988).

There had been little need or opportunity for traffic engineering prior to the car. All vehicles traveled at relatively low speeds, had short stopping distances, and did not routinely require their drivers to have quick reactions. Thus, vehicular priority did not have to be predefined and was usually by might rather than by decree. These circumstances also meant that the behavior of horse traffic was much less disciplined and less predictable than that of modern car traffic, which largely explains the poor accident record of urban horse traffic. Nevertheless, even prior to the car the increasing volumes of horse traffic had required some action in traffic management. Pedestrian safety islands in the middle of streets were first introduced in Liverpool in the precar days of 1862. The initiative came from a saddler called Hastings whose advocacy of a refuge at a dangerous crossing was only heeded after a leading citizen was killed at the very location.

When specific controls were needed, horse-drawn traffic was controlled by policemen's hand signals. This was often far from effective, and pedestrian crossing signals were introduced in London in 1868 at Bridge Street and New Palace Yard outside the British houses of Parliament in order to provide safe crossing for the members. The signals used manually activated semaphores, with red and green gaslights at night to indicate Stop or Caution to "all persons in charge of vehicles and horses" (fig. 6.3). The caution light meant "go, with care." The meanings associated with the red, yellow, and green displays used in these signals came from the color convention already in common use in maritime and railway signal technology (Wilson 1855).

The device's first major mishap occurred when it scared a troop of cavalry into confusion. Two policemen were killed in a subsequent event. Then a gas explosion in 1869 killed the police operator lighting the lamps. These events resulted in an initial prejudice against the wider use of mechanical signals. Nevertheless, an improved device was installed at the location in 1872. Progress was slow and, by the First World War, only a few hand-controlled semaphores and arms carrying red kerosene lamps were in use (Institute of Transportation Engineers 1980).

The few precar traffic control regulations that were introduced, such as the United Kingdom's Red Flag Act, were largely aimed at preventing a few from being a nuisance to the majority and not at giving useful guidance to the majority. The car changed all this passivity and also brought a whole new set of traffic problems.

Consequently, the practice of traffic engineering was born. The birth was at times brutal. In 1908 a semicomic Farmers' Anti-Auto Protective Society made some tongue-in-cheek suggestions that many took up with serious intent (Robinson 1971). For example, one proposed American traffic bill would have required any car ap-

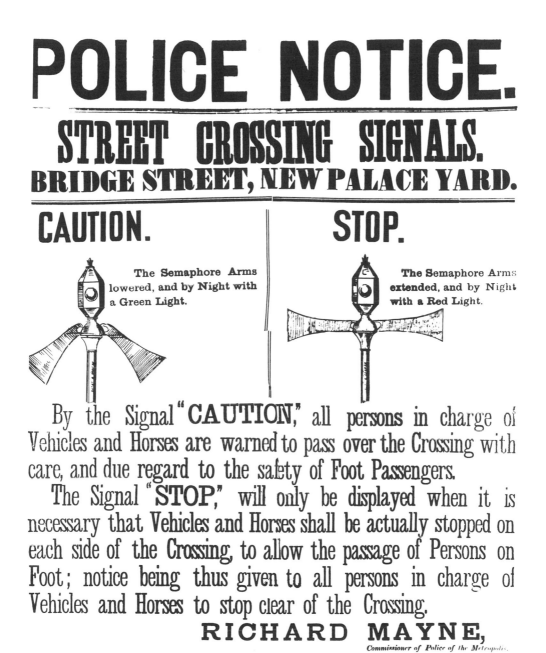

POLICE NOTICE.

STREET CROSSING SIGNALS.
BRIDGE STREET, NEW PALACE YARD.

CAUTION.

The Semaphore Arms lowered, and by Night with a Green Light.

STOP.

The Semaphore Arms extended, and by Night with a Red Light.

By the Signal "CAUTION," all persons in charge of Vehicles and Horses are warned to pass over the Crossing with care, and due regard to the safety of Foot Passengers.

The Signal "STOP," will only be displayed when it is necessary that Vehicles and Horses shall be actually stopped on each side of the Crossing, to allow the passage of Persons on Foot; notice being thus given to all persons in charge of Vehicles and Horses to stop clear of the Crossing.

RICHARD MAYNE,
Commissioner of Police of the Metropolis.

Figure 6.3. The police notice used to advise of the first traffic signals. *Copy with the permission of the Science Museum, London.*

proaching a crossroad to completely stop, its "engineer dismount and examine the roadway thoroughly, sound his horn vigorously, then halloo loudly or ring a gong or discharge some explosive device," before finally crossing the intersection. Similarly, early Pennsylvania motorists were instructed than "if the driver of an automobile sees a team of horses approaching, he is to stop, pulling over to one side of the road, and cover his machine with a blanket or dust cover painted or colored to blend in with the scenery" (Pettifer and Turner, 1984). Another Protective Society suggestion required vehicles to be painted in accord with the season of the year (Robinson 1971).

In a more serious mode, some parts of Switzerland banned cars altogether, or on Sundays required them to be towed through urban areas by horse or ox. At the extreme, motoring was totally prohibited in the canton of Graubünden between 1902 and 1928 (Krampen 1983).

The message had also filtered through to motorists. The famous French impressionist artist Henri Matisse was a motoring enthusiast. When asked at the turn of the century what he would do if he encountered a car coming in the opposite direction, he replied: "Should the inconceivable happen, I should of course halt my motoring car, descend from it and take shelter in a nearby field until the other had gone by."

It was soon established that most drivers would not stop in the Matisse manner, and the requirement for drivers to use hand signals to indicate their steering intentions was first introduced in Berlin in 1902. Turning continued to be hazardous, and center-turn bans were first introduced in Buffalo in 1916.

By 1910 it was clear that right of way had to be clearly defined in order to accommodate the fast new vehicles. The awareness was such that the International Law Association established a committee on road traffic regulations. Two years later the committee recommended that drivers faced with cross traffic give way to the vehicle on the driver's curb side (i.e., yield to the right for driving on the right and yield to the left for driving on the left). This followed German rules established in 1906 (Kincaid 1986) and has remained the official international recommendation to the present day, as in the 1952 United Nations convention.

However, a number of left-hand drive countries, such as the United Kingdom and Australia, also adopted give way to the right, which meant center-side rather than curb-side priority. Thus, the practice of giving way to the right effectively became universal. The reason for this fortunate perversity is that Australia simply copied the American give way to the right rule, even though the two countries drove on opposite sides of the road. The United Kingdom had consistently had trouble accepting international traffic conventions. Although the U.K. Automobile Association had been advocating center-side priority since 1921, the rule was only formally adopted in 1966. In the interim, Britain relied largely on a major/minor road system and the use of stop signs. Center-side priority subsequently proved effective in roundabout operations, because traffic circles function poorly with curb-side priority, which requires the circulating traffic to give way to entering traffic. This explains why, during most of the twentieth century, the roundabout has only developed in countries with center-side priority and which coincidentally also drive on the left. France, recognizing this problem, introduced center-side priority rules for roundabouts in 1984.

Traffic circles had long been a feature of grand urban design, providing a fine site for civic monuments, as evidenced by the 1821 Monument Circle built in Indianapolis. Traffic movement past them was relatively random. To curb this disorder, gyratory traffic flow within the circle was proposed by Holroyd Smith in London in 1897 and by Eugène Henard in France in 1903. The concept was first applied at the Columbus Circle in New York in 1905, followed in 1907 by Parisian applications at place de l'Etoile and place de la Nation. Henard, a famous urban architect-engineer, in 1906 also proposed the grade-separated interchange in the course of developing a plan for Parisian development.

The deliberate use of gyratory flow as an intersection control measure gave rise to the special-purpose roundabout, first installed by Barry Parker and Raymond Unwin at Sollershott Circus in Letchworth in Britain in 1910 (Laurence 1980). It was built with a 16 m diameter circle circumscribed by a 7 m ring of pavement. Traffic signs were only added after twenty years of operation. There was a spurt of roundabout development in the United States in the 1920s, but installations were generally underdesigned for the rapidly increasing traffic volumes and suffered from the curbside priority rule. These circumstances subsequently gave roundabouts a bad name in the United States, and they came to be little used. In 1925 an English committee determined that "the Circular System of Control, as adopted in New York, was not suitable for this country." However, in yet another demonstration of committee myopia, the devices were progressively introduced from 1926 on, and England subsequently became replete with roundabouts (Todd 1989).

A further new innovation was the traffic control tower, which allowed a police officer to exercise intersection control from a centrally located and very visible position. It was introduced in Paris in 1912 when an ornate 4.5 m high bronze structure was erected at the intersection of rue Montmartre and boulevard Montmartre. A policewoman sat in a glass enclosure at the top and manipulated a boxlike device carrying the instructions for stop and go in red and white. The tower operated for twenty-two days, but was then abandoned because Parisian drivers ignored the signals. Nevertheless, it was subsequently copied in many cities.

The first modern-style traffic signals were installed in Salt Lake City in 1912 by Lester Wire, a city policeman, although the signal's red and green electric lamps were still manually operated (Sessions 1976).[5] Five years later, an electrician called Charles Reading helped Wire to link six of his signals together to provide signal coordination for traffic traveling at 30 km/h (Mueller 1970; Armstrong 1976).[6] This move to new technological solutions reflects the growing and insatiable demands that the car was beginning to place on urban areas.

Detroit obviously saw its Motor City vehicle-production role as including traffic safety obligations and in 1909 formed a traffic division of thirteen policemen who used hand signals to direct traffic. However, the incentive was not to control the IC car but to reduce horse-drawn vehicle accidents, which were occurring in Detroit streets at a fatalities per distance traveled rate that was 50 percent above the American average (Borth 1969). In 1915 the Detroit Traffic Division produced the first stop sign; in 1917 it introduced the traffic control tower to the United States, with red and green lamps on the ends of semaphore arms at the corner of Woodward and Michigan avenues. In 1918 it introduced the first three color traffic signal. A 1920 innovation showed the colors in up to four traffic directions. The color changes were still manually activated and were commonly assisted by gongs, whistles, and bells. An example of early American traffic-signal design is given in figure 6.4.

Figure 6.4. An early American traffic signal. *From Sherrard (1930).*

The traffic signals were not always placed on roadside poles. New York State tried at least two alternatives, both using policemen as the lantern bearers. In 1921 New York City equipped its officers with vests to which traffic-signal sets were attached. In Syracuse, the policemen were equipped with an umbrella with a stop/go sign on its top. Overhead lane controls were first used in 1927 on the New Jersey approach to the Holland Tunnel.

Vehicle actuation is the very useful technique whereby the signals change in response to the traffic using the intersection. The key is vehicle-detection technology. Vehicle activation was introduced in Baltimore in 1928 via sound-sensitive signals; the signal changed when an approaching driver blew the horn or loudly revved the engine (Institute of Transportation Engineers 1980). Other devices, activated by either light or tire pressure, were introduced at the same time. The current inductance loop method for detecting vehicles at signals was in spasmodic use prior to 1930 but was first widely used in 1934. More recent advances in traffic-signal technology are discussed in chapter 9.

Until the mid-1920s traffic control and management were very much the business of police departments. The traffic engineering profession perhaps began with the publication in 1924 of engineer Arthur Tuttle's conventionally optimistic view that "the traffic relief problem has now assumed such proportions that the engineering service must take the lead in providing a solution. The recognition of this responsibility by the profession should bring about the freeing of the streets for the traffic needs of today and the days to come."

In that same year Burton Marsh was appointed by the city of Pittsburgh to be the world's first traffic engineer. However, the first codification of traffic engineering practice did not occur until 1932, with the work of Arthur Tuttle and his colleague Edward Holmes.

Traffic Signs

Routes along early paths were often marked with broken twigs, sticks, or stones. More elaborate waymarkers were signs carved on rocks, pieces of vertical stone called *stoops,* stones piled into cairns, and marked trees (Hey 1980). One variant encountered on the prehistoric English ridgeways was to line the way with the burial mounds of local chieftans (Addison 1980). The practice was also common among the Greeks and Romans, who often buried their dead in graves beside the approach roads to their towns and cities (Mumford 1961).

The Romans used marble shafts about 2 m high at 1.5 km spacings as distance markers or milestones. Within about 200 km of Rome, these milestones showed the distances to that city. The centerpiece in Rome was the golden milestone erected by Emperor Augustus in 20 B.C. at a corner of the Forum. It used bronze plaques and, later, gold letters to show the distance to the various parts of the empire. Elsewhere, milestone markings were based on the distance to the nearest large town. The word *mile* comes from the Latin *milia passum,* "a thousand steps."

Smoke was always a daytime sign of a nearby town. Later, a number of towns lit lights or blew horns at night to guide travelers to their facilities.

Early in the seventeenth century, the duc de Sully and Cardinal Richelieu introduced a system of signing at French crossroads using either cairns or posts. Mathew Simon's 1635 *Directions for English Travillers* notes that directional signs were found in England "in many parts where wayes be doubtful." Ole Rømer in Denmark instituted an extensive network of mileposts at 1.8 km spacings in 1697.[7] In 1698 an English law required each parish to place guideposts at its crossroads. Finger posts, often shaped like a pointing hand, and stone mileposts set in the ground became more common from the eighteenth century. The developments of toll roads and the postal service in the eighteenth century led indirectly to great improvements in signposting, and some signs even included travel times.

Road signs came under police protection in the 1800s. In the 1760s Ben Franklin actively promoted milestones as an aid to his postal service (Sessions 1976). In the United States in the eighteenth century, signposts were often funded by public subscription. Modern road signs began with the efforts of the cycle clubs at the end of the nineteenth century. The clubs erected many road signs, of which the commonest were To Cyclists: This Hill Is Dangerous and Caution for Cyclists. As cycle clubs developed into automobile clubs they often took their signing charter with them. Between 1895 and 1896 the Italian Touring Club erected 140 cast-iron signs carrying directional arrows, mainly on the coast road near Senigállia. In 1901 the Automobile Club of America began signposting roads with cast-iron strip signs.

The first attempt at international traffic signs occurred at a meeting of the international League of Tourist Associations held in London in the late 1890s. The proposal was based on Italian arrow signs. Although the matter remained topical among tourist bodies and automobile clubs, formal government action did not occur until 1908 at the initial meeting of the international road group, PIARC. The resulting PIARC signs were first used for a road race in Austria in 1909 and ratified at an international convention later that year (Krampen 1983). The four standardized signs were for crossroads, railway crossings, sharp curves, and gutters across roads.

Symbolic signs were introduced in Austria in 1910 and consisted of paintings of wagon-wheel brake shoes on roadside rocks at locations where a steep slope demanded cautious braking. The incentive for the signs was a then famous road death at Brennbuhel near Imst in Austria in 1854. King Frederick Augustus II of Saxony had been killed when his coachman failed to use his brakes in just such a situation (Krampen 1983).

Pressures from the 1921 International Traffic Congress and the 1923 PIARC to amend the 1909 convention led to a League of Nations study beginning in 1923 and ending with a convention being passed in 1926. Nations progressively endorsed the convention between 1929 and 1939, with Britain being one of the last signatories. The British were initially separated from the League of Nations convention by a desire to accompany their signs with written messages. Standard signing in Britain began with the Local Government Board in 1904, but uniform signing was not introduced until 1921.

A further league convention in Geneva in 1931 was devoted to the further standardization of road signs and expanded the set to include the Attention! and right of way signs and the abstract No Entry sign using a horizontal white bar on a red disc. The latter drew on Swiss experience, which in turn was based on the practice adopted in some earlier European states of marking their boundaries with their formal shields and of tying a blood-red ribbon horizontally around the shield when they did not wish visitors to enter. The large blue and white freeway-style direction signs with their distinctive lowercase letters naming each direction were introduced during the 1930s on the German autobahnen.

The first American signing manuals were produced by Ohio and Minnesota in 1921. Signing practice was codified nationally by AASHO in 1924 and 1927, drawing on existing state systems. The octagonal stop sign arose out of this process. A *Manual of Uniform Traffic Control Devices* was first produced in 1935, with a second edition in 1949.

Many European signs were destroyed during the two world wars to avoid assisting the navigation of foreign armies. After the Second World War, the United Nations took over sign standardization and continued the practice of producing signing conventions, beginning with a convention in Geneva on road and motor transport in 1949. The outcome was the Protocol on Traffic Signs and Signals. It retained all the old warning signs except the caution triangle and added fourteen more. In 1968 the United Nations convention in Vienna proposed an international system based on an acknowledgment that the European and American systems were of equal value. Thus, the number of warning signs increased from twenty-four at Geneva to forty-one at Vienna. The world still lacks an international system of traffic signs.

By the time of the 1949 United Nations protocol there were effectively three signing systems in use around the world: the European system, which was close to the United Nations protocol; an African system, which was a combination of English and European practice; and a Pan-American system founded on the *U.S. Manual of Uniform Traffic Control Devices* and relying more on words than on symbols.[8] The reliance on words reflects the fact that the U.S. system had to accommodate only one official language. Although called Pan-American as a result of decisions at Pan-

American Road Congresses in 1939 and 1949, its English basis meant that it had little impact in South America and was effectively the U.S. system. Fifty years later there are still two major systems in use in the world, the European and the American, with European symbols gradually becoming the world standard.

Reflective lenses were used on railway crossing gates in Switzerland from 1918. Reflectorized road signs were introduced in 1921 with the use of ribbed glass panels. The panels were supplanted by glass-bead technology following the development of the beads by Rudolph Potters in 1914 and Micheli's application of the technique in 1925. Luminous paints have been in use since World War II.

Systematic route marking began with a 1704 Maryland law requiring trees beside a route to be marked with an elaborate system of notches, letters, and/or colors (MacDonald 1928). The system could indicate whether a road led to a ferry, church, or courthouse. One south Maryland road between the Potomac and Patuxent rivers has retained the name Three Notch Road. Another Three Notch'd Road ran from Richmond, Virginia, to the Shenandoah Valley from 1730 to 1930, after which it became part of U.S. 250 (Paulett and Lay 1980). In the early days of the cycle and the car, major American routes were often distinguished by coded bands of color painted on roadside poles and trees.

In 1914 the Swiss formed a society for route numbering, and by 1915 had erected several thousand signs (Krampen 1983). Route numbering began in the United States in Wisconsin in 1918, and national route numbering was introduced by AASHO in 1925 (Sessions 1971). This effort was associated with the definition of a system of interstate highways. Intense political and commercial debate preceded these discussions (Paxson 1946).

Pavement Marking

The use of centerlines marked on the road surface to separate opposing traffic possibly began with the Roman emperor Trajan, who used a row of elevated stones to create two lanes on the road he was building north from Ezion-Geber to Petra and beyond in the Middle East (see fig. 3.1). A more recent centerline marking was the use of light-colored stones on a road built in about 1600 between Mexico City and Cuernavaca, which was still in existence in 1949 (fig. 6.5; White 1939). When the same method was used on the English Bridge over the Severn at Shrewsbury in the 1820s, wagoners frequently indicated their dissatisfaction with the arrangement by throwing the stones into the water (James 1964a).

Modern pavement line marking began on the new Brooklyn Bridge in New York in 1883. The first pavement application was possibly in 1907 when stop lines were used in Portsmouth, Virginia (Sessions 1976). Edward Hines in Wayne County, Michigan, introduced centerlines on bridges and road curves in 1911, following a near collision between a car and a horse and buggy on a bridge. At around this time it was quite common, but usually illegal, for small roadside businesses to advertise by chalking advertisements on the new asphalt road surfaces. One of the more graphic demonstrations of the effectiveness of centerlines occurred accidentally in Maryland in 1920 when black bitumen was used to seal the longitudinal joint down the center of a new concrete road.

PIARC delegates in 1913 opposed painting curbs and obstructions white, as this would be "only of service to persons who are the worse for liquor." Their advice

Figure 6.5. Oldest existing (1600–1949) centerline marking, on a road between Mexico City and Cuernavaca. *From White (1939).*

was not followed—indeed, even better delineators were demanded. Bullseye reflectors (involving a glass bead set in a rubber block and then mounted on signs and posts) were introduced in 1923 (James 1964a). This idea was extended in 1935 by P. Shaw when he developed the now common catseye reflector, which was glued to the road pavement rather than to a post. It was subsequently wiped clean of accumulated road grime by the tires of passing vehicles.

Road Maps

Road maps date from very ancient times. The oldest known map is dated at 2200 B.C. and is of the ancient Mesopotamian city of Lagash. It is carved on the lap of a statue of the city's ruler, Gudea. The Greeks introduced the science of cartography and, consequently, improved maps. The next advance came late in the first century B.C. when Marcus Agrippa, at the request of Emperor Augustus, spent twenty years mapping the Roman Empire. The master copy of his work was engraved in marble on a wall of the Forum in Rome. In the second century A.D. Claudius Ptolemy published his influential eight-volume *Guide to Geography*. The work of Agrippa and

Ptolemy served as a model for the famous Roman itinerarium, which was basically a route map giving distances between stations or post houses, together with other data such as road conditions, rivers, and shortcuts. The most famous survivor, the Antonine Itinerary, compiled about A.D. 300, simply lists Roman placenames with distances between them. Another well-known Roman map was the one Caesar kept, showing the Roman road network in hammered gold and the principal cities represented by precious jewels.

Ibn Khurdadhbih, a ninth-century Arab geographer of Persian descent who worked in the al Jibal (Media) area near Babylon, translated Ptolemy's geography into Arabic. In 846 he published a work called *On Geography,* which included a "Book of Roads and Provinces" containing detailed and accurate itineraries for the road systems of the known world (Ronan 1983). The itineraries were centered on Babylon, with the four main arms (see fig. 3.1) north along the Caspian Sea, east to Asia, south to Arabia and west to North Africa and Spain.

A few pilgrims' route guides were produced between the eleventh and sixteenth centuries. The best known were those of Saint James of Compostela and a woodcut map—Das ist der Romweg—made by Erhard Etzlaub in Nuremberg between 1492 and 1501. The first comprehensive book of road maps was Charles Estienne's famous 1552 *La Guide des Chemins pour . . . Royaume de France,* which passed through seven editions by 1560 and provided a model for other books, which were soon to cover most of Europe (Fordham 1924). They were all collections of itineraries and strip maps rather than sets of full-page maps. The first English guide was Richard Rowland's *Posts of the World* published in 1576.[9] When John Ogilby produced his popular Britannia collection of strip maps and associated commentaries in 1675, the products were without doubt useful, reliable, and well used. He was helped in his task by the indefatigable Robert Hooke, who gave advice ranging from the principles of cartography to the design of a measuring device for road distances, which was called the "chariot way-wiser" (Ogilby [1675] 1970).

Road guides in the United States began with *A Survey of the Roads of the United States of America,* produced by Christopher Coles and published in 1789. The first useful modern maps were produced by the cycle clubs in the early 1880s, and Michelin began producing its famous guides in 1900. The German army subsequently used Michelin maps to plan its successful invasion of northern France in 1914.[10]

Distance-measuring devices (odometers) were known to the Romans; Vitruvius in about 25 B.C. described a machine consisting of a series of notched wheels mounted on a carriage (Vitruvius 1960). The first comprehensive road inventory was undertaken in the United States in 1914 (Robertson 1985).

Speeds and Speeding

Before the appearance of the bicycle and the car, road vehicle speeds seldom exceeded 15 km/h, with 25 km/h as the maximum possible speed. Nevertheless, even these speeds were often considered excessive. In 1487 Paris banned trotting and galloping and punished any offenders with a flogging. The law was not very effective, and in 1540 it was extended to require horses to be led by hand. In 1652 New York, or New Amsterdam as it then was, followed suit and banned wagons from being driven at a gallop within the city. Note has already been made of the early nineteenth-century belief that traveling at over 50 km/h would be physically damaging. A century later the speeding car caused general public disquiet and on most roads also produced unwanted noise and large clouds of antisocial dust. Armed with the earlier precedents, the public still thought 50 km/h to be excessive and demanded speed controls.

In Britain the Red Flag Act of 1865 had kept speed limits to walking pace and was not relaxed until a 20 km/h limit was introduced in 1896 and raised to 33 km/h in 1904. In response to public pressure, the 1910 PIARC recommended the following speed limits to prevent road damage: cars, 18 km/h; trucks with axle loads under 4 tonnes, 16 km/h; trucks with axle loads between 4 and 7 tonnes, 10 km/h. The PIARC recommendations represented merely pious hope and reaction; for the previous seven years, such cars as the 45 kW Mercedes 60 had been capable of traveling at speeds of well over 130 km/h. Not surprisingly, there was a great deal of pessimism about controlling car speeds. As Logan Page asked quite prophetically in the same year as the PIARC recommendations, "If we cannot control the speed of motor traffic in the infancy of the motor era, how can we expect to cope with it in the future?"

As a physical response to the growing public concern, German and French towns built sharp-sided, 200 mm deep ditches across their main village roads to limit car speeds (Carey 1914). Deviously located and enforced speed signs were used to trap speeding motorists. Heavy local fines for the often innocent victims became commonplace. In Britain the Automobile Association was formed by motorists in 1905 to outwit such measures. Such clubs used officials wearing club colors to warn of approaching speed traps without being seen to do so deliberately (Flink 1974). The confusion and discontent that this situation caused was a major factor leading to uniform traffic laws and regulations.

By 1918 American speed limits ranged between 25 and 50 km/h, except for one state, which permitted 70 km/h. By 1928 the common limit was 60 km/h (Good 1978). Meanwhile, and seemingly independently, the top speed of even the average car increased steadily from about 90 km/h in 1925 to 140 km/h in 1940.

A major stage in the development of traffic engineering occurred in both Germany and the United States in about 1930 with the acceptance of the concept of using a particular design speed to design roads to be compatible with vehicle behavior. By 1934 common design speeds were 80, 100, and 130 km/h, depending on the class of road. The alternative approach of designing the road and then calculating the safe speed that eventuated, remained British practice for some decades (Good 1978).

It is not coincidental that this stage occurred at the time of the first autobahn and parkway construction and just a year or two after the development of reliable speed-measuring equipment. Before this, speeds had commonly been established by timing vehicles over a measured distance, using either an electric buzzer or a dropped handkerchief as a signal. Ropes, or even wires, were then stretched across the downstream road to force the detected speedsters to stop. Some circumvented these measures by installing scythes on the front of their cars to cut the impeding ropes. A novel alternative used in Ontario was to throw a plank studded with nails a measured distance in front of an oncoming car. If traveling at the legal speed, the car would be able to stop in time. Otherwise, it would be brought to an uncomfortable and costly stop.

Road Layout and Geometry

The early factors determining road alignment were discussed in chapters 1 to 3. The relatively low power of horse-drawn vehicles had resulted in all the emphasis in precar times being placed on the grade of the road. Where possible, grades were limited to 5 percent, and the horse worked best on grades of 3 percent or less. However, there are reports of wagons successfully negotiating 12 percent grades (Rose 1952a).

An advantage that the car and the truck brought to the road builder was that they were far less demanding on vertical grades than were animal-drawn vehicles and could manage 13 percent grades. Nevertheless, conservative attitudes kept the new road grades to well below 10 percent (Cron 1976b).

Precar speeds of below 15 km/h led to road curves being frequently no more than kinks between pieces of straight alignment. The distance that a driver could adequately see ahead, called the *sight distance,* was an irrelevancy at low operating speeds and, with the coming of the bicycle, was solved more by bell ringing than by realignment of the road to eliminate hidden curves and crests. The horizontal alignment, or curvature, of the road was only important insofar as generous curves were necessary to enable long animal teams hauling wagons to negotiate corners. On minor roads a 15 m radius was adequate, and a 25 m radius was ample for major highways (Cron 1976a). However, by the turn of the century the self-powered vehicles had raised that minimum to 50 m. This uncharacteristically favorable response of the road designer to the needs of the car driver was partly induced by the need to limit the tendency of the faster vehicles on tight curves to shift large amounts of loose surfacing material from the inside to the outside of the curve.

A more characteristic response to the problem of tight curves in the early years of the century was the erection of warning signs and the placement of large mirrors to allow the driver to see around the curve.

Minimum curve radii gradually increased from the 50 m endorsed at the first (1908) PIARC to 100 m in the early 1920s, to 150 m in the late 1920s, to today's 500 or so meters. The major technical debate of the time was the link between safe curve speeds and curve radius. Nevertheless, the chosen speed was not a safe speed at all but a discomfort speed, derived from railway alignment practice. It related the speed, somewhat fictitiously, to a lateral slipping—or side friction—factor derived for passengers in railway carriages in a manner that has persisted in some design codes to the present (Good 1978). A sound early derivation of the discomfort criterion for road travel was produced by Thomas Agg in the early 1920s and was formally introduced into design codes in 1932. The 1920s also saw curve design introduce centerline markings and minimum sight distance requirements. It was not really until this time that the emphasis in road design finally switched from vertical to horizontal alignment and the car—as opposed to the wagon and the truck—was at last formally recognized in the road design process.

Not surprisingly, most road geometric requirements continued to derive mainly from railway practice. For example, the layout of the transition curves between a straight section of road and the curved section at a bend was based on criteria developed to control the swing of rail carriages on curves.[11]

In the 1930s, autobahn designer Hans Lorenz went beyond these transitional curve concepts and used smooth, continuous, and mathematically defined road curves that could be driven at constant speed. This method thus linked curve design to the design-speed approach discussed above. Slowly, the world also learned the lesson that highway curves were desirable ways of avoiding the accident-causing monotony of long, straight lengths of road.

Pavements had been continuously and relatively steeply sloped transverse to the traffic direction on both straight and curved road lengths in an often futile attempt to drain water from the road surface. However, the cross slope encouraged vehicles to keep to the center of the road in order to straddle the longitudinal ridge or crown. As late as 1879 a technical writer was to comment that "the arc of a circle is practically the best cross section for street pavements. If not from choice, at least from necessity, the bulk of travel is along the center of the street [which] is almost level" (North 1879). A few road makers complicated the issue further on straight lengths of road by preferring a continuous cross fall, rather than a crowned cross section, in order to improve surface drainage.

There was one area where a continuous cross fall did have its place. The banking of curves was commonplace in railway practice but was uncalled for on roads when peak road speeds were only 15 km/h. The car soon changed this situation. In order to encourage motorists to stay on the correct side of the road as they concerned, the old reverse cambers had to be removed and the cross sections super-elevated at curves (Cron 1976a). The road builders also believed—reasonably correctly—that banking reduced the road damage caused by speeding vehicles (PIARC 1910). Banking was first used on roads in 1906 on the Long Island Motor Parkway and racetrack, where the provision of banking met stiff opposition from the owners of the slower horse-drawn vehicles, whose lack of centrifugal force when cornering led to them sluing down the banking.

Banking is now designed to help a car negotiate a curve with minimal steering-wheel manipulation or tire squeal. The speed at which this can be done with no input at all from the steering wheel is called the "hands off" speed (Lay 1990). Banking is another example of a road technology borrowed from the higher-speed railway practice.

In the United States, William Eno founded the Eno Foundation for Highway Traffic Regulation in 1921, and the Studebaker Corporation funded the Bureau for Street Traffic Research at Harvard University in 1926 (Montgomery 1988). Another major step was the creation of the U.S. Highway Research Board Committee on Highway Traffic Analysis, which began its contribution by analyzing the capacity of a single traffic lane (McLean 1980). Traffic capacity of a road is the maximum vehicular flow rate, usually measured in vehicles per hour (veh/h), that a traffic lane or an entire road can accommodate. It is the critical factor in determining whether a road can meet the traffic demand imposed upon it.

In 1927 George Hamlin of the Pennsylvania Highway Department gave an accurate prediction of the maximum capacity of a lane as 2,000 veh/h at 34 km/h. It soon became obvious that such predictions were very sensitive to the assumptions made in their derivation, and so experimental data were obtained in 1928 using aerial photography. The next major contribution came with Bruce Greenshields's use of photogrammetry in 1934, followed in the next year by spot speed surveys of some twenty-two thousand Ohio vehicles. These data led to his now famous linear relationship between vehicle speed and traffic density (i.e., vehicles per kilometer). This pioneering work led to a maximum capacity prediction of 2,200 veh/h.

Greenshields's effort was followed by an extensive series of empirical and analytical studies by Olav Normann of the Bureau of Public Roads, introducing the concept of the practical capacity of a traffic lane; that is, that the theoretical estimate is affected by such real-life factors as road alignment and curbside parking. Normann also chaired the committee that produced the first and second editions of the widely used (U.S.) *Highway Capacity Manual,* which provided a basis for most subsequent developments and new road designs around the world. He later applied his extensive speed-density data to more complex studies of passing (overtaking) and truck behavior (McLean 1980).

Driving on the Left or on the Right?

One road rule of some general interest concerns traveling on one side of the road or the other. The first such regulation was introduced by the Chinese bureaucracy of the Western Zhou dynasty, whose 1100 B.C. *Book of Rites* (Zhou Li) stated: "The right side of the road is for men, the left side for women and the center for carriages." The rule applied to the wide official roads and, because of the low travel speeds, was more concerned with protocol than with head-on collisions.

Although few countries were blessed with such wide carriageways, during most of transport history left- or right-side driving requirements were irrelevant. Relatively few roads could take wheeled vehicles, and when they could, they were rarely wide enough for two vehicles to pass with ease. It was normal to keep in the ruts and as close to the center of the road as possible (see fig. 3.7). The center of the road was the high ground. Lead horses in packhorse strings and stage wagon teams frequently wore bells to proclaim their possession of the center. Some Roman bridges did permit two-way traffic, with each bridge having its own priority arrangements. Of course, vehicle speeds on such bridges were so low that drivers did not require any advance knowledge of whether they had right of way.

In such circumstances, the key rule was the right and might convention, which told oncoming traffic which vehicle had the right and might to stay on course and which had to pull aside. A German traffic law of about 1200, the Sachsenspiegel, specifically required pedestrians to give way to horse riders, horse riders to wagons, and empty wagons to full wagons. Such requirements subsequently operated in practice in many parts of the world; as late as 1930, an Australian government report recommended a law defining who had right of way by requiring "on narrow paved roads [under 3.6 m] that the heavy vehicle have right of way in the event of it being necessary for one vehicle to partly pull off the road" (Kemp and Crawford 1930).

For foot soldiers the left hand was normally used for shield carrying and the sword worn on the right (Day 1984). Thus, more troubled times led to a defensive tendency to walk on the right and to keep the shield facing the oncoming traveler. When firearms came to be carried, the right-to-left direction of the barrel across the body also encouraged moving to the right as a defensive gesture.

Traveling on the left in Japan is said to stem from the courtesy of keeping the sword-side away from oncoming travelers. In particular, Japanese Samurai swords were large and trailed along the ground. Even an accidental clash of two swords could be seen as an aggressive act. Nevertheless, Japan's continued use of the left side of the road for driving was the result of strong pressure from Britain during the nineteenth century. Britain was also responsible for the course of traffic in China, at least until 1946, when that country moved to the right, in traffic if not in politics.

Keeping to the right was also favored by the tendency to lead a horse by the right hand, with the handler therefore walking to the left side of the animal. The left was thus called the "near side," but the far side was called the "off side." These two terms are still sometimes used technically in defining traffic priority (Kincaid 1986). Similarly, those in charge of carts and packhorse strings held the lead horse by the right hand and thus found keeping to the right easier for passing. Indeed, it remained common for led animals to keep to the right, even in countries where keeping to the left became the norm for other traffic. The British army attempted to manage the situation by requiring animals to be led by the left hand rather than the right in order to keep such travel naturally on the left (Kincaid 1986).

Thus, keeping to the right was common until an increase in horse riders brought new priorities. Right-handers mount a horse from the left, which means that they normally begin on the left side of the road. This tendency was reinforced by the use of roadside steps to help riders reach the saddle. Stepped mounting stones on the left side of the road were particularly common prior to the introduction of the stirrup and were a feature of formal Roman roads, where they were placed at 12 m intervals along the roadside. Somewhat more romantically, the need to leave the right hand free for both salutation and aggression and the fact that right-handed horsemen wore their swords on the left led horse riders to travel on the left when safe passing was required.

These factors led naturally to left-side travel for horse riders, as opposed to the right-side tendency of foot traffic (Hopper 1982). The difference was probably beneficial for the safety of both because it allowed early evasive measures to be taken. An attempt to clarify the situation was made by Pope Boniface VIII, who suggested in 1300 that pilgrims and other travelers to Rome for a jubilee should keep to the left when crossing the Ponte Sant'Angelo in Rome. For five hundred years this minor suggestion gained much unjustified notoriety as a papal intervention into traffic regulation (Hopper 1982). The jubilee had been planned as a fund-raising substitute for the recently finished Crusades. Two million pilgrims descended on Rome. Ironically, Boniface is historically famous for the precedent-setting rejection of his papal powers by the French king.

The need for a clear rule of the road grew as horse-drawn wheeled vehicles became more common in the seventeenth century. On vehicles with one line of horses, the use of the right hand for the whip led to the driver favoring the right side of the horses, and therefore keeping to the left so that he could judge clearances to oncoming vehicles. This was further emphasized with the introduction of four-in-hand coaches at the end of the eighteenth century. The lead horses in this harnessing system could only be controlled by a skillfully wielded whip.

However, with more than one line of controlled horses, the driver tended to the left side of the vehicle to permit his right hand to manage the whip and the reins, which were naturally located between the lines of horses. This favored driving on the right for passing. Such multiple-line harnessing arrangements and the associated larger vehicles were more common in continental Europe than in England. European practice also tended to postilion riders seated on the horses rather than to drivers seated on benches. This practice also favored driving on the right because the postilions always sat on the left-hand horse both for ease of mounting and to allow their right hand to control both lines of horses. The bench allowed more flexibility.

Perhaps, then, the different sides of the road chosen on either side of the Channel can be traced to both the large expanse of flat terrain on one side and to the dominance of the coaches and ridden horses of the aristocracy on the other. Although the move toward more elaborate vehicles for the wealthy also led to keeping to the left being the obvious mode during the nineteenth century, the revolutionary counter had already been cast.

A major impetus for right-hand driving in the United States came from the design of the Conestoga wagon, which had led to the winning of the West. The wagon was operated either by the postilion driver riding the left-hand near horse—called the wheel horse—or by the driver walking or sitting on a "lazy board" on the left-hand side of the vehicle. He kept to the left in both cases in order to use his right hand to manage the horses and operate the brake lever mounted on the left-hand side. Passing therefore required moving to the right to give the driver forward vision.

When not actually passing an oncoming vehicle, drivers usually preferred to move to the side of the road opposite the side they used for passing, so that they could judge how close they were to any dangerous road edges. The natural situation for passing traffic can be summarized as follows: walkers and horse riders kept left; led horses and carts kept right; postilion riders and drivers of European- and American-style wagons kept right; and drivers seated on benches on English-style carriages or ones with a single line of horses kept left (Kincaid 1986).

Road traffic began to increase in volume and speed during the eighteenth century, and the above mix of practices led to a clear need for some legal rule-making. In response, in 1722 the lord mayor of London appointed three men to ensure that traffic kept to the left and did not stop on London Bridge. They were possibly the world's first traffic police (Sessions 1976). The London Bridge carriageway was at best 5 m wide and was then the sole road across the Thames in London. As wheeled traffic increased, the need for some decision on who had right of way on the bridge became imperative. The wording of the order is instructive: "The Court being sensible of the great inconveniences and mischiefs by the disorderly leading and driving of Cars, Carts, Coaches and other Carriages, over London Bridge . . ." The wording also indicates that a number of previous unsuccessful orders had been made (Home 1931).

The first national decree was introduced in Saxony in 1736, requiring travelers to keep to the right when crossing a new bridge over the Elbe. In 1756 the first relevant English legislation supported the lord mayor of London's earlier decision and required traffic on London Bridge to keep to the left (Rose 1952a). The rule was extended to Scottish towns in an act of 1772 and to Ireland in 1908, and it became widespread in Britain after the Highway Act of 1835. In 1787 New York City and County introduced a passing priority rule based on whether or not the traffic was entering or leaving the city. Incoming traffic had priority. A similar rule had applied on some of the English roads leading to London. Indeed, the London Bridge ordinance quoted above refers more to the direction of travel in and out of London than to a general keep-to-the-left philosophy.

Following the French Revolution in 1789–1792, Robespierre's administration introduced a law requiring travelers in Paris to keep to the right. This satisfied both the technical and political needs of the times; the poor in France had traditionally walked on the right to permit proper deference to the left-keeping nobility as they rode in their carriages with their drivers seated on benches. It was, of course, also safer to walk into oncoming wheeled traffic rather than be overrun by it. With the Revolution, citizens henceforth thought it wise to align themselves with those on the political left. Napoleon extended the keep-right practice to heavy military vehicles with their postilion drivers, and the rest of the traveling public had little option but to follow suit (Hopper 1982). An 1852 decree made it compulsory to keep to the right while passing. Although a government commission recommended left-hand driving for France in 1909, its advice was ignored and the right-hand law was finally codified in 1921. The French rail network, SNCF, was initiated by a British engineer and still operates partially on the keep-to-the-left principle.

Napoleon used his usual administrative method to spread the right-side rule throughout Europe, with driving on the left subsequently being a mark of countries that had managed to resist or avoid his onslaughts. The success of traffic control regulations in managing French troop movements during World War I, widely reported in the news media, strengthened the commitment of many administrations to instituting formal traffic regulations and led to many left-side countries changing over in the decade after the war.

Perhaps the most curious situation occurred in Austria where, for most of the nineteenth century, half the country drove on the left and half on the right. The situation was eased but not eliminated in 1921, and the two alternatives remained until the coming of Adolf Hitler in 1938. Hitler conducted similar traffic engineering realignment exercises in Czechoslovakia and Hungary in 1939. Benito Mussolini resolved similar issues in Italy in a similarly autocratic way. The problem in Italy had been that, while the countryside and some cities drove on the right, a number of the major cities required left-hand driving. Mussolini soon decreed national uniformity.

The common American practice of right-side driving was encouraged by regulation on the new Lancaster Turnpike in Pennsylvania in 1792, with the first general laws occurring in New York in 1804 and in Canada in 1812 (Labatut and Lane 1950). These dates were undoubtedly influenced by the French precedents mentioned above. Curiously, left-side driving persisted in the United States on the National Road until the 1850s (Rose 1952a). Governor Lachlan Macquarie introduced a left-side regulation in Australia in 1810 (Lay 1984). Despite the 1812 date of the first Canadian edict, left-hand driving was still the norm in parts of Canada in the 1920s, and Newfoundland did not change across until 1947, two years before it became part of Canada.

From a practical viewpoint, however, the high longitudinal crowns and deep roadside ditches that were an inevitable part of most rural roads meant that rural travelers still commonly used the center of the road and only had a left- or right-side decision to make when passing the occasional oncoming vehicle. The old might and right rule generally dominated.

With the growth of traffic, the roadside ditches also led to a growing tendency in the United States in the late nineteenth century for drivers of light horse-drawn vehicles to both drive on the right and sit on the right to avoid the greater evil of the ditch. It was also common practice with bench-seated drivers of single-line horse-drawn carriages, where the need to accommodate the whip in the right hand predominated.

The first cars had central-tiller steering and so did not need to commit themselves as to the driver's location. However, with the introduction of the steering wheel in 1898, a central location was no longer technically possible. Car makers usually copied existing practice and placed the driver on the curbside. Thus, most American cars produced before 1910 were made with right-side driver seating, although intended for right-side driving (see fig. 6.2; Rose 1952a). Such vehicles remained in common use until 1915, and the 1908 Model T was the first of Ford's cars to feature a left-side driving position. In France, cars with drivers seated on the right were common until well into the 1920s. There are, incidentally, good technical reasons—such as ease of curbside parking and pedestrian safety—for curbside driver placement as opposed to the current center-of-the-road location of the driver (Kincaid 1986). On the other hand, there are no technical reasons for preferring driving on either the left or the right side of the road.

Summary

The previous chapter described how the technology of powered transport took some enormous steps forward during the nineteenth century, steps so large that they often left behind the much more slowly developing road infrastructure. The next chapter explores the response of pavement technology to these pressures. By the end of the nineteenth century there had thus been a congruence of events seemingly stage managed to produce the modern motorcar with its pneumatic tires, gasoline-powered IC engine, and ability to satisfy the new, universal desire for private travel. No flanged wheels restricted the driver's choice of route. The one missing factor was a good road system. Indeed, the roads of the day were such that it was wise to use large-diameter wheels to provide sufficient clearance over rocks, stumps, and other obstacles.

The practical gasoline-powered internal-combustion engine was soon supplanting all other mobile power sources. And these new engines were so light and effective that the new motorcars could—well, almost could—operate on the old ways. Rail's sixty years of dominance were destroyed by the inflexibility of the permanent way—which had made its initial use so popular. Nevertheless the car, with an admirable sense of history, drew heavily on railway technology for its traffic engineering basis.

The car presented a new and desirable power source and fitted neatly into the technological jigsaw of the day. In addition, its individuality meant that it also fitted into the emerging acceptance of social democracy, which was greatly changing the way society of the time operated. The bicycle had heralded this technico-social revolution, and the car provided the means by which it could be expressed in daily life. However, the fit into social democracy was not immediate, as the car was initially a plaything of the wealthy. It took Henry Ford to realize that every citizen was a customer and so market forces, rather than social forces, democratized the car. The truck, on the other hand, was always the honest workhorse of the road. Both vehicles also had different impacts on the infrastructure. The car demanded better road surfaces in order to protect those near the roadside from its excesses; the truck simply smashed through all but the best of surfaces.

Much of the rest of this book explores how society has subsequently adapted to the car and truck and then examines the social and technological consequences of our third quantum jump in travel capacities.

Pavements

Chapter 7

*As a branch of Civil Engineering, street paving has
never occupied a very high place.*

FREDERICK DEACON 1879

*T*his chapter discusses the development of road pavements in the
nineteenth and twentieth centuries, following the widespread adoption of the methods of Telford and McAdam. The two centuries brought their different technical
problems. In the nineteenth century, horse-drawn vehicles provided the pavement
engineer with a dilemma. The horse required good surface friction and resilience to
accommodate its hooves; on the other hand, the hard, iron-tired wheels of the time
required a smooth and rigid surface. Yet both hoof and wheel used the same portion
of the pavement surface. In the twentieth century, the speeding car and powerful
truck greatly increased the demands placed on the road surface as a whole new set
of horizontal loads were created and as vertical loads increased. The technical tactics
used to win these battles provide a fascinating story.

The Problem with Macadam

The nineteenth-century techniques of McAdam and Telford provided a running surface of compacted but unbound broken stone, usually with the individual
pieces of stone restricted to about 20 mm in size. It may have crossed the reader's
mind that this would not have made an ideal urban pavement surface, even for the
solid iron-tired wheels then in vogue. Indeed, although the surface was blinded by
working fine stone and sand into any openings between the individual pieces of
broken stone, these materials had little or no long-term cementing effect. The blinding process was thus far from effective, particularly when McAdam's strong advice
was ignored, as was often the case, and organic material was used for blinding. This
short-term glue soon turned to either mud or dust, depending on the season.

Furthermore, the stone at the surface was regularly dislodged by the hooves of passing horse traffic and usually replaced by horse excreta.

Thus, urban macadam surfaces, although a great improvement on their predecessors, were frequently abraded, dusty in summer, muddy in winter, slippery, slimy, and malodorous. A technical writer in 1845 stated of macadam: "Each morning the devastations of the day before are repaired, but when the stream of traffic again covers the street, the dirt recovers its rights, and during 300 days of the year the roadway becomes an ocean of mud or a desert of infected dust" (Boulnois 1919).

In dry weather regular watering was required to allay the dust, and in winter regular shoveling was needed to remove the mud. A nineteenth-century device for removing mud from the road surface is shown in figure 7.1. However, the process was rarely so organized. In urban areas people made a living as street sweepers (or street orderly boys or crossing sweepers), using straw brooms to clean 200 m or so of road or a particular crossing, or to prepare paths for pedestrians wishing to cross the road or enter premises without soiling their shoes and clothes. Some two thousand sweepers found employment in London alone. Writing of sweepers between 1840 and 1850, Henry Mayhew ([1850] 1968) commented: "We can scarcely walk along a street of any extent, or pass through a square with the least pretensions to 'gentility', without meeting one or more of these private scavengers. Crossing sweeping seems to be one of those occupations which are resorted to as an excuse for begging."

Street cleaning is an old profession, with records of its role in ancient Rome, in Paris from 1154 (Parsons 1939), and in London as early as 1280. The titles scawageour, scavengeour, rakyer, skawager, and street surveyor were in use by the end of the fourteenth century. Despite the pretensions of the last title, it was still a street cleaning position (Salusbury 1948).

Many street-cleaning laws were passed in Paris from 1154 on. They have been described as "a long series of ordinances and edicts which were by their very terms unenforceable and so were not observed, and which were accompanied by the . . .

Water Cart.

Figure 7.1. Nineteenth-century device for cleaning the road surface. *From Delano (1880), with the kind permission of the Institution of Civil Engineers.*

unattractive details of incompetency, graft and personal favouritism" (Parsons 1939). Note the similarity between this view of street-cleaning regulations and the Webbs' opinion in 1913 of the weight of load edicts (chapter 2).

The timetable for some of these Parisian edicts is illuminating:

1154	Locals to share carts used to dump street refuse
1349	The dumping of rubbish and the running of pigs in the street made illegal
1374	Abutting owners must pay for street cleaning
1389	The nobility must also pay for street cleaning
1399	The religious must also pay for street cleaning
1405	Throwing street rubbish into the Seine made illegal
1476	A poll, or per head, tax levied to pay for street cleaning
1517	Street cleaning made a task for prisoners
1540	Street-cleaning standard issued

Street refuse was thus a long and troublesome problem. In the nineteenth century, London was apparently slower than most cities to act to improve its streets. Commenting on that city's use of macadam pavements in 1879, visiting Leeds engineer Deacon wrote that "to the unsophisticated provincial the manner in which, on a hot July day, fashionable London rolls over her tainted macadam pavements, apparently without even smelling them, is a mystery almost as great as [the poor London water supply]." Nor did the macadam impress the street sweepers. One of Mayhew's characters, Billy, commented to him that he "couldn't abide this mucky-dam [macadam] . . . its sloppy stuff and goes so bad in holes. Give me the good solid stones as used to be" ([1850] 1968).

It was widely believed that many in authority in London favored macadam because it kept traffic volumes down and provided a favorable, somewhat roughened surface for their own horses. Indeed, on any hard, smooth surface a horse cannot travel at much over walking pace.

The ever-present mess on the street pavements was one of the reasons why nineteenth-century women favored ankle-length clothing (Thompson 1970). Certainly, trousers did not replace breeches and stockings until routine street cleaning became commonplace.

The mess was not due solely to problems with macadam. Untended urban streets had always been either messy or dusty. An English act of 1662 had required Londoners to sweep the streets outside their houses twice a week and parishes to provide a street scavenger (Pawson 1977).

In particular, the sanitary logistics of the horse were enormous. The car alleviated these problems but rapidly created an insurmountable new set as its pneumatic tires sounded the death knell for unbound urban surfaces. Indeed, the 1910 PIARC resolved that "macadam carried out following the methods of Trésaguet and McAdam, causes dust and mud, is expensive to maintain, and is suitable in large cities only for streets where the traffic is not very great or heavy" (355).

In the countryside, macadam was still greeted with great joy for providing affordable roads that were passable in all weather. The horrendous alternative to macadam was illustrated in figure 6.2. Macadam's main problem was that it created a great deal of either dust or slush. Coach travelers and motorists on macadam roads often had to wear goggles and linen dustcatchers. In the summer the London–Bath road was watered daily to lay the dust (Jeffreys 1949).[1] In California, oil was sometimes spread on the road as a dust palliative, though it occasionally either caught fire or was eaten by hungry cattle (Robinson 1971).

A cheaper alternative to macadam for low traffic volumes was the careful use of mixtures of sand and clay to provide a reasonable surface. Indeed, in many areas it was impossible to find suitable stone for use in macadam. If clay alone was available, it was sometimes burnt in roadside fires to produce an inert, bricklike material. An alternative in situ technique involved placing logs on the road clay and then covering them with further clay. The logs were then burnt and the surface subsequently rolled. This technique developed in the 1930s into a mobile machine in which logs were burnt in a furnace whose heat was directed onto the road surface. The technique required a plentiful supply of trees.

For major rural roads the only serious alternatives to macadam and telford construction were various direct uses of timber. Frequently, particularly in the New World, timber was in more plentiful supply than stone. In the simplest method, logs were placed longitudinally in the wheel ruts. The resulting running surface was difficult. An alternative was the corduroy road (fig. 7.2), which was built with the logs placed transverse to the direction of travel, often on a base of two longitudinal logs. The rough running surface was, at least, navigable and was often improved some-

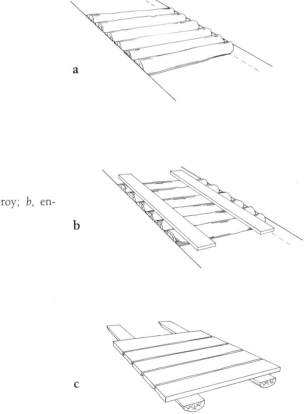

Figure **7.2.** Wooden roads: *a,* corduroy; *b,* enhanced corduroy; *c,* plank.

what by adding sand between the logs or by attaching longitudinal running planks to the top of the logs (fig. 7.2b).

Corduroy was demanding of timber, uneven, and often short lived. Nevertheless, the practice was widespread. As the Novgorod example (p. 60) illustrates, one advantage of the technique in soft ground is that as the timbers settle more can be stacked on top. Another advantage is that the method is simple to apply in treed regions. Consequently, corduroy roads are still in use in remote areas.

The related plank road—or farmers' railroad—was developed in Toronto, Ontario, in 1835 by Darcy Boulton, who used it on Yonge Street. Longitudinal timbers were placed with a flat upper surface slightly below ground level. Transverse 3.6 m planks were pounded into place until they bore on the longitudinal timber (fig. 7.2c). The surface was then covered with sand (Wixom 1975). There is a suggestion that the concept had been brought to Canada from Russia (Newlon et al. 1985). The plank road was strongly marketed as a private venture and became quite popular in North America, where it was first used at Syracuse in 1846, and in Australia. However, plank roads were not durable and rarely lasted more than ten years. A very similar technique to the plank road has been found at Glastonbury in the Meare Heath track, which has been dated at about 900 B.C.

The short life and rough ride provided by wooden roads made it clear that timber was not the answer to the growing nineteenth-century demand for better quality roads.

Native Asphalt and Mastic

The technological push for better road surfaces during the nineteenth century came not from rural roads with their relatively low traffic volumes, but from the demands of city life and busy city traffic. In retrospect, the most significant result of this push was the development of asphalt roads.

There is much confusion over the use of the words *bitumen* and *asphalt*. This book maintains the precise, consistent, and traditional definition that asphalt is a mixture of bitumen, sand, and pieces of stone. Nevertheless, North's 1879 remark that "dictionary and encyclopaedia makers, as well as chemists, seem to use these terms [bitumen and asphalt] interchangeably" remains true today. *Webster's* unhelpfully permits both interpretations, and the *Shorter Oxford English Dictionary* seriously confuses the two words and probably means bitumen in a number of references to asphalt. Misunderstandings concerning asphalt are indeed widespread. One work on the history of road transport (Paterson 1927) described asphalt as "a kind of liquid lava found between Italy and Switzerland."

The roadmaking use of asphalt required advanced technology and did not occur until the nineteenth century. Technically, asphalt comprises pieces of stone, sometimes called "aggregate," bound together by a bituminous adhesive such as bitumen or tar. The four theories for the origin of the word *asphalt* are that it came from (1) Sumerian or Greek roots meaning stable, secure, or nonslippery; (2) the Hebrew word *shatel,* meaning "low lying," used to describe the Dead Sea, which had been an early source of bitumen; (3) the name of a Mesopotamian tribe expert at using bitumen for waterproofing houses; or (4) the Greek word *sphallo,* meaning "to split," referring to its brittle nature when cold (Forbes 1936).

In nature, asphalt occurs when bitumen mixes with the local rock formation, most commonly with relatively porous limestone. To distinguish this asphalt from the manufactured product, we will call it *native asphalt,* although it has also been called natural asphalt, rock asphalt, bituminous rock, and asphalte. We will use the term *natural asphalt* to describe paving asphalts made directly from native asphalt. Deposits of native asphalt existed throughout the Middle East, including sources at Ur, Hit, Nineveh, and the Dead Sea (see fig. 3.1; Abraham 1960).

In Europe, native asphalt quarries operated on the Greek island of Zakinthos and at Zante in Albania at about the time of Christ, and the Romans worked mines in the Val de Travers in the Swiss canton of Neuchâtel and at Seyssel beside the Rhône in France. Native asphalts have since been found in many parts of the world, ranging from various oil sands, to bituminous deposits at Buton in Indonesia, to European-like deposits in West Virginia and California. Scottish native asphalt deposits were called boghead shale (Peckham 1909).

Spanish writers in 1526 first mention the Trinidad Pitch Lake at La Brea, a seemingly bottomless 60 ha pit of native asphalt with a very high—40 percent—bitumen content. In 1595 Sir Walter Raleigh used the material for caulking his ships, because it was less temperature sensitive than the Norwegian wood pitch then in common use (Forbes 1958b). California Indians similarly had a long history of using bitumen to caulk canoes and to fix arrowheads. The mineral matter in Trinidad Lake native asphalt is fine sand and colloidal clay, and it is thus closer to a mastic than to a conventional asphalt. The famous demonstration of the viscous nature of the lake's contents was to use a sharp hammer blow to shatter its seemingly brittle surface.[2] The hammer was then left lying on the nearby unshattered surface, through which it slowly sank from sight. After Raleigh, Trinidad Lake was effectively forgotten until 1864, although the governor of Trinidad had tried using the contents of the lake in the 1820s to keep down grass in Port of Spain, but found that it only "added to its fertility" (Merdinger 1952).

The Val de Travers and Seyssel deposits were both in the Jura Mountains and were mostly limestones mixed with about 12 and 7 percent bitumen by mass, respectively. The Seyssel native asphalt also contained "marine shells and shark's teeth" and was sometimes called Seyssel molasses (Malo 1866; Delano 1880). Other native asphalts in the area were mixtures of quartz and bitumen. The Jura product was described as a Swiss confection: "the colour is a rich chocolate; it is hard and easily broken, giving an irregular fracture" (Ellice-Clark 1880). It could thus be mined as a solid and then crushed to a manageable size or sawn into cubes.

Georgius Agricola, in book twelve of his *De re metallica* in 1556, described one of a number of methods then in use for removing bitumen from native asphalt (Forbes 1936; Abraham 1960). In the seventeenth century a German adventurer, W. Jost, published a study of the effects of heat on native asphalt and later unsuccessfully tried to exploit his findings commercially by applying them to Val de Travers material (Merdinger 1952). In 1711 Jost met a physician named Eirini d'Eyrinys, who was in the Val de Travers (Neuchâtel) region on an official mission for the Swiss government.[3] Neuchâtel was then under the control of the king of Prussia. Dr. d'Eyrinys was incidentally seeking bitumen for medicinal uses, and his interest in native asphalt was aroused by Jost. He subsequently carefully investigated and documented the major European deposits: in 1712 at Val de Travers; in 1730 at Hannover/Braunschweig, and in 1737 at Seyssel.

To meet the needs of the medical and fuel markets, d'Eyrinys removed the bitumen from the native asphalt by placing it in stirred, boiling water for an hour, after which time the bitumen floated to the top and the rock settled to the bottom.

As a sideline, d'Eyrinys powdered some of the native asphalt and then heated the material to produce a compound for use as a waterproof covering and for caulking ships (Broome 1963). However, it proved to be too viscous to be practical. In about 1720 he solved this problem by adding about 15 percent by mass of distilled bitumen to the powdered native asphalt. This produced a material which, when warmed, could be applied to surfaces and then worked with a wooden hand-float to produce a smooth finish. An appropriate early name for this product was *poured asphalt*, but it is now called *mastic*. In pavement technology, mastic now covers all mixtures of bitumen, sand, and fine stone particles.

He formed a company to exploit his native asphalt, bitumen, and mastic and published a glowing treatise on their properties (d'Eyrinys 1721). Dr. d'Eyrinys's admiration for his products had no limits, judging by the treatise, which painted native asphalt as a "panacée universalle" and compared his mastic with the bituminous mortars used in Babylon in biblical times (Malo 1866). The mastic clearly had its merits and some floors and steps built using it were said to be in excellent condition a hundred years later in 1838 (Forbes 1958a). Despite the company's efforts, the use of its products was not widespread, with little more to show than some ship caulking and the sealing of the water basins at Versailles Palace in 1743 (Simms 1838). A Swiss banker named de la Sablonniere, who been marketing the materials in Paris since 1720, slowly pushed d'Eyrinys aside. The change in company management had little impact on the market.

Mastic was first used to make a road pavement in 1796, when it was employed as a waterproof surface on the timber deck of the 72 m span Sunderland iron bridge in England; it incidentally provided a running surface for traffic.[4] Urban footpaths came into common use at about this time and there was a clear market need for a smooth and cleanable walking surface. Mastic had the potential to meet that need.

In 1797 the French government gave the rights to mine the native asphalt deposits at Seyssel to Secrétan, who subsequently made a mastic from the powdered ore by mixing it with bitumen from nearby natural seepages (Abraham 1960). This product was used for footpath construction in Paris, Lyon, and Geneva in 1810 (Broome 1963). However, it was "often promoted for purposes for which it was not fitted," and the serious exploitation of mastic for footpaths did not begin until the Société Eyquem began its marketing efforts in 1827.

One frequently encountered confusion is the long-running terminological overlap between the glue (bitumen) and the glued mixture (asphalt). One prime example of the confusion surrounded Secrétan's mining rights. Did he have a concession to mine the asphalt (i.e., the bitumen-limestone mixture) or just the bitumen? Litigation to determine what was actually meant by his concession began in 1797, and was not decided until 1845 (Chabrier 1875). The issue was complicated by the fact that in 1802 the rock strata mined at Seyssel had changed dramatically toward a pure limestone. This book consistently uses the terms in parentheses above, whereas many confusingly refer to both the glue and the mixture as asphalt.

The mastics proved highly effective in waterproofing decks and footways, but were often unacceptably slippery. This was overcome by adding sand to the surface, in the manner developed for tarmacadam in 1824. The technique was first used with mastic on Pont Morand in Lyon.

The Seyssel mine was taken over by H. Grafen, comte de Sassenay in 1833. He set about reversing the previous over-zealous marketing strategy by systematically establishing and then documenting the correct way to use mastic. He particularly insisted on the need for skilled workers.

Successful tests of mastic footpaths were undertaken by Jean Partiot on the Champs-Elysées and Pont-Royal in 1835 and 1837, and in 1838 the city of Paris adopted mastic for making footpaths. The first mastic used commercially in the United States was imported from Seyssel in 1838 for use on a footpath in Philadelphia (Abraham 1960). One motive for Partiot's tests was to show that various bitumen sources could be used for producing mastic. However, the so-called bitumens derived from other sources did not perform well, so practice reverted to the original d'Eyrinys technique of using bitumen derived from the same native asphalt as was used in the mastic (Malo 1866). Ductility tests to help improve performance prediction were introduced by Quirit de Coulaine in 1850.

The high bitumen content needed to make the mastics workable meant that they were either too brittle in cold weather or too soft in hot weather to be able to resist cracking, rutting, and deformation under the action of the iron-tired wheels and horseshoes using the road. There was therefore little or no use of mastic for roadmaking.

The Val de Travers deposits in Neuchâtel were not worked commercially until 1837, when de Sassenay bought into the mine, buoyed by the success of the Seyssel mine and its market potential in footpath construction and excited about the quality of the native asphalt he found at Val de Travers. For instance, its higher bitumen content made it easier to work than the Seyssel product. In 1870 the mine was bought by the English-owned Neufchâtel Bituminous Rock Company. German native asphalt deposits, notably at Limmer near Hannover, began to be used for pavements in 1843.

As experience with footpath construction increased, it was observed that when the sand or grit used to roughen the surface accidentally penetrated into the hot mixture, it significantly stiffened the finished mastic layer. Therefore, after 1850 it became common to add less expensive sand or grit to the mastic during mixing to produce a cheaper and stiffer material called gritted mastic. By 1866 some 30 percent by mass of grit was being used for Parisian footpaths (Malo 1866), and by the 1880s about 40 percent of the mastic was sand or grit. The current German *gussasphalt* developed from this mastic technology. This addition of a foreign material such as sand to the native asphalt was an important yet controversial step toward a solution to the paving problem.

Mastic was delivered to construction sites in 25 kg "cheeses," whose particular shape was the trademark of the mastic producer. For instance, the Val de Travers cheeses were hexagonal. The cheeses were then inefficiently heated on site until the material was spreadable. The use of mastic became much more widespread in the 1860s as a result of a handling system, developed by Léon Malo, a French civil engineer, which permitted the material to be heated and prepared at the depot rather than at a construction site (Malo 1861). The hot mastic was then delivered to the job site in a heated container. The situation further improved in the 1870s when the availability of heavy compaction machines eliminated the need to heat the mastic until it was soft enough to hand trowel into place.

At the same time a better understanding had developed of the way in which a pavement functions. Engineers now accepted that mastic could provide a good, wear-resistant surface course and be carried by a strong underlying structural course. No longer was the one material expected to perform two separate tasks. In addition, construction equipment capable of producing such courses now existed. The first such job was the 1869 Threadneedle Street trial to be discussed below (Broome 1963).

However, the potential for pavements to meet the growing nineteenth-century demands for better road surfaces required some quantum jumps beyond mastic technology. Before exploring those jumps, it is necessary to look at the direct use of native asphalts and then at two very competitive alternatives to the use of bitumen as the glue to hold pavement mixtures together.

Natural Asphalt

Despite the developments in mastic technology, there was still an unsatisfied need for a surface that could be kept intact, in shape, clean, dustless, free of mud, and waterproof while providing adequate friction for the tractive needs of iron-shod horses' hooves. Surfaces had to be waterproof because many underlying structural courses and natural formations lose strength and stiffness if they are moist. It is usually far more economical and practical to design to for wet conditions and make no provisions to inhibit water entry. This lesson was demonstrated by the Romans and has been a hallmark of good road design ever since.

A natural extension of mastic technology was to use native asphalt in some other way. By the middle of the nineteenth century, native asphalts were in plentiful supply from Switzerland (Neuchâtel), Germany (Limmer), France (Seyssel), Sicily (Ragusa), and Trinidad. Some commercial encouragement for the development of these mines had been received from the use of mastic for footpath construction. Another incentive in the mid nineteenth century was that the cheap freight rates now available from the new railways were making it economical to ship native asphalt from the mines to the cities.

One possible technique for using native asphalt directly in road construction is to cut it into bricklike cubes. There is serious question as to whether this method was ever used; its appearance in the literature may be the result of subsequent writers misinterpreting the early descriptions of the compressed-block method.

In the compressed-block method, native asphalt was powdered, sometimes mixed with further bitumen, and then heated until it could be pounded into cubical molds to produce asphalt blocks (Simms 1838). These were first used for paving in Paris on the Champs-Elysées in 1824 and for larger works on the same street in 1837. The latter trial was not successful; the blocks were too brittle in winter and too soft in summer.

A major trial of paving products was conducted in 1838–1839 on Oxford Street between Tottenham Court Road and Soho Street near Soho Square in London. It was probably the world's first significant pavement experiment, comparing Seyssel compressed blocks with mastic, stone setts, and wood blocks, with and without various bituminous glues and coverings (fig. 7.3). The tests favored wood blocks, and the compressed asphalt blocks performed poorly (James 1964b). The Soho Square blocks measured 300 × 300 × 75 mm. As the drawing shows, they were

Figure 7.3. The 1838 pavement trial in Oxford Street, between Tottenham Court Road and Soho Street, in London. The contemporary sketch, by George Scharf, shows the compressed asphalt blocks being placed and then finished with a bitumen mortar. *With the permission of the Hulton Picture Company and from their Hulton-Deutsch collection.*

placed on a sand bedding layer with the 75 mm dimension vertical. Regardless of their inherent properties, the blocks were probably too thin to ever have survived heavy iron-tired traffic. In 1840 Paris had some success with compressed blocks set on concrete, but a subsequent trial in 1848 led to early failure (Tronquoy 1868).

The largely unsuccessful initial experience with compressed asphalt blocks saw a major decline in interest in asphalt surfaces for roads in the thirty years from 1840 to 1870. By 1850 London had no asphalt paving (Darcy 1850). Writing of this situation, Henri Darcy, the inspector general of bridges and roads, commented: "The English have little confidence in asphalt. Originally their engineers too easily accepted all the promises of imprudent speculators and suppliers. Experience contradicted those promises and the reaction remains today [1850]. They don't even use it for footpaths—they reject all new approaches with an explicit but unjustified defiance."

Malo developed and patented an improved method for the manufacture of compressed native asphalt blocks, and Edward de Smedt introduced it into the United States in St. Louis in 1873. The new blocks were only partially successful, but seven years later the development of a powerful new compressing machine in the United States finally led to asphalt blocks of adequate density (Richardson 1905). A major initial application of these new blocks occurred in Baltimore in 1885.

The new blocks commonly had pieces of broken stone added to the powdered native asphalt, following the earlier experience with mastic. They were about 200 × 100 × 50 (deep) in Europe and 300 × 60 × 125 (deep) in the United States. The use of much deeper blocks in the United States was partly the result of conservative design practice and partly because they were often placed on unbound macadam rather than on concrete, in the European manner (Tillson 1897). The failure of thin asphalt blocks in the Oxford Street trials has already been noted. Thinner hexagonal blocks were used for footpath construction, where the loads were lighter. Asphalt blocks found wide use on residential streets and on steep slopes, where the joints provided additional friction. However, they tended to wear excessively on heavily trafficked routes. They were commonly used in the Americas into the 1950s, although they had long before fallen from favor in Europe (Merdinger 1952).

The second method of using native asphalt for roadmaking followed the mastic technology. The native asphalt was ground to a fine powder, heated to about 140° C, raked into place, pressed, tamped with hot 10 kg rammers, and rolled and then finished with a hot iron until the surface was resonant and glazed. Grinding the native asphalt to a powder was appropriate because some of the limestones in European native asphalts were unconsolidated and soft. The resulting material was effectively a low-bitumen-content mastic. It only became practical as construction equipment able to place and compact the stiffer mixture became available. Improved equipment also made it possible to do the grinding at the job site.

Unfortunately, the pressing and tamping process led to such asphalt often also being called compressed, and there is frequent confusion in the literature between the compressed blocks and the compressed in situ layers. We will call the latter product *natural asphalt*.

The natural asphalt method originated in 1849 when André Mérian, who was chief engineer of the Swiss canton of Neuchâtel, visited the Seyssel asphalt mine in France, where he observed horse-drawn trucks taking native asphalt from the mine to a mastic factory. Mérian noted that pieces of native asphalt falling from the trucks were being powdered by the iron-tired wheels and then cold-rolled by later wheels into a useful natural asphalt road surface. He returned home and immediately conducted trials aimed at transferring this serendipitous technology to his local material. As a demonstration he cold-rolled powdered Neuchâtel native asphalt into a macadam pavement on a road, between Val de Travers and Les Verrières, leading to Pontarlier in France (Malo 1866; Merdinger 1952). (It will become apparent that the material would have resembled asphaltic concrete.) After twenty-five years the trial road was still providing good service (Forbes 1958c; Broome 1963).

The message soon spread to Paris and, in the same year, Henri Darcy conducted a similar trial on the Champs-Elysées. The road was still performing well when Darcy reported on it a year later. De Coulaine also tested the technique, using a 40 mm layer of powdered native asphalt at Saumur on the Bordeaux to Rouen road, with similar initial success (de Coulaine 1850). A year of successful performance was all the enthusiasts needed. The method was extended to other roads at Saumur, to roads in the further, Loire-side towns of Angers, Tours, and Briare (Meyn 1873) and to the avenue de Marigny in Paris. Chapter 3 noted that this job probably used the first self-powered road roller; it would have been much needed.

Malo's 1866 book describes another 1850 French trial, which was an immediate failure. He claimed it was spoiled when the experimenters unsuccessfully added various additional material to the asphalt. This comment needs to be taken with a grain of salt. Malo was probably referring to bitumen not derived from French native asphalt, in which he had a commercial interest, although various resins and coal tar were certainly often added in a random way to spice the mix.

Darcy visited London in 1850 to review English paving practice, and his unfavorable comparison of the pavements of macadam, wood setts, and stone setts with the new Parisian natural asphalt led him to recommend the further use of the last named (Darcy 1850; Malo 1861).

Although mastics were placed hot in order to make them workable and provide a good footpath surface, Darcy—emboldened by the initial success of the trials—advocated the cold placement of natural asphalt in what was called the Dufan method. The method was cheaper than hot placement and gave a sufficiently rough surface to avoid the slipperiness associated with hot-placed mastics. However, cold-placed layers lacked adequate ductility, since asphalt will only meld into a coherent layer at temperatures well over 60° C. Predictably, then, the trials of de Coulaine and Darcy produced short-term successes but long-term failures: severe cracking and crazing began to occur under service conditions. The Dufan method was abandoned, and Darcy might well have wished to rewrite his confident report of 1850. The long-term success of Mérian's initial trial is unexplained, but was probably due to the material's melding into the underlying open macadam and to frequent maintenance.

The consequences of overselling in order to exploit the new native asphalt sources had been dramatic. The Parisian asphalt industry of the time had also been involved in a number of financial scandals, which turned people against the material (Malo 1866). Partly as a consequence of this, by 1869 the great majority of the streets of Paris were paved, not with asphalt products, but with stone setts ("Public Works in Paris," *Engineer* 27, no. 5 [1869]: 107).

The industry needed a savior, and that man was Léon Malo, manager and superintendent of the Seyssel mine in the 1850s and 1860s. His work *Guide practique pour la fabrication et l'application de l'asphalte et des bitumes* first appeared in 1861 and went through several editions before being published as a book in 1866. It made a major contribution to late nineteenth-century natural asphalt technology. Malo was very insistent on the use of the correct definitions for bitumen and asphalt, definitions that have stood the test of time and are used in this book. Unfortunately for readers, they do not coincide with current American practice, which uses the words *asphaltic cement* or even *asphalt* where Malo would use *bitumen*.

From his experience making and using Seyssel mastic since 1838, Malo had no doubt that the powdered native asphalt had to be placed hot for the natural asphalt pavement to be a success. In 1854 he and H. Vaudrey were able to put their ideas into practice in Paris to produce natural asphalt pavements on a small (800 m²) scale on the rue Bergère near the rue de Conservatoire, and then four years later on a larger (8000 m²) scale on the rue Faubourg St. Honoré and the place du Palais Royal (Malo 1866; Earle 1974).

The rue Bergère pavement lasted for sixty years (Abraham 1960). However, natural asphalt subsequently used on the rue de Rivoli was placed on a poor natural formation and soon failed. Over half a century was to pass before the often deliberate commercial misinterpretation of this failure by a somewhat shady industry was to be forgotten. In the interim, it was conventionally and incorrectly assumed that asphalt lacked the strength to act without a strong supporting structure, and asphalt was therefore commonly used as a 50 mm thick surface course on a 100 to 150 mm thick concrete pavement. In addition to industrial sabotage, natural asphalt had been typecast as a material for making footpaths and was then the major material used in Europe for that purpose (Malo 1866).

Practically, the problems encountered with placing asphalt prior to the mid-1860s were identical to those experienced with mastic and were similarly resolved by the advent of improved field equipment. As a consequence of all these factors, between the 1854 success on the rue de Bergère and the mid-1860s, Paris placed only 10,000 m², or less than a kilometer, of natural asphalt a year. With the new construction equipment, the situation improved somewhat with 100,000 m² in 1866 and 280,000 m² in 1869, mainly in the quarter between the rue de Rivoli and the boulevards de la Madeleine, Poissonierè, and de Sebastopol, not far from the earlier trials.

However, the success was short-lived and natural asphalt continued in disfavor in Paris during the 1870s and 1880s, "owing to failing contractors on the maintenance—and its slipperiness." The construction rate dropped to less than a kilometer a year, giving a total of only 20 km in 1885. The reference to "maintenance" in the above quotation arose because of the doubts of the Parisian city fathers about the durability of asphalt. These doubts led them to pay their asphalt contractors over a lengthy period, during which time the contractors had to maintain the pavement. By 1885 this approach was accepted in both Paris and London, with pavement contracts including construction and maintenance and with obligations and repayments being typically over five years but sometimes extending over twenty years (Greene 1885).

A major trial of Val de Travers native asphalt was undertaken in 1869 on 400 sq m of Threadneedle Street in London, where 50 mm of natural asphalt were laid on a 200 mm concrete base (Peckham 1909; Earle 1974). Even though the asphalt was placed in very wet conditions and its surface had to be leveled with up to 10 mm of mastic, it was still performing relatively well in 1880 and encouraged the wider use of natural asphalt in London (Haywood 1871).

Unfortunately, the underlying problems remained. The Threadneedle Street natural asphalt was just as slippery and abradable as both its predecessors and its current competitors. It was said in 1874 that asphalt was "most cruel and dangerous to horses" (Engineering Record 1890). In 1885 the major streets of London were still paved with wood blocks, which were considered more suitable for horse traffic, and—as in Paris—natural asphalt paving was being added at a rate of less than a kilometer a year (Greene 1885).

The application of native asphalts in California began prior to 1865 using local materials (called *brea,* a word derived from the Spanish word for *pitch*), possibly near Santa Barbara, and spread to San Francisco and Los Angeles by 1894, although it is possible that the material being used was bitumen and not asphalt (Peckham 1909). There are also suggestions that the Spaniards and Mexicans were using California native asphalts for paving prior to 1838 (Baker 1903). These asphalts, largely mixtures of bitumen, sand, and clay, would have been masticlike and easy to handle (Tunnicliff, Beaty, and Holt 1974).

In 1870 de Smedt conducted an unsuccessful trial of European natural asphalt in Newark, New Jersey, on a section of William Street bravely opposite city hall (Broome 1963). Subsequently, generally unsuccessful trials of natural asphalt were placed in Philadelphia in 1871, in Washington and New York in 1872, in Chicago some years prior to 1879, and on Pennsylvania Avenue in Washington in 1876 (FHA 1976). As with many prior asphalt pavements, the last trial performed successfully for ten years but was condemned as excessively slippery.

There were signs of change, however. The more popular wood blocks were far from perfect, and asphalt advocates claimed that detailed observations showed that asphalt caused less damage to horses when they fell, was more resilient under hoof impact and therefore generally easier on the horse, and led to lower vehicle operating costs (Richardson 1905).

As surface-preparation methods improved, natural asphalt was to provide a century of successful road surfacing. A small amount was still being hand laid on British streets in 1960 (Road Research Laboratory 1962). Nevertheless, it always lacked structural strength, scored easily, and was too slippery to use on roads with grades of over 2 percent.

The bitumen content of native asphalts varied from deposit to deposit, ranging between about 5 and 40 percent by mass. This variation and the variations in the inherent rock type led to much, sometimes acrimonious, debate about the effectiveness and application of particular native asphalts. The facts—which never seemed to be directly stated at the time—are that the Trinidad native asphalt performed like a mastic and could not carry heavy traffic; the Val de Travers native asphalt contained a little too much bitumen for hot climates and thus tended to soften noticeably and be pushed and shoved by traffic in warm weather; the Seyssel native asphalt had rather too little bitumen and was hard and awkward to place. Not surprisingly, mixtures of Val de Travers and Seyssel native asphalts were sometimes used (Malo 1885).

In retrospect, it might seem obvious that the solution to the inadequacies of the native product was to add bitumen "to taste" or as needed. There were to be major consequences of the realization that an asphalt could be manufactured, and that it was not necessary to rely on the happenstance of nature, as was the case with the natural asphalts. The idea of modifying asphalt properties by additions of bitumen or stone was usually dismissed as moving away from the natural product. This dismissal was based on past unsuccessful experience with the use of various additives, as in the attempt by de Coulaine to produce cold asphalt and in the use of soft stone in early road trials. This unfortunate but strong reluctance to tamper with nature delayed the introduction of a satisfactory road surface by half a century.

The earlier story of the discovery of natural asphalt at the Seyssel mine has the ring of a technological legend, appearing in a number of guises. For example, a related sighting was an equivalent quarry road trial conducted at the Seyssel mine, mirroring the competitive success at Val de Travers (Malo 1866). A third form ascribed the earlier discovery of mastic to the falling of bituminous material from a truck in an otherwise similar sequence (Earle 1974). A similar nineteenth-century story in Australia had ships' asphalt (mastic?) falling from a truck in Adelaide and serendipitously proving to be a good paving material (Lay 1984). The story reappeared a fifth time in 1901 when Purnell Hooley observed that a barrel of tar which accidentally burst over slag at England's Denby Iron Works produced an admirable road surface. After some further research, he founded the Tarmac Company in 1903 to market the new product (Earle 1971). Tarmac was to prove extremely successful in Britain and received significant official endorsement. Mixing tar with natural stone was by then a well-established practice, so Hooley's total serendipity in producing tarmacadam based on slag might be doubted. Copper-making slag had been used for roadmaking in the eighteenth century (Albert 1972) and iron-making slag was a common component of nineteenth-century structural courses.

Further outbreaks of the legend occurred in Santa Barbara in southern California, in Mexico, and in Trinidad, thus bringing the total to eight. In a letter to

Scientific American in December 1898, M. Meigs of the U.S. Corps of Engineers reported sightings nine to twelve: in Pennsylvania, where a leaking oil pipe accidentally improved a clay road; in Austin, Texas, where oil from leaky drums in a depot was compacted to a hard, firm pavement; in Missouri, where the inevitable barrel of oil fell from a wagon onto a road and subsequently and accidentally produced a good pavement, and from the Pennsylvania Railroad. All this led Meigs to reach thirteen with U.S. Engineer Office experiments in Indiana, which were directly sponsored by Rockefeller of Standard Oil.

Tar and Tarmacadam

Tar is commonly obtained as a byproduct of timber or coal processing; the major source prior to the nineteenth century was from charcoal-making using Scandinavian timber. This product, known as Stockholm tar, was used primarily for caulking ships. Coincidentally, McAdam had a commercial interest in a tar works, but does not appear to have used tar for road construction. Tar and bitumen are very similar; both are viscous liquids at normal temperatures, although tar is somewhat the harder and less durable of the two. Like bitumen, it had some popular uses outside roadmaking. For example, in the eighteenth century tar water rivaled soda water in popularity and the presence of tar was thought to improve the taste of wine.

An increasing interest in coal was a harbinger of the Industrial Revolution. Coal tar was patented in England in 1681 and coal gas in 1691 (Abraham 1960). Coal gas was soon recognized as a very convenient source of fuel and lighting, and from 1823 it was used for powering IC engines. As a consequence of the growing production of coal gas, tar became increasingly available as a byproduct of the process.

The market for the gas was partly for street lighting. Street lighting itself was not a new idea. Prominent Roman citizens had hung lamps before their street doors in A.D. 200, and lights had been introduced in the streets of prosperous Ephesus and Antakya (see fig. 3.3) in about A.D. 400 (Mumford 1961). Extensive street lighting probably began in A.D. 850 with the use of oil lamps in the well-paved city of Córdoba. It is said that the lighting was so extensive that a person could walk for 17 km in a straight line lit by public lamps. Much later, the city of London in 1416 and then an English act of 1662 required householders in some areas to provide lamps outside their homes to light the street. Subsequently, street lighting became relatively common in London by the end of the eighteenth century. Coal gas was first used to light streets in 1811, and by 1824 it was being reported that tar could "now be obtained from the gas-works at a very cheap rate."[5]

Given the concurrent increase in the use of macadam pavements, it was an innovative and synergistic step to try the surplus tar from the gas works as a replacement for the sand, stone, dust, and soil then in use as blinders for open macadam surfaces. The tar was poured on the surface and filled the open surface pores of the macadam. The step certainly had not been taken with the rarer and more expensive bitumen distilled from native asphalts. According to a letter in *Mechanics Magazine* (3 July 1824, 270), tar was first used in this manner in 1822 for footpath construction on Margate Pier. The tar was also allowed to cover the surface to a depth of about 5 mm to achieve a level surface, which was then sprinkled with sand to provide adequate friction. Subsequent applications in Gloucester and Cheltenham in 1832 drew on paving precedents and supplemented the tar with earth and fibrous vegetable matter mixed with the tar (Law 1962).

It was soon realized that using tar alone was inadequate because the surfaces lacked strength and stiffness. This need gave rise to Alexandre Happey's concept of precoating individual pieces of broken stone with tar so that they would stick together when placed in the pavement course. The method was first used on a 3 km length of road between Nottingham and Lincoln. The product came to be known as precoated or coated macadam. As a more traditional alternative, in the late 1830s tar was used as a mortar for stone paving blocks (Earle 1974).

As is bitumen, tar is heated to lower its viscosity so that it will flow during construction. A problem with many of the tars was that they then set very slowly. This perhaps aided the serendipitous use of the tar in the early 1840s, not only for precoating stones for new work and as a blinder of pores on an existing macadam surface, but also as a binder gluing together all the stones in an existing macadam course. The fluid tar seeped down and penetrated through the interstices between the stones of the naturally porous macadam courses, which were made of single-size stones. The product came to be known as penetration or grouted macadam, and its effectiveness depended on the extent to which the tar was able to penetrate. Penetration was initially aided by sweeping, and later by the application of surface pressure. The technique was used for surfacing and as a structural alternative to portland cement concrete and was popular until well into the 1930s (Deacon 1879).

Even when used accidentally, the tar was clearly effective as a binder. Within a few years, the new composite tar-stone material, known as tarmacadam, came to be widely used in the United Kingdom. The shortened term *tarmac* is still colloquially used to describe black-topped pavements, particularly on airfields. *Tarmac* has also had a proprietary meaning in the United Kingdom, where for many years the word described a commercial product made of tar and blast furnace slag.

The above tar-based processes worked well on rural roads. On the other hand, various trial applications of tar in urban streetworks in Britain and France were often unsatisfactory. A relatively successful London tar trial in 1869 led to its wider use, although tar was still considered unsuitable for heavily trafficked streets (Earle 1974; Deacon 1879; Peckham 1909).

Early tarmacadam street trials were unsuccessful, because the tars used had a deservedly poor reputation for setting slowly and being excessively hard and brittle once they had set. In the United States "tar roads" were mainly tar mastic surfacing courses rather than tar-stone mixes. A Washington, D.C., report in 1878 described them as "lacking in vitality." A Philadelphia report of 1884 discouraged the use of tar, for "gradual oxidation takes place, by which the tar loses its cementing properties and becomes inert" (Griffin 1987).

Competition from other binders increased late in the nineteenth century, and to circumvent tar's slow setting, fluid behavior, and poor durability, the tar by-product was distilled to produce more viscous and durable tars. These more marketable materials came to be known as road tars and led to a much improved product. For instance, the data to be presented in figure 7.4 will indicate that in 1891 penetration macadams rated more highly as structural courses than did portland cement concrete. As late as the 1960s, such tars shared the United Kingdom market equally with bitumen (Road Research Laboratory 1962). However, bitumen possesses a number of technical and cost advantages over tar and is now by far the more common binder (Lay 1990). Nevertheless, many citizens still believe that most roads are built of tar.

One of the important lessons that the tar developments taught was that it was not essential to rely on unmodified, natural products. Paving materials *could* be manufactured. Clearly, the innovations with tar had taken the world a long way along the route to a successful road pavement.

Cement and Concrete

To be successful, a road pavement needs more than a good running surface for traffic. Great care has to be taken with the drainage of the natural formation. And between the running surface and the formation is the structural course, which gives the pavement its load-carrying strength. During the time that mastics and tarmacadam were being used to provide better running surfaces, major advances were also occurring in the technology of the structural courses.

In particular, cements were having a rebirth. Cements are binders that play a very analogous role to bitumen; indeed, the two materials are sometimes respectively called cementitious and bituminous binders. Continuing the comparison, the addition of water to cement is analogous to the addition of heat to bitumen; a mixture of cement, water, and sand produces mortar, which is analogous to mastic; and a mixture of mortar and pieces of stone produces concrete, which is analogous to asphalt. The word *concrete* comes from the Latin *concretus,* meaning "compounded."

Cements are usually made from burnt limestone or dolomite, and hence could be discovered serendipitously at campfires. They have been used since 6000 B.C. and were extensively employed by the Egyptians, Cretans, Greeks, and Romans and then largely forgotten. The first roadmaking records are of roads built on Crete in about 2600 B.C., partly with flagstones and partly with a concrete made from a gypsum mortar. Many of these roads also featured major retaining walls. There is evidence of roads in India built with cement bound surfaces in 1000 B.C. The Greeks were using lime-cements for building by about 300 B.C., and their knowledge was certainly passed on to the Romans, perhaps via southern Italy. In about 200 B.C. the Romans greatly enhanced its strength by adding pozzolan, which is volcanic ash from Pozzuoli on the Bay of Naples. They then mixed it with gravel to produce the strong structural concrete used throughout their empire. It was a major innovation. Vitruvius (1960) correctly described their cement mixture as "a kind of natural powder which from natural causes produces astonishing results."

The Romans used concrete extensively, it had had some use in Saxon England in about A.D. 700, and the Normans used it for major building works. However, the art of burning lime, grinding it, and then adding other materials to produce high-strength and water-resistant concretes and mortars was slowly lost, and was effectively nonexistent by the beginning of the eighteenth century (Stanley 1979).

Hydraulic cements are cements that can set in damp places or under water and are thus particularly valuable for building bridge piers. Many of the Roman cements were hydraulic, and this aided the construction of their magnificent achievements in bridge building. Cement made from pure lime is not hydraulic; the property comes from impurities in the lime or from additives. After a series of empirical experiments on different mixtures of natural lime and clay, John Smeaton rediscovered hydraulic cement and used it in the construction of the third Eddystone lighthouse in 1756.

The relative weakness of lime-cement restricted its widespread use. Its roadmaking application was more often for waterproofing than for strength. A more useful cement was discovered in 1796 by a vicar and wealthy lime burner, the Reverend James Parker, using clay-lime nodules found in London clay on the Isle of Sheppey. Parker called his product Sheppey Stone or Roman cement.

The reintroduction of cement led to a rebirth of concrete. Telford strongly advocated its use as a pavement structural course, which was not surprising, for it was a logical extension of his Telford block method (see fig. 3.8d; Forbes 1953). Indeed, its first use was in 1828, when Telford and Macneill employed concrete as a trial on a 2.5 km stretch of Telford's Holyhead Road at Highgate Archway in north London. They selected concrete because the site was poorly drained and the pavement structure had to be able to resist water damage. The concrete used was 150 mm thick and 5.4 m wide and was topped by a macadam running surface keyed to the structural course by pieces of broken stone set in cement mortar. The trial was successful and yet apparently unconvincing; it led to few further applications of concrete (Committee on Science and the Arts 1843; Paxton 1977).

Modern cement was invented in England in 1824 by Joseph Aspidin, a Wakefield bricklayer, who called it portland cement on the optimistic but commercially wise assumption that it resembled in color and texture the building limestone quarried on the Isle of Portland.[6] In Aspidin's process, limestone was finely ground and burnt. Following Parker's lead, it was then mixed with clay, reburnt, and then finely ground to produce cement. Aspidin began production in 1843 and discovered the significant virtues of burning at higher temperatures some seven years later. Portland cement did not take over from natural cement until about 1890.

In 1846 an existing lime-concrete structural course supporting wood blocks was removed from the Strand in London and replaced with one made of portland cement concrete supporting granite setts (Bone 1952). It was the first application of this now common roadmaking material. Construction practice on this job was illustrated in figure 3.11.

There was some unconvincing use of 150 to 200 mm thick concrete structural courses in France in the 1840s, but it appears that the mixes used had a very low stone content and were prone to excessive cracking (de Coulaine 1850). In addition, they proved quite expensive and, in the absence of large mixing plants, difficult to construct. A common method was to place alternate layers of mortar and stone, beating the stone into the mortar with the backs of spades (Deacon 1879). There was also considerable municipal objection to constructing concrete slabs, because they subsequently made many underlying services inaccessible, given the equipment of the time (Gillespie 1856).

Indeed, the use of concrete for roadmaking did not become firmly established until after 1865, when a student of Telford's named Joseph Mitchell applied the established tar-penetration method to concrete roadmaking, allowing cement mortar to penetrate downwards into the voids of an existing open-graded macadam layer. Penetration concrete became a popular form of construction, particularly for surface courses. The process was often aided by rolling the grouting material into the voids (Kennerell 1958). The method was still in use in the 1930s, although its relevance was reduced as modern concrete mixing plants became available.

Mitchell constructed three penetration concrete test pavements (Mitchell 1867):

1. A 45 m trial in 1865 near the Inverness railway station. This trial was initially successful, but it was later reported that "the surface was very good, but when the road commenced to break, it went to pieces very fast" (North 1879).
2. A 100 m trial later in 1865 on the mall in St. James's Park, at the foot of Green Park in London. This pavement failed, probably because a heavy roller was taken over the concrete soon after it was placed, causing cracking.
3. A 45 m trial in 1866 on the George IV Bridge in Edinburgh. The concrete proved half the cost of the stone block alternative, the trial was successful, and the pavement provided good service for some forty years (Paxton 1977). Surface wear was reduced, although there were some complaints about traffic noise. It is not clear whether the concrete carried stone setts or 60 × 70 mm concrete blocks (Merdinger 1952).

Mitchell suggested the George IV trial showed that concrete was a potential structural course for supporting stone setts and only advocated the use of concrete without surfacing "for secondary streets" (Deacon 1879, 96).

When the design of concrete mixes came to be well understood in the late 1860s, concrete became a strong and reliable product. The process was aided by the development of new construction equipment such as introduction of continuous concrete mixers in 1875. It soon became common practice to support asphalt on 150 mm or so of structural concrete. The question that this poses is, What purpose did the asphalt serve? Before the pneumatic tire, asphalt was primarily used to provide adequate traction for horse traffic. Concrete was rarely used for surfacing in the nineteenth century, because its smooth, impervious finish made it slippery and unsuitable for use with iron-shod, excreting animal traffic. In addition, the concrete of the day was rapidly abraded by horses' hooves.

The disappearance of animal traffic and the appearance of pneumatic rubber tires were dual boons to the cement industry, which began a vigorous advertising campaign in the 1920s. Despite such efforts, up to 100 mm of asphalt was placed on top of the concrete, which was typically about 150 mm thick. Experts of the time said that the asphalt had more resilience. Certainly the stiffness of asphalt, as measured by its elastic modulus, is only about 1,000 MPa compared with about 40,000 MPa for concrete (Lay 1990). But was the lower stiffness of value in the motorcar age?

The factors that contributed to the continued, but seemingly unnecessary, use of asphalt on top of concrete were probably (1) a reluctance to deviate from the status quo; (2) the slippery surfaces of the uncovered concrete; (3) a marked tendency for the early concrete bases to crack badly due to concrete's low tensile strength (see, e.g., Delano 1880; Agg 1916); (4) problems due to the concrete of the time being relatively slow setting; and (5) low concrete strengths, as the key factors influencing strength had not yet been identified.

In this sense, the asphalt was a form of readily repairable flexible carpet or overlay. Nevertheless, by the 1920s the concrete industry had begun to realize the full competitive potential of its product. Technically, from 1922 the industry learned to use joints to control cracking, and in 1926 Westergaard introduced design methods for controlling stresses, strains, and deflections. The subsequent history of concrete pavement design is traced in Ray (1964).

Competition between asphalt and concrete then began in earnest. At least in rural parts of the United States, concrete soon had the strong upper hand, as shown by the lengths of different rural road types in use in 1930 (table 7.1).

An interesting public perception of the problem in the 1920s was given by John Steinbeck in his book *The Wayward Bus:*

> The highway to San Juan de la Cruz was a black top road. In the twenties hundreds of miles of concrete highway had been laid down in California, and people had sat back and said, "There, that's permanent. That will last as long as the Roman roads and longer, because no grass can grow up through the concrete to break it." But it wasn't so. The rubber-shod trucks, the pounding automobiles, beat the concrete, and after a while the life went out of it and it began to crumble. Then a side broke off and a hole crushed through and a crack developed and a little ice in the winter spread the crack, so the resisting concrete could not stand the beating of rubber and broke down.
>
> Then the county maintenance crews poured tar in the cracks to keep the water out, and that didn't work, and finally they capped the road with an asphalt and gravel mixture. That did survive, because it offered no stern face to the pounding tires. It gave a little and came back a little. It softened in the summer and hardened in the winter. And gradually all the roads were capped with shining black that looked silver in the distance.

Steinbeck's explanation as to why the concrete failed and the asphalt succeeded was based on views widely held at the time by both technologist and layman. They were, nevertheless, wrong. The concrete failed for the reasons given above. Specifically, measures were not taken to either enhance its inherently poor tensile strength or shield it from tensile stresses. The asphalt succeeded because it had relatively good tensile strength and not because it "softened in the summer and hardened in the winter." Note that Steinbeck used the word "asphalt" where we would use bitumen.

There is another important method for using cement in roadmaking. Because

TABLE 7.1
Lengths of Different Rural Road Types in Use in the United States in 1930

Road type	Length in Mm (1000 km)
Gravel	650
Sand-clay	150
Concrete	106
Bituminous penetration macadam	50
Asphalt	16
Bituminous surface treatment	8
Brick on concrete base	7

SOURCE: HEB 1930.

many soils, such as clays, have electrically charged microsurfaces and are thus surface reactive, a small percentage of an additive with free ions, such as cement, can have a dramatic and beneficial effect on the properties of the soil. The process is commonly called cement or soil stabilization (Lay 1990). It is particularly useful for improving the physical properties of the natural formation of a road. The amount of cement used is far less than would be needed to make a conventional cement mortar. Stabilized mixtures of lime and clay were used for roadmaking in ancient China, India, and the Roman Empire.

In more recent times, stabilization experiments began in the American South in 1906 with cement being worked into the soil using plows and tilling devices. The Amies Cement Company in Philadelphia was granted American patents in 1910 for a stabilized material they called soilamies. Postwar stabilization trials were conducted in the United States in the 1920s, and formal testing began in South Carolina in 1932 and was taken over by the Portland Cement Association in 1935. The impressive results saw the rapid adoption of the method around the world (Catton 1959). It was included in U.S. Army specifications by 1937 and was used with such success by German army engineers building airfields during World War II that their engineering manuals were made secret documents. Despite this secrecy, the better publicized concurrent successes of the U.S. Corps of Engineers in their airfield work led to the widespread adoption of the method in many Allied countries.

Paving Blocks and Rails

The preceding discussion has shown how the world was moving toward a paving solution based on producing a single, coherent pavement structure. Nevertheless, some of the older systems based on building the pavement from discrete elements were continuing to develop. The three most fruitful of these will now be examined.

Stone Pavements

The first pavements were of beaten earth. When something better was needed, inventive men naturally turned to timber and rock. Occasionally, convenient flat pieces of stone would have been available; however, uncut and often waterworn stones called cobblestones (or pebbles, catsheads, or rubble), usually at least 150 mm in size, were the traditional stone paving material. In 1533 an act was passed requiring cobblestones to be brought to London from the seashore, probably from Kent (Merdinger 1952). Such an ordinance was necessary because stone other than small river gravel is scarce in the vicinity of London. The cobblestones were usually placed on a bed of sand and then wedged shoulder-to-shoulder into place. However, their smooth, rounded shape and the lack of any binder to hold them in place meant that the practice was often unsatisfactory. In the nineteenth century the availability of mastics as mortars greatly improved the behavior of cobblestone roads.

The longest side of a cobblestone was placed vertically to prevent the stone from tilting when a wheel bore on only a portion of its running surface. This placement orientation also favored the use of cobbles whose running surface was small relative to the contact areas of wheels and hooves. For maximum effectiveness, the cobbles were set on a thick, sand bedding layer to allow them to be readily located and to reduce the concentrated, localized loads that they applied to the underlying natural formation. Nontechnical reviews of urban paving in the nineteenth century have referred to such sand layers as fragile, claiming that sand was less effective than concrete as a structural course (McShane 1979). This is a misconception as there is no reason why an adequately compacted sand layer should not provide more than adequate strength and stiffness. Indeed, sand layers are still used in modern road construction and are far from fragile. Sand possesses the advantages of being free-draining, inert, and strong.

Cobblestones produced a surface that was uneven, noisy, and slippery, and they were never regarded as a satisfactory paving solution. Despite the rough and noisy ride, cobblestone paving was quite extensively used—even in recent times. In 1876 there were 150 km of cobblestone road in both New York and Cincinnati (North 1879). Nevertheless, alternatives were in some demand. One relatively expensive option was to use large, hand-cut stone slabs known as flagstones.

It was difficult to maintain an even surface with flagstone paving, and a common hazard was the muddy water that squirted from the joints when the flagstones were stepped upon. In addition, heavy wheel loads toward one edge of a flagstone could easily cause it to tilt or crack. Therefore, the Romans and their predecessors frequently used relatively small flagstones for street paving. The technique was reintroduced in 1184 when large (2 sq m) flagstones were used in response to Philip II's order to pave the street outside his castle at the Louvre in Paris. The use of large flagstones ignored Roman experience, and the surface soon became uneven under traffic. Consequently, in 1296 the city leaders downgraded the size requirement to a more practical 250 mm × 250 mm.

The widespread advent of wheeled urban traffic caused a general move toward the use of smaller, flat-faced stone blocks, although flagstones remained the appropriate material for footpath construction. Such cubical blocks were known as setts and produced a flatter traffic surface than either flagstones or cobblestones. By virtue of their more frequent joints, they also provided a less slippery surface than did flagstones. Many pavements were hybrids between the random roughness of cobblestones and the organized flatness of the stone setts.

Setts about 125 × 125 × 150 mm deep were introduced into Paris in 1415, and by 1720 had gradually increased in size to 230 × 230 × 230 mm (Moaligou 1982). They were placed on a bed of sand and drained to a centerline gutter. Because many roads built for royalty used setts rather than rough cobbles, they came to be known as *pavé du roi*.

In 1824 St. George's parish in London asked Telford to advise them on the best system of paving. He considered the request of such importance that he referred it to the Institution of Civil Engineers, of which he was president. The decision was made to recommend whinstone, or granite setts, placed "on a sound bottom." Recall that "bottom" was the term then used for the underlying layer of Telfordian pitchers that provided the pavement with its structural strength (Moore 1876). The pitchers were also made of granite, which split readily into cubical shapes. This pavement proved effective in carrying heavy freight carts and wagons heading to ports and warehouses.

By 1850 common European sett sizes had diminished again to about 160 × 160 × 100 mm deep as narrow-rimmed iron-tired vehicles became more common.

Belgian granite became famous for sett-making, providing Belgian blocks to many cities; for example, North commented in 1879, "The City of New York is largely paved with Belgian pavement." Scottish granite blocks were also well regarded, and in 1865 were exported as far afield as Montreal (St. George 1879). In addition, many port cities had plentiful sett supplies based on discarded ships' ballast. Other cities passed ordinances requiring all incoming vehicles to bring some paving material with their payload.

Typical sett plan dimensions eventually dropped to about 75 × 75 mm, with the large 150 × 150 mm setts restricted to special circumstances. The smaller, mosaic setts were placed on structural courses that commonly consisted of 150 mm of concrete covered by a 50 mm sand bedding layer (Kennerell 1958). The joints between setts were filled with sand, mortar, pitch, or tar. Urban horseshoes often had small protuberances called *calks,* which were intended to improve traction by catching in these joints. For the same reason, it came to be accepted that "the width of [a sett] should not exceed the breadth of a horse's foot" (Moore 1876). A photograph of an 1890s sett pavement under horse traffic is given in figure 5.4.

Laying setts in circular arcs—or fantails—permitted looser tolerances on sett size, because the joints did not need to line up (a good review of set practice is given in Simon 1970). Stone sett streets were constructed in much of Europe up until 1940 and in Paris and Liverpool for some years after the end of World War II. They have recently returned to be used in the restoration of old areas.

Pavements using segmental blocks such as flagstones, bricks, or setts had little inherent strength and thus required a strong structural course. However, their major drawback was that they commonly resulted in uneven, slippery, and noisy surfaces, particularly under the action of iron-shod hooves and wheels. The noise alone was enough to lead many cities to remove them from heavily trafficked routes (Greene 1885). Many block pavements also abraded easily, although thirty-year lives were achieved with the more expensive granite setts. As a maintenance measure, it was common practice to take up worn setts, redress them, and then re-lay them on minor streets. Such salvage procedures were needed, for the initial cost of a sett pavement was about ten times the cost of macadam. Setts were thus an expensive paving solution.

A different objection to setts arose in nineteenth-century Paris when they were used by insurgents constructing street barricades. They were so used in the 1830 revolution. Victor Hugo wrote in *Les Misérables:* "The barricade was built with setts. . . . Not a stone out of line, it might have been made of china. . . . There were corpses scattered around and pools of blood on the pavement." The issue is discussed at some length in an 1850 technical paper by de Coulaine, who used the example of barricades build in June 1848 from setts and wooden blocks as a strong argument for using asphalt for urban pavement surfaces. His message was not heeded, and in 1871 rioting members of the Commune again used setts for barricades. Discontented Irishmen had similarly used setts for missiles in 1864. For such reasons, there was a strong official preference for a more coherent and less throwable pavement surfacing. Nevertheless, reports indicate that in the 1870s over 75 percent of the streets of Paris were paved with setts ("Public Works in Paris," *Engineer* 27, no. 5 [1869]: 107; Moore 1876).

Supplies of roadmaking stones are scarce in many parts of the world. For some four millennia, burnt clay bricks have often been the only available paving material in areas of the Middle East, although their expense restricted their use to important avenues. In addition, unless they had been particularly well laid, brick pavements often proved excessively noisy under iron-tired horse traffic. Brick paving was recently brought to near-perfection by the Dutch, and this has led to a world-wide resurgence in its use for street construction. Concrete bricks or blocks were first used in Edinburgh in 1866 and have also had a contemporary street revival, due mainly to their decorative appearance.

Timber Blocks

Timber blocks were first used for paving in fourteenth-century Russia. Recall, incidentally, the discussion in chapter 3 of timber-plank paving being used in Novgorod from the tenth century. The technical advantages of the block over the plank are that it is easier to maintain a level surface and the much tougher end grain of the timber can be presented to the traffic. Wear is thus reduced by about 75 percent. A common, but ineffective, early construction technique was to place the blocks on two layers of 25 mm planks in order to overcome any differential settlement of the underlying sand and natural formation.

Timber blocks were applied with some technical skill by M. Gourieff in St. Petersburg in 1820 when he set hexagonal blocks on a structural course of crushed rock and sand. He filled the joints between the blocks with sand, poured boiling tar on the surface to overcome the rotting problem, and then sprinkled sand on the cooling tar to provide a surface with adequate friction (Merdinger 1952). The technique was successful and led to 175 mm hexagonal softwood blocks of St. Petersburg fir being exported to England. They were used in Manchester in 1838 and, in the same year, "won" the Oxford Street pavement trials discussed earlier. These applications gave wood-block paving such a fillip that a number of companies were established in London to apply the new method. However, the enthusiasm was short-lived, because the "successful" trial pavements had all rotted away within seven years. In the harsh light of operating experience, granite setts returned to a position of dominance (Moore 1876).

Due to the widespread reliance on wood in many areas of construction, the prevention of rotting was a major topic of technological debate. Various, and largely unsuccessful, patented methods of rot-proofing wood blocks were proposed from 1792 onward (Committee on Science and the Arts 1843). Nevertheless, wood pavements rarely lasted more than five years. Tar- or creosote-impregnated blocks were introduced commercially in 1867 and brought some improvement.[7]

However, timber blocks were still far from the perfect pavement solution. They wore quickly into wheel ruts, their surfaces deteriorated, they smelled when damp, and they still slowly rotted. It was widely believed that gaseous vapors from the rotting wood pavements were a source of malaria, typhoid, and other fatal diseases (Lay 1984; see also St. George 1879 and Moore 1876). The blocks were less satisfactory in this respect than was asphalt, as table 7.2 indicates.

In one memorable setback, the creosote-soaked wooden streets of Chicago burnt vigorously in the Great Fire of 1871. Earlier, major street fires had occurred in San Francisco in 1850 and in New York after an election celebration in 1870. Nevertheless, wood-block paving continued to be widely used in the absence of any serious competition (see Mountain 1903 for a comprehensive review). It also pro-

TABLE 7.2
Hygiene Problems Associated with Various Road Surfacings

Pavement type	Relative number of loads of mud produced	Filth rating
Macadam	1.0[a]	very poor
Setts	0.7	poor
Wood blocks	0.2	good
Asphalt	0.1	very good

SOURCES: Delano 1880 and Mountain 1903.

[a]A typical surface produced about 20 mm of mud per week (Rochester Executive Board 1886).

vided a useful technique for land speculators looking for a cheap pavement that would last until the land was sold.

Four factors helped wood-block paving counteract the above difficulties. The first was the use of the Gourieff overlay to reduce slipperiness and seal the blocks against moisture (Law and Clark 1907). The second was the use of creosote to preserve the blocks, and the third was the use of Australian hardwood blocks, which were imported into a number of countries beginning in 1888. The hardwood blocks possessed much-improved durability and weather resistance (Lay 1984). Sydney was regarded as "one of the first [cities] to arrive at perfection in wood paving" (Mountain 1903). The fourth factor was the widespread belief that wood-block paving was safer than any of the alternatives. For example, in the 1890s the accident rates in London on various street types were calculated in terms of the distance traveled in kilometers by a horse between accidents as 220 for asphalt, 320 for granite setts, and 550 for wood blocks (Haywood 1871). Using these data, the owners of fifteen thousand horses petitioned London authorities in 1873 to discontinue the use of asphalt because of its greater propensity to cause accidents.

By the turn of the century, wood-block paving was the internationally preferred option. In 1906 it was reported in England that "wood laid on concrete seems to have finally had the mastery over granite and asphalt" (Civil Engineering 1981). Reports in 1923 indicate a still vigorous debate over the type of wood to use to produce a successful pavement (Wood 1923). The last wood-block pavement in Paris was constructed in 1938 (Moaligou 1982).

Cartways

An interesting amalgam of road and railway practice occurred with the cartway, which was formed of long, narrow flagstones placed in the wheelpath in order to provide a smooth wheel track (Law and Clark 1907).[8] Cartways built of high-quality stone both aided the passage of the common iron-tired carts and wagons and restricted the abrasive effect of their wheels on conventional stone paving (Hadfield 1934). An additional advantage was that the horse pulling the cart could walk on a better traction surface between the smooth tracks (Lay 1984). The economic value of a cartway was that it permitted a horse on the level to haul 12 tonne, which was four times its capability when working on a conventional road (see table 3.1). Cartway construction ceased in the 1870s as general paving quality began to improve (Deacon 1879).

Early cartways were installed in Pompeii and Cornwall. Their more modern use originated in Pisa and, by 1826, they were also in service in such other Italian cities as Florence, Siena, and Milan (Forbes 1958a). In the 1840s, Milan street paving was regarded as the best in Europe and its cartway tracks were each 1 m wide (Committee on Science and the Arts 1843). Lord Palmerston, having observed the effectiveness of cartways in northern Italy, "despaired of anything so good being introduced in England" (Gillespie 1856). In London the first cartway was placed by James Walker on Commercial Road in 1825 and provided a route to the East and West India docks. It was made of granite blocks 1.5 m long, 400 mm wide, and 300 mm deep, with a track gauge of between 1.2 and 2.1 m. Raised curbs were used to provide guidance to the cart wheels (Deacon 1879).

Some of the cartway networks were quite extensive. For instance, there was a 20 km stretch in New York State between Albany and Schenectady, which operated between 1834 and at least 1901 (Gregory 1931). On the Holyhead Road, Telford and Macneill used a granite cartways on hills with slopes of 5 percent or more and proposed constructing a cartway from London to Birmingham to carry road locomotives and horse carriages.

Prefabricated flat steel sections were sometimes used instead of flagstones to create a metal cartway (or wheelways or plateways). The method was first used in English coal mines in 1785 and, in the outdoors, in Glasgow at the turn of the century on steep hills and, in 1816, alongside the Forth and Clyde Canal (Dyos and Alcroft 1969). Of course, at about the same time a related technique was being used by the tram with its flanged wheels and iron rails set into the road surface. The availability of steel rolling mills gave the technique an end-of-century resurgence, despite decreasing demand. In Melbourne a 15 km stretch operated between vegetable farms and their city markets. It employed specially rolled steel plates and was based on a system used for some hundreds of kilometers of plateway in Germany (Lay 1984; Coane, Coane, and Coane 1908).

Some interesting secondary surfacings were tested. On the Holyhead Road, Macneill developed and patented the addition of 25 mm square iron cubes embedded in the surface at 100 mm centers in order to increase its strength. The total use of iron for paving began in London where it was employed to pave a footpath at the corner of Old-Street and City Road in 1834 and for the deck of Blackfriars Bridge in 1836. Subsequent trails occurred in Glasgow and Boston in the early 1850s. In 1856 interlocking, gravel-filled circular iron cells 300 mm in diameter and 150 mm deep were introduced on Leadenhall Street in London and then into New York and Boston later in that same year (*Engineer* 1 [25 Jan. 1856: 48]). In 1857 cellular iron slabs were tried on the roadway outside Paddington Station in London and in 1859 on King Street in Westminster. The cells were smaller than a horse's hoof. Unfortunately for the idea, the surface wore unevenly, fractured, and severely injured horses that fell on it (Dawson 1876). Another trial on Olive Street in St. Louis proved to be impossible to maintain and buckled into ridges in hot weather (Moore 1876). Iron blocks were unsuccessfully used on Cortland Street, New York, in 1865 (Merdinger 1952). Nevertheless, a trial based on earlier English practice was conducted in Paris in 1924, and a further 9000 m^2 were placed during the 1930s.

Iron plates were spasmodically tried in general road construction, for example, in Berlin in 1877 on Unter den Linden. They suffered the same evenness problems as had once plagued flagstones. Further unconvincing tests of wide steel plates as surfacing and structural courses and for enhancing the properties of the natural formation were conducted in the 1920s (Hadfield 1934).

Glass was also used for roadmaking, presumably as broken pieces used in constructing structural courses although there are reports of glass blocks being used in Lyon (Tillson, Haylow, and Richardson 1907). Likewise, there are reports of the use of leather as a surfacing (Merdinger 1952). One of the more novel paving materials was compressed sewage sludge, which was produced by a factory near Manchester (Malcolm 1934).

Manufactured Asphalt

By the middle of the nineteenth century, courses providing a strong and stiff pavement structure could be obtained using either conventional unbound macadam, Telford pitchers, concrete, or tarmacadam. For pavement surfaces, penetration macadam, natural asphalt, wood blocks, and stone setts were in common use by the 1870s.

The next and final stage in this sequence of pavement development was the production of manufactured rather than natural asphalts. The term *manufactured asphalt* (or synthetic or imitation asphalt) implies that the product is made by mixing selected sand and stone with bitumen or tar. The method offers the now obvious advantage that the stone size and strength and the bitumen quality and proportion do not depend on chance, on equipment size, or on nature's geological whims.

Recall, however, that there was a strong nineteenth-century belief that nature's mix should not be tampered with. Thus, the new material was commonly and depracatingly called artificial asphalt. In addition, there was the considerable commercial vested interest of the native asphalt suppliers. Thus, it took some considerable time for people to realize and accept that nature could be improved upon. The timing of this realization in the latter half of the nineteenth century has many parallels in the history of science and technology.

European Beginnings

The story begins in 1837 when compressed asphalt blocks were produced from a mixture of mastic and quartz fragments in order to meet a demand for patterned black-and-white paving at the entrance to the place de la Concorde in Paris (Simms 1838). Thus, the quartz pieces were added to alter the color rather than the strength of the mix. The trial was initially successful but failed after six months in service due to rutting and unevenness. It was recognized at the time by de Coulaine and others that the failure resulted from a very high mastic content preventing particle-to-particle contact of the pieces of stone, thus giving a mix with inadequate strength and stiffness. In the technical terms of our later discussion, this first manufactured mix was open-graded and more like hot-rolled asphalt than asphaltic concrete (Committee on Science and the Arts 1843). It is also probable that only relatively thin (40 mm) asphalt blocks were used, mainly to permit easy access to underground services (de Coulaine 1850).

This failure gave manufactured asphalt a bad reputation in Europe, as the following words by Ellice-Clark (1880) illustrate:

> In 1838 a trial pavement was laid in Paris, and the matter was taken up by financiers, who, not content with honestly giving the numerous uses to which it might be applied, attributed to it qualities it never possessed, exaggerated the importance of its virtues, and soundly claimed for it an adaptability to almost every conceivable object in constructive works. Speculation in asphalte became so great as to reach fever heat; panic and collapse followed, and this retarded the proper use of the material for a considerable period. It was therefore not until 1854 [that the situation improved, when Malo used heat to produce natural asphalt].

The view is confirmed by a report of the time (Committee on Science and the Arts 1843), which refers to "the various asphalts [being] made very much the objects of speculation . . . , some of them fraudulent."

Further small trials were conducted between 1840 and 1843, beginning with a section at the place Louis XV in Paris (de Coulaine 1850). In these tests the quartz was powdered to a fine size, so the material was closer to a mastic than to an asphalt. Some of the trials were unsuccessful because the bitumen used soon lost its ductility. However, many others were still performing well after nine years. These had used a relatively ductile bitumen as a spinoff from the earlier, futile attempt to employ cold placement with natural asphalt.

The United States Takes Over

The impetus for the development of manufactured asphalt was to come from the United States, which had few useful deposits of native asphalt and only the masticlike Trinidad Lake native asphalt to draw upon in any significant quantity. It also had no entrenched industry groupings. Some usage of American native asphalt for roadmaking did occur in California, Kansas City, and Ardmore, Oklahoma, in the 1890s (Peckham 1909). The first American manufactured-asphalt trials may have been in 1868, and manufactured asphalt made from crushed stone and Trinidad Lake native asphalt was used to produce asphalt blocks in San Francisco in 1869 (Merdinger 1952). In 1871 N. B. Abbott tested manufactured asphalt in Washington, D.C., using tar and creosote as binders. The surface of the poorly proportioned product soon failed. Abbott persisted and, despite never solving the surfacing problem, within a couple of years was producing useful 300 mm thick asphalt structural courses. Although these early American trials were not completely successful, they had broken away from the restraints of Europe.

Consistent success with manufactured asphalt could not occur until the principles behind the operation of asphalt were understood. This understanding began with the work of Edward de Smedt, a Belgian chemist who had emigrated to the United States in 1861 to work at Columbia University on coal dust problems. It has been suggested that de Smedt had previously learned of mastic technology in Paris

(Peckham 1909). Other, stronger, evidence indicates that his attention was diverted from coal to asphalt by the success of portland cement concrete.

De Smedt took up the challenge with gusto to produce a paving asphalt and began in 1870 with field tests in New Jersey of imported natural asphalt. When these trials proved unsuccessful, he realized that strength and stiffness would be achieved if the mineral particles in the mix were in point-to-point contact. He therefore selected the size of all the particles in his mix to produce minimum air voids prior to adding the bitumen; that is, smaller particles were selected to fit inside the spaces remaining when larger particles were in point contact (Forbes 1953). Today, his mix would be called a well-graded, close-packed, or maximum-density mix.

With this understanding, de Smedt then developed and patented manufactured asphalt for trials in Battery Park and on Fifth Avenue, New York, in 1872–1873 and on Pennsylvania Avenue (near the Capitol) in Washington, D.C., in 1876–1878. His mixture used sand, a small amount of finely crushed limestone, and 10 percent by mass of bitumen and was thus very similar to the European native asphalts. He used bitumen refined from West Virginian natural asphalt for New York and from Trinidad Lake asphalt for Washington (Halstead and Welborn 1974). He also adventurously used a little bitumen from petroleum refining in his mixes. Patenting of mastic mixes was already common, and various asphalts and binders continued to be patented and given magic names. One such material of de Smedt's time was Barnell's Liquid Iron asphalt, which used iron ore or iron sesquioxides instead of sand.

Manufactured-asphalt trials were also conducted by C. E. Evans on Connecticut Avenue in Washington, D.C., in about 1873. These failed "after a short time" (Richardson 1905). Nevertheless, with half the streets in major American cities still unpaved, there were sustained, strong market pressures for good urban paving (McShane 1979).

De Smedt went to Washington, D.C., in 1876 to take up the newly created post of inspector of asphalts and cements, following President Ulysses Grant's desire "to make the City of Washington a Capital City worthy of a great Nation" (Robertson 1958). Apart from altruism, there were other reasons for the president's desire.

A Board of Public Works had been established in Washington in 1871 under Boss Alexander Shepard; however, its efforts in roads and other works were such that it bankrupted and discredited the District of Columbia and ended local government in Washington. Grant and Congress then appointed a commission of officers of the Army Corps of Engineers to supervise Washington roadmaking—with specific priority given to Pennsylvania Avenue, which then had a surface of rotting wood blocks. The president was probably prompted by a Washington Board of Health report, which noted that although 80 percent of new paving was in wood blocks, the poorly laid blocks rotted within three years and were a source of "typhoid, malaria and intermittent fever, dysentery, diphtheria, etc" (St. George 1879). Public works were also favored to overcome the effects of a depression in 1873. There was thus an anxious market for new pavement technology.

The terms of the commission required it to lay trials of the best-known pavement systems. It advertised for proposals, putting no restriction on the materials that could be offered, and forty-one responses were received. Two pavement systems were selected, and the commission set the trial for Pennsylvania Avenue in 1877; the plan adopted used 40 percent Val de Travers natural asphalt—which was still a far from perfect technology in Europe—and the remaining sixty percent was 60 to 75 mm of the new manufactured asphalt placed in two layers on a 150 mm concrete structural course (Warren 1901; Robertson 1958). The manufactured asphalt was about 14 percent bitumen, 76 percent sand, and 10 percent pulverized limestone, mixed at about 150° C in a box containing revolving agitators. The bitumen came from Trinidad Lake native asphalt. The mix thus followed the de Smedt practice and was much more of a mastic than a modern asphalt (Collingwood 1891).

The trial was a success. The sections of manufactured asphalt and natural asphalt lasted fourteen and ten years respectively. Twenty years later an expert commentator wrote of them: "A commission appointed under the authority of Congress reported in favor of paving Pennsylvania Avenue with asphalt, using the European bituminous rock for one portion, and the Trinidad asphalt for the remainder. From the completion of that work dates the entire success of asphalt pavements in the United States" (Tillson 1897). Note, however, how even he downplays the techniques used and concentrates on the material sources. This belief that the source determined performance was, of course, encouraged by the suppliers and was a major dampener on asphalt development.

In this climate, commercial and technical feelings were intense and the trial pavements were not universally seen as unbridled successes. For instance, one expert of the time commented that the pavements were "poorly laid," were placed on "damp concrete," and had badly cracked after two years. He doubted that they would last another three. Discussing the tendency of manufactured asphalt "to disintegrate in a rotten manner," he commented on "the inherent difficulty of making an artificial mixture equal in quality to the natural product" (North 1879). He must have found difficulty with his views during the trial's subsequent twelve successful years.

Of course, manufactured asphalt, with its well controlled and designed for the purpose properties, had to be the eventual winner in the contest with natural asphalt. But such logic was not so apparent at the time, and it is salutary to see the counter arguments in the context of modern product rivalry.

The technology of asphalt roadmaking advanced under de Smedt's lead and the construction management skills of the Corps of Engineers. A further key reason for the success of manufactured-asphalt pavements from 1876 on was that they were laid by contractors working to a good margin of profit and required to financially guarantee the performance of their product (Warren 1901). The manufactured-asphalt technique had been exported to Australia by 1890, with contractors there offering a three-year guarantee (Mountain 1894).

A schoolteacher named Amzi Barber became convinced of the need for street improvements through his dabbling in Washington real estate. In 1877 he made a major entrance into the asphalt industry when his Barber Company gained the franchise to import Trinidad Lake asphalt for the Pennsylvania Avenue project. An English group had the Crown concession from 1888 to 1908 for the entire lake, which was then the world's major bitumen source (Broome 1963). Barber's business had prospered, and he was able to form an agreement with the English group to give him a monopoly for twenty years (Borth 1969). Incidentally, North (1879) complains that the confusing use of the word *asphalt* for bitumen was introduced into the United States by traders in Trinidad Lake bitumen.

A series of mergers in 1902 gave the Barber Company effective control of the American asphalt industry through the General Asphalt Company of America, which

TABLE 7.3
1897 Survey of 122 American Cities
with Paved Streets and Populations over 10,000

Paving type	Paved length in km
Asphalt	230
Granite setts	1,920
Wood block	1,220
Brick block	1,160
Rock (macadam?)	910
TOTAL	**5,440**

SOURCE: Armstrong 1976.

was commonly known as the Asphalt Trust and was a common target for scandal-mongering journalists (Hveem 1971; McShane 1979). Barber later withdrew from the trust and formed a new Barber Company, using Bermudez asphalt from Venezuela. Descendants of the Barber Company remain prominent in the asphalt industry today.

As table 7.3 clearly shows, asphalt was still a minor paving solution at the turn of the century, although George Tillson's 1897 data indicate that 230 km may be a very low estimate for asphalt paving. Tillson also claimed that Buffalo, with 3.12 km², had more asphalt paving than any other city in the world. Its nearest contenders in his list were Berlin at 1.36 and Paris at 0.34. The Barber Company had obviously sensed a market well short of saturation.

The survey also indicates that there was significant urban paving taking place by the end of the century. The work made the streets more usable for pedestrians, easier for vehicular traffic, and more satisfying for the strong sanitary movement that arose in about 1880. This pressure group deplored streets replete with stagnant water and mud and advocated asphalt paving. Most American urban streets were paved by 1924 (McShane 1979).

The Richardson Era

In 1887 de Smedt was followed as inspector of asphalts and cements in Washington by Clifford Richardson, who had been with the Department of Agriculture and had traveled extensively in Europe. He was another chemist, but with a strong geological bent. His charter as inspector was to overcome the empirical nature of existing asphalt design methods (Engineering Record 1907).

Richardson's initial impact on the technology can be seen in figure 7.4, which shows Washington street paving specifications after four years of his influence. Each component in the cross sections now served a specific, planned purpose. Traffic loads were carried by strong, stiff structural courses of concrete or penetration macadam. The cushion and binder courses were innovative. The cushion course was a mastic rich in bitumen; its threefold purpose was to cover and level the rough surface of the structural course and to provide a load-carrying key between the structural

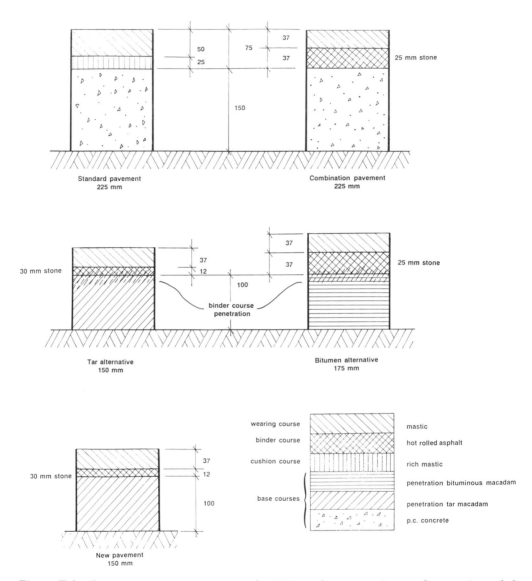

Figure 7.4. Alternative pavement sections in the 1891 Washington, D.C., specifications (compiled from Engineering Record 1891a). The shading illustrates that the binder course keyed into the penetration macadam structural course, leaving only 12 of its 37 mm proud of the base.

course and the wearing course. Not particularly successful, the cushion course produced misshapen surfaces due to its susceptibility to being pushed and rutted by traffic (Tillson 1897). The alternative to the cushion course was the binder course.

The binder course was inspired by the effectiveness of the stony structure of the penetration macadams, which suggested that mastics might be strengthened by the deliberate addition of a small proportion of broken stones larger than 30 mm in size. However, this produced a stiff material that was difficult to place into its final position. To overcome this drawback, extra bitumen was added, raising the mass percentage to 16, and the fine stone particles normally found in mastic were omitted. Technically, this absence of stones of intermediate size made it a bitumen-filled open-graded asphalt. The proportions used meant that the large stone particles were not necessarily in contact. Large stones were finally being added to the manufactured mixes developed by de Smedt. The move undoubtedly reflected the availability of improved construction equipment as well as a better understanding of pavement behavior.

The resulting product, which came to be known as hot-rolled asphalt, represented a major step forward in asphalt technology. We will subsequently use the initials HRA (for hot-rolled asphalt) to describe this material, although in the early days the material may have been tamped into place rather than hot-rolled. When HRA was occasionally used as a major pavement course in the United States, it was called stone-filled sheet asphalt or Topeka asphalt. The former term related to its history, and the Topeka connection is described below.

Nevertheless, the binder course was unnecessary in this situation, and from 1915 it slowly gave way to a simple priming coat of bitumen applied to the top of the structural course (Broome 1963). This provided the same covering and shear transfer at a fraction of the cost. However, the binder course had played its role in introducing a major new component to pavement construction.

The wearing surfaces in figure 7.4 were easily worked mastics based on the de Smedt manufactured asphalts. They had little inherent strength and were shoved and rutted with relative ease. Richardson was also to find a solution to this problem. In 1894 he resigned as inspector of asphalts and cements to become a director of the Barber Company. The company knew that Europe was unaware of the new manufactured asphalts, and one of Richardson's first jobs was to follow up this market opportunity during visits to London, Paris, and Berlin. Richardson was succeeded as inspector by Allan Dow, who had been a chemist with the Barber Company. Dow occupied the post, now called inspector of pavements, until 1906 when he went into private practice. His invention of an improved penetrometer will be discussed below.

During these visits, he advised on the use of the American surface courses in pavement trials on King's Road and Pelham Street in the London suburbs of Chelsea and South Kensington (Attwooll 1955). The demonstrations were "not an entire success" and suffered from "scaling" (Richardson 1905). Richardson hypothesized that the scaling was due to the effect of London fog on the Trinidad Lake asphalt used; indeed, moisture can affect whether bitumen adheres to stone. In 1895 the Trinidad Lake asphalt was replaced with material from Bermuda, and the pavements failed even more quickly. Success finally came in 1896 when cement powder and fine, angular sand were added to the mix. Technically, these are fillers used to increase the stiffness of the binder and reduce the air voids and, incidentally, the cost of the mix. The type and grading of the filler is critical (Hoiberg 1965). Cement also improves adhesion between bitumen and stone. The additives certainly led to a sound, impervious wearing course, which came to be called sand carpet in Britain and sheet asphalt in the United States.

Pavements carrying horse-drawn traffic had long been plagued with a slipperiness problem. Although sheet asphalt was quite smooth, its flat imperviousness meant that it could be readily washed and swept, and this regular maintenance dramatically diminished the danger. In addition, the inherent slipperiness was often reduced by following the long-established technique of incorporating a covering of coarse sand or grit into the still sticky asphalt surface.

The concurrent demands of the cyclists further emphasized the need for the smooth surfaces that only sheet asphalt could then provide. Thus, technology—with impeccable timing—finally produced an affordable pavement surface that was durable, smooth, and of adequate skid resistance. It was just what would be needed for the new pneumatic-tired motorcar. Well, not quite; the surfaces were still quite slippery when wet. Indeed, in 1933 Parisian taxi drivers struck over the slipperiness of asphalt paving in that city. The solution, first introduced in Surrey in England, was to roll precoated pieces of broken stone into the surface of the sheet asphalt.

Richardson's encounter in London with a new set of materials and problems forced him to think through and consolidate his approach to manufactured asphalt (Engineering Record 1907). The beneficial effects of adding fine material soon flowed through to the HRA mixes. Given the success of HRA, it was perhaps inevitable that Richardson would experiment with further increasing the proportion of large stones in the mix.

However, it was not simply a matter of adding more stones. Experience with macadam had shown that maximum density was obtained by using a continuous gradation of stone sizes, each fitting into the other's interstices. Called a well-graded mix, its many points of interparticle contact make it stiffer and stronger in compression than a traditional macadam mix. Adding too much bitumen to a well-graded mix loses the benefits of interparticle contact. One of Richardson's innovations was to add just enough bitumen to coat the particles with a thin bitumen film to act as a glue but not as a separator. Such an effect can also be produced by filling the voids in an open-graded mix with a stiff well-graded mastic.

This new material was known as asphaltic concrete (AC) and had the additional advantages that it (1) was cheaper because it used about half the bitumen needed for HRA, requiring 8 rather than 16 percent by mass; (2) did not require all the pieces of stone to be ground to a fine size, as did the de Smedt mixes; and (3) was inherently stiffer, which meant that softer—and therefore more durable—bitumens could be used (Warren 1901). Even when heated, AC was much stiffer than HRA and therefore only became feasible as construction equipment improved.

The term AC dates from 1861 when Léon Malo defined it more generally as a mixture of bitumen, powdered native asphalt, sand, and stone. By 1880 mixtures of bitumen, sand, and stone were known alternatively as asphaltic or bituminous concrete (Delano 1880)—dual terms that still exist today. To confuse the issue, penetration tarmacadam was sometimes also called bituminous concrete (Deacon 1879). Malo's definition did not distinguish between AC, in which the voids are stone-filled, and HRA, in which they are largely bitumen-filled. Richardson in 1905 noted that the term AC had been in use for "many years" to describe a material using bitumen rather than portland cement binder. AC was introduced into England in about 1920 and for some time was called bituminous macadam. It is now referred to there as dense bituminous macadam rather than as AC.

In 1896 Richardson tested his first AC on a footpath outside the Barber office on West Avenue at Sixth Street in Long Island City. The American market slowly responded to the new asphalts, and the area placed rose from 5 sq km in 1904, to 6 sq km in 1905, and then to 7 sq km in 1906 (Engineering Record 1907). Unless surfacing was used with AC, it tended to unravel under the action of incessant iron-shod hooves and wheels. Therefore, Richardson continued to recommend the use of a 25 to 40 mm thick sheet asphalt wearing course on roads where traffic was heavy. Others less confidently used courses up to 75 mm thick. The Barber Company's combination of sheet asphalt placed over AC was known at the Standard Sheet Asphalt Pavement.

In 1905 Richardson published a textbook containing the world's first reliable manufactured-asphalt specifications. It is said that when the Barber Company realized this, it bought as many copies as possible to prevent disclosure of too many "trade secrets" (Halstead and Welborn 1974). Certainly, for some time Richardson was able to successfully patent his approach to manufactured asphalt. His method for optimizing his asphalt specification was empirical. Over twenty years he inspected 3 Mm of asphalt road in one hundred cities in the United States, Britain, and France. He observed pavement performance under traffic, taking asphalt samples so that he could determine the asphalt composition. From this data base of successful and failed mixes he developed a final AC specification, which was to stand the test of time remarkably well.

A Variety of Recipes

AC is not particularly strong in tension and fatigue because the binder is only a thin film between the pieces of stone, and HRA is a far better material in this respect. They therefore play two quite different roles within a pavement. Nevertheless, over most of asphalt's history there has been vigorous rivalry—usually based on national preferences—between the two products. For instance, a British view (Road Research Laboratory 1962) was that Richardson's work provided "the basis of the modern rolled asphalt surfacing, which has long held the preeminent position as a heavy duty surfacing . . . although in the United States itself the 'asphaltic concrete' commonly used nowadays is not based on Richardson's principles." Did Richardson recommend HRA for the British and ignore his fellow Americans? No, Richardson developed the three asphalts that underpin today's practice: sheet asphalt, HRA, and AC. National practices and a lack of understanding of the underlying principles led to the selective adoption of his ideas. Clearly there is a role for both types of asphalt, as reflected in recent British practice which even advocates the use of both HRA and AC together in a composite design (Powell et al. 1984).

There have been numerous examples of communities strenuously arguing the merits of their own selection of sheet asphalt, HRA, AC, or some intermediate product. The reality is that the availability of local materials, rather than technical theory, often determined the original local choice. For example, if well-graded, angular, and low-cost sand was available, then sheet asphalts would be more effective. On the other hand, the availability of good broken stone favored the ascendancy of AC, even as a surface course. Good sand and good stone are more likely to be found apart than together. Of course, each of these three asphalt products has a distinctly different role to play and should be used in association rather than in competition. Only recently has it come to be realized that both AC and HRA were only part of the ultimate solution.

At the turn of the century, a variety of proprietary manufactured asphalts were being marketed in the United States with such trade names as Willite and Romanite. The most common of all was Bitulithic, an AC introduced by the Warren Company in Park Place, Pawtucket, Rhode Island, in 1901. The Warrens had been associated with the tar business since 1845, and two of the family, Earl and Fred Warren, had been associated with the Barber Company from the outset. Fred left Barbers in 1900 to form the independent Warren company in association with his six brothers.

Warrens realized that AC provided an adequate running surface for rubber-tired traffic without the need for a sheet asphalt surfacing. This insight provided Bitulithic with a unique commercial advantage. But tradition died hard and, in the words of Hoiberg (1965), "the introduction of this pavement for city streets was an uphill fight, was much ridiculed, and at first met with little success." The fight was not aided by the occasional remaining iron-tired vehicle continuing to fracture and dislodge the large pieces of stone in the AC. In these circumstances the company supplied a mastic surfacing and ingeniously marketed this old solution as a new product called Warrenite-Bitulithic. Indeed, AC only became a major surfacing competitor for sheet asphalt in the early 1920s when there was a clear preponderance of cars with gentler pneumatic tires (Hoiberg 1965). For rehabilitation work, existing pavements of macadam or Telford pitchers were primed and then covered with 50 mm thick layers of AC.

Prior to 1910 most of the proprietary asphalts were patented, a situation that lasted until a landmark patent infringement case when Warren unsuccessfully alleged that the city of Topeka had infringed his Bitulithic patent.[9] The case failed because the maximum stone size in Topeka AC was only 12 mm, whereas the Warren patent specified a maximum stone size of between 37 and 75 mm. Until it expired in 1919, the Warren patent continued to stultify the American use of mixes with pieces of stone larger than 12 mm. Indeed, the industry honored the patent for many years after it had legally expired. In 1929 Bitulithic was still being widely used, and an equivalent of 100 sq km of Bitulithic pavement had been placed around the world.

Success with the new manufactured asphalts still did not come easily. Writing in 1907 one American expert, Tillson, observed that "few engineers of the country are qualified to design asphalt pavements for different conditions. That is one reason why such pavements have so often failed." Tillson had published a well-received textbook on pavements in 1901, and at about the time of the above remark he became chief engineer of the Bureau of Highway in Manhattan. Richardson welcomed the appointment, noting that New York at the time did not possess "a single well constructed pavement" and that the pavements that did exist were a disgrace to such a large city" (Tillson, Haylow, and Richardson 1907).

Despite the presence of Bitulithic, the conservative combination of sheet asphalt surfacing, AC intermediate course, and portland cement concrete structural course found ready acceptance in New York and Washington. By 1910 it was the most commonly used paving design in American cities (Merdinger 1952). Indeed, in contrast to Bitulithic, the AC layer was sometimes omitted completely, with all structural strength obtained from the concrete course. Writings of the time sometimes create confusion by referring to the whole multilayer composite structure as sheet asphalt (Agg 1916).

In 1904 there were only 30 km of bituminous roads in rural America. By 1914 this length had grown to 17 Mm, comprised mainly of surface treatment or penetration macadam. These two pavement types reached their American zenith at about this time. Their subsequent decline occurred because they were unable to manage the solid tires of the heavy new motor trucks just appearing on the roads

(MacDonald 1928). Indeed, the increase in motor truck traffic after World War I had a major destructive effect on pavements. Trucks also made it possible to bring good roadmaking materials over much larger distances than had been feasible with horse and cart, a major benefit. For all these reasons the length of the more expensive asphalt rural roads went from zero in 1914 to 16,000 km in 1924.

Such an increase in rural asphalt usage also required an increase in mechanization. The initial asphalts had been produced by the penetration method or by in situ mixing, often aided by precoating the stones with the binder. The first portable asphalt mixers were in use by 1909, but the precoat method was still common in the 1920s (Wellborn 1969). Charles Pope in California in 1924 developed an asphalt-spreading machine based on a similar device for concrete, and machines for laying and finishing asphalt were introduced in 1928 (Scherocman and Martensen 1984; Hveem 1971).[10] However, effective mechanical spreading of premixed asphalt did not occur until 1931. Nevertheless, hand placement of asphalt remained commonplace throughout the 1930s. Curiously, machines for pavement recycling were introduced in the 1920s and thus predate machines for pavement spreading.

The asphalt layers were commonly supported on a 150 mm to 200 mm thick portland cement concrete course. It is interesting to speculate why the asphalt industry was so slow to realize that asphalt had its own inherent structural strength and could replace concrete entirely. Why did the portland cement concrete courses continue to be used under structurally strong asphalt; why the belt and braces?

Although Malo in the 1860s had advocated placing asphalt directly on a dry and well-drained natural formation, his advice was rarely followed, for many mistakenly blamed the rue de Rivoli failure on the use of this approach.

In 1891 Richardson permitted the use of penetration macadams as an alternative to portland cement concrete (see fig. 7.4). However, the option was rarely exercised because asphalt was usually much more expensive than concrete; trials of asphalt in 1890 on an unbound macadam base had been failures, and this memory lingered on for many decades (Tillson 1897); and Richardson personally favored concrete, and such was the respect for him that this was close to an official command (Civil Engineering 1981).

The best reported successful application of an asphalt structural course was on the Thames Embankment in 1907, but even in this case it was somewhat disparagingly noted of the underlying unbound macadam that, "owing to the compacting of the material under long traffic the [macadam] was in a hard, dense condition" (Civil Engineering 1981). The first inklings of the emergence of a positive view of asphalt as a structural course did not occur until the early 1920s (Agg 1916).

As late as 1927, textbooks stated that the resilience of AC—which made it such a good road surface—meant that, by virtue of its associated lack of rigidity,[11] it had to be supported on a "solid, substantial" foundation (Coane, Coane, and Coane 1908). In 1930 it was still common for AC to be placed on a 150 mm portland cement concrete course. The use of thicker layers of AC to counteract its alleged resilience produced a product that was either much more expensive than conventional concrete or liable to *wave* (corrugate), particularly during construction rolling. A less conservative view was adopted in suburban streets where authorities in the early twenties were recommending a 50 mm thickness of asphalt placed on top of an unbound course (Agg 1916).

An important obstacle to the structural use of asphalt was the absence of any viable method for the engineering design of either the asphalt or the underlying natural formation. The prediction of formation behavior only became possible after an investigation of Californian pavement failures in 1928–1929. This work pointed to the need to know formation properties before selecting pavement thicknesses. To meet this need, in 1930 the California Division of Highways introduced the California Bearing Ratio (CBR) method for assessing formation strength, basing it on an earlier Bureau of Public Roads technique called the Bearing Power Determinator (Hogentogler 1923). The CBR method measures the deformation caused by a load applied to a flat disc placed on the formation being tested. Design curves relating the pavement thickness to this deformation were first produced in 1942, based on field observations over the preceding twelve years (Porter 1949). The design method was widely promulgated during the Second World War by the U.S. Corps of Engineers.

Finally, in the 1960s the results of the AASHO Road Test showed that thin asphalt was ineffective under heavy traffic but that a thick layer of asphalt placed directly on the natural formation performed adequately. Faced with increased competition from concrete, which had also received good AASHO results, the American asphalt industry moved quickly to promote thick layers, which they called deep lift asphalt.

Engineering textiles were first used to enhance asphalt in about 1935, when woven cotton fabrics were spread between asphalt layers in South Carolina. The cotton fibers provided only temporary benefits due to wear and rotting. Bitumen-impregnated hessian was used with some success by General Slim in Burma during World War II. Plastic fibers are currently showing considerable promise as additions to modern road construction.

Competition in the paving industry has always been strong and, sometimes, unscrupulous to the extent that it was often difficult for customers to discern the true technical merits of the products being offered to them. A particularly interesting review of the merits of the various pavement types was given by Tillson in his 1901 book and is included as tables 7.4 and 7.5. The final decision clearly still rested on the engineer's judgment and preconceptions.

TABLE 7.4
Merits and Costs of Various Pavement Types

Quality	Maximum possible points	Granite setts	Asphalt	Brick	Macadam	Cobbles
Durability	21	21	15	13	7	15
Cleanability	15	11	15	12	5	2
Low vehicle operating costs	15	7	15	12	6	4
Cheapness	14	2	4	3	7	14
Sanitariness	13	9	13	11	5	2
Ease of maintenance	10	10	6	6	3	2
Good surface friction	7	6	3	6	7	5
Favorableness to travel	5	3	5	4	5	0
TOTAL	100	69	76	67	45	44

SOURCE: Based on Tillson 1901 and Agg 1916.

TABLE 7.5
Comparative Costs of Pavement Surfaces, Normalized to Unity for Gravel

Costs	Gravel	Best gravel	Water-bound macadam	Bituminous penetration macadam	Portland cement concrete	Sheet asphalt and AC	Setts	Wood block
Pavement	1	2	7	10	16	18	10	17
Car operations	1	0.92	0.94	0.90	0.85	0.85	—	—
Annual maintenance	—	—	2.8	—	—	1.4	1.0	1.6

SOURCES: Based on Moore 1876, Rochester Executive Board 1886, Carey 1914, and Agg 1916.

Bitumen

First, it is necessary to clarify some definitions. *Bitumen* is the viscous, heavy component of petroleum and is a strong, durable glue. The word is relatively recent and probably comes from attempts to Latinize the older term *pitch* into *pitchumen* (Abraham 1960). This *pitch* is the same word as in the phrase "to pitch a tent—both uses originally meant "to cover"—and links to widespread use of bitumen as a waterproof covering (Buttrick 1962). The term *mineral pitch* was synonymous with bitumen, but *pitch* later came to be the common term for both wood resin obtained by boiling or distilling the turpentine obtained from coniferous trees and for compounds made from animal fats. Later still, *pitch* was used to describe the relatively hard residue left over when coal tar was distilled.

Alternative terms for bitumen have been "Jews' pitch," "slime," and, in India, "earth-butter." In early writings the words *bitumen, pitch, molasses,* and *naphtha* were often interchangeable. A nineteenth-century French term for bitumen was *pissasphalte* (Malo 1866). For practical reasons, many of the bitumens used in those days were actually *mastic,* which is a mix of bitumen and fine granular material.

Chapter 3 described the pre-Roman use of bitumen as a mortar for flagstones and paving bricks and noted that the technology disappeared at about the time of Christ. Nevertheless, bitumen was still widely used by the Romans, not for roadmaking, but as a building mortar, lamp fuel, waterproofing agent, and paint and, in warfare, as boiling tar.

For Medicinal Purposes Only

After the Romans, bitumen was also used for exorcising evil spirits, as a disinfectant, for embalming, and as a medicine for treating cuts and wounds, skin disorders, swollen feet and hands, sore eyes, rabies, tumors, swellings, gout, and, when mixed with beer, diarrhea, and other stomach upsets (Forbes 1936). The embalming use was widespread, and the word *mummy* comes from the Persian word for bitumen, *mumia.* Until the eighteenth century, European apothecaries sold desiccated mummies as a medicine. Whatever virtues the product had probably came from the bitumen and not from the ancient flesh. Bitumen was not always good for the stomach. The Apocrypha in Daniel 3:23 records how Daniel defeated a dragon by feeding it a mixture that included bitumen. The dragon, with legendary fire in its belly, burst open upon swallowing the combustible offering. In the sixth century, Byzantine friars mixed together naphtha and bitumen and, later, saltpeter to produce Greek "wild fire," which was used for setting fire to enemy ships.

From at least the fifteenth century, various bituminous ground seepages were collected in western Europe and sold for medicinal purposes, for example, as St. Catherine Oil, St. Quirinus Oil, and Blood of Thyrsus (Forbes 1958b). The latter came from Bavaria, but there were also sources in Miano, near Parma in Italy. In 1555 Agricola in the twelfth book of his De Re Metallica described how bitumen could also be obtained by using heat to separate it from native asphalt (Forbes 1958b).

In 1690 bitumen was "discovered" at Pitchford in England's Severn Valley, some 15 km west of Coalbrookdale. It appears to have been used as medicine, for caulking, and as lighting fuel (Ironbridge Gorge Museum Trust 1979). The medicine was sold as Betton's British Oil. The name of the town suggests that bituminous material had been known in the area in much earlier times (Forbes 1958b).

By the eighteenth century bitumen was being widely promoted in Europe as a medicine for treating chilblains, itching, toothache, boils, sheep pox, ringworm, gout, epilepsy, and blindness (d'Eyrinys 1721; Abraham 1960). It was also widely used as a beauty aid; indeed, coal tar soap remains a popular item in today's supermarkets.

Similarly, the Seneca used oil found in the Oil Creek in Pennsylvania to make a medicine named Seneca Oil. It was widely promoted for rheumatic relief and gained fame in the Revolutionary War, when soldiers used it to treat the rheumatic complaints from which many suffered. Another bituminous medicinal product, this time from the Pittsburgh area, was Kier's Rock Oil and Petroleum Butter, which was sold for treating burns and bruises (Welty and Taylor 1958).

Down to Earth

The gradual development of asphalt in the nineteenth century created a steadily increasing demand from the paving industry for an effective binder. Many of the tars then in widespread use were less than adequate. A potential natural competitor arose in 1857 in the form of bitumen, which is produced as a "heavy end" residual byproduct of commercial oil refining—gasolines and kerosenes are the "light end." The California petroleum was particularly rich in bitumen, which was initially used commercially for making building mastic, roofing, and fiber pipes.

However, from a pavement viewpoint the initial bitumens available to the large American market were little better than tar. The gritty Trinidad bitumen tended to be quite hard. This not only caused pavement problems, but there had also been a number of instances where solidified Trinidad bitumen had had to be chopped from the holds of ships. Many of the East Coast bitumens were high in paraffin and froze when cold. The early Californian bitumens were also sensitive to temperature. This led to pavements rapidly cracking and deteriorating. Most bitumens were too viscous for the then-current pavement equipment to handle, and different viscosities were needed for different applications.

Another step away from the natural solution occurred when de Smedt realized that these problems could be overcome by blending different bitumens together in a process that came to be known as *oiling*. Initially he used California bitumen as a blending addition to his native asphalt binders. Eventually, the more fluid West Virginia bitumens proved more valuable softening additions and usually constituted about 20 percent of the blend (Krchma and Gagle 1974).

A typical Washington, D.C., blend was "a mixture of one part distilled heavy petroleum oil with four parts Trinidad bitumen" (Mountain 1894). Blending to produce desirable bitumen properties became common, and up to 102 different grades were available in the United States until the Department of Commerce stepped in in 1923 and cut the number of grades back to 9.

The confusing American practice of using the term *asphaltic cement* instead of *bitumen* apparently arose because the word *bitumen* came to be restricted in that country to straight petroleum bitumen and the term *asphaltic cement* was used for bitumen blends. As blending died out and petroleum bitumen took over, the terminology unfortunately remained unaltered. The 1987 edition of *Encyclopaedia Britannica* (volume 10, 98c) refers to bitumen as "an asphalt substance."

To determine the suitability of a blended bitumen for use in making asphalt, the tester would chew a sample as vigorously as possible in order to assess its viscosity. A mixture of bitumen and chicle gum was used as a chewing gum in Mexico, and some bitumen suppliers provided a small box of chewing samples so that a foreman could calibrate his jaw (Abraham 1960; Hveem 1971). This oral tradition was put under notice in 1888 when Professor H. C. Bowen of Columbia University, a consultant to the Barber Company, introduced the first penetration tester for determining bitumen stiffness cum viscosity. The device employed a "no. 2 cambric sewing needle" and measured the way it penetrated into a bitumen sample. Bowen used the method to actively advise Barbers on the appropriate native asphalt to import from Trinidad (Engineering Record 1891c). Its descendants, little altered, still control the selection of most of the world's bitumen (Halstead and Welborn 1974).

Bowen conducted his tests in a room kept at 25° C, and Richardson suggested the procedural improvement of keeping the specimen in a water tank at the desired temperature. The idea was taken up by Dow, whose device permitted careful temperature control of the specimen. Viscosity was first measured directly in 1898 using an instrument developed in Germany. Such devices were to prove invaluable for controlling both bitumen blends and bitumen sources. Nevertheless, the chewing test was still in use in 1910 (Krchma and Gagle 1974).[12]

As an alternative to blending, tars and bitumens were often made more fluid either by a process known as *cutting,* in which the bitumen was mixed with a less viscous but more volatile petroleum fraction, or by adding an emulsifying agent to allow the bitumen to be mixed with water. Cutback tars were used in Washington prior to 1879, and cutback bitumens were in use from the turn of the century (North 1879; Krchma and Gagle 1974). Following a commercial path similar to that of blended bitumens, some 119 grades of cutback bitumen were being marketed in the United States in 1929 before official action dropped the number down to 9.

In 1896 the Union Oil Company sent the first shipload of California bitumen to the East Coast. This was an important move because it broke the monopoly the Barber Company had gained on the Atlantic coast as a result of its control of the Trinidad deposits. The monopoly was further broken when the Warren brothers set up in competition to Barbers in 1901. Trinidad material was not made available to them, so they turned to the new California bitumens. The final blow occurred with the widespread availability of bitumen-rich Mexican Panuco crude oils in 1915. From then on, petroleum bitumen dominated the American binder market (Hoiberg 1965).

A major boost to bitumen availability came as the new IC engines consumed more and more of the light end of oil refinery output. Until that time, petroleum producers often had to rely on sales of heavy-end products to support their operations. From about 100 tonnes of bitumen in 1880, by 1902 some 20,000 tonnes per year were being produced (Halstead and Welborn 1974). However, the widespread use of petroleum bitumen did not occur until refining capabilities increased in the 1920s.

The early petroleum bitumens were "all more or less carelessly manufactured, without laboratory control," and their viscosities varied enormously (Engineering

Record 1907; Richardson 1905; Krchma and Gagle 1974). The superior refinery practice of "bitumen blowing"—in which the bitumen is hardened by oxidation with blown air—was patented in 1893 and in commercial use in 1904. This new technology permitted production of a range of controlled bitumens that could be used unblended (Krchma and Gagle 1974; Lay 1990).

Unblended petroleum bitumen was first used for roadmaking in the early 1880s, when it was mixed with in situ materials around oil fields and refineries to improve local service roads. The success of this technique led to the practice of pouring bitumen and tar over roadside windrows of loose material, then spreading them into place on the pavement surface to produce an oiled road (Wellborn 1969). The first formal record of an oiled road is in 1894 at Ortega Hill in Summerland near Santa Barbara, California (Peckham 1909; Krchma and Gagle 1974). Oily sands from oil drilling were used to fill ruts and potholes and later spread across the entire road. The first large job was a 10 km trial conducted by Los Angeles County in 1898.

The method is most effective with mixtures of sand and stone and relies largely on increasing the effective size of the smaller particles in the surface layer. For the first day or two after application, there is an unfortunate tendency for the slowly hardening bitumen to soil vehicles and clothing. A popular commercial extension of the process was a material called petrolithic, which was a compacted mixture of bitumen, earth, and water. The material lacked strength, support, and coherence, so the treatment rarely lasted more than a few months in wet climates and a year on lightly trafficked roads in dry climates. The result was often little better than a black, sticky mud. To make the petrolithic last longer required much more effort to be devoted to preparation and placement and more control to be exercised over application rates. A move to place the bitumen by spraying rather than pouring assisted in the development of the spray and chip seal method to be discussed below.

A technically sound oiled-road design method based on the surface area of the stones was developed in California in the 1920s and 1930s (Hveem 1971). The method showed that, to prevent the mixture from becoming too plastic, it was essential to limit the amount of bitumen used. This consideration led in 1925 to Prevost Hubbard and Frederick Field's development of their stability test for measuring mix stiffness. The test is now applied to asphalt design generally.

Bitumen was delivered to job sites in wooden barrels. The barrel staves were often burned to heat the bitumen and lower its viscosity. Steam-heated transport containers were first used about 1900.

Solid rubber roads were tested in the Admiralty Courtyard in London in 1840. Rubber slabs, usually on a concrete structural course, were then successfully used in London outside St. Pancras Station from 1870 to at least 1934 and similarly outside Euston Station. The purpose was to keep the noise of vehicles servicing the busy stations from disturbing the occupants of nearby buildings (Hadfield 1934). Likewise, there are reports of the use of leather as a surfacing (Merdinger 1952). Rubber was also used as a protective cap on wood-block paving in trials on Southwark Street, London, between 1912 and 1915 and in block form elsewhere in that city in 1926 (Boulnois 1919; Hadfield 1934). Cork was also added to asphalt to reduce traffic noise outside hospitals; the product was quiet but lacked durability. Molasses was tried as a bituminous binder in Newton, Massachusetts, in 1908 but tended to dissolve in water. During pavement trials beginning in 1929 in Holland and Indonesia, the Dutch successfully mixed rubber with bitumen in order to enhance its elasticity (Road Research Laboratory 1962). The technique is widely used in modern pavement practice.

The main competitor for bitumen in the twentieth century has been portland cement. The rivalry between the two materials continues to this day, with economics and local custom usually determining which is used. The rivalry is between cement concrete and asphalt for high-cost roads, but at the low-cost end of the spectrum, asphalt competes with bituminous spray and chip sealed roads. We will now discuss these remarkable sprayed surfaces.

Sprayed Surfaces

Although tar had been sprayed on macadam surfaces since the 1830s, it took a well-publicized trial in 1880 near Bordeaux in France to give credence to the idea. Subsequently, bitumen and tar were made fluid by emulsification or heating and then brushed, painted, sprayed, or squeezed onto urban macadam surfaces to suppress dust and to provide waterproofing (Gregory 1931; Tunnicliff, Beaty, and Holt 1974). Sand and grit were spread on the surface to reduce its stickiness and improve its traffic friction. The process was far from perfect. Initially it lasted but a few months and adhered tenuously to bystanders, nearby vehicles, and adjacent houses. Public outcry was extreme (Boulnois 1919). One major cause of these early problems was the widespread European use of fluid tar emulsions, which often took over a week to harden in cool conditions (Hughes, Adam, and China 1938). It was some time before more suitable road tars were produced.

In the face of a continuing need, the technique was gradually made more useful and reliable. A significant leap forward occurred in 1902, when there was a severe loss of casino business in Monte Carlo due to the dust arising from the limestone macadam surface of the Grand Corniche Road, an old Roman way reconstructed by Napoleon. A Swiss physician, Ernest Guglielminetti, in Monte Carlo to give a lecture on mountain sickness and associated devices that he had invented, impressed Prince Albert I of Monaco, who asked him to help solve the road dust problem. Albert had been actively involved in science, both as a practitioner and as a patron. Guglielminetti was led to try tar through his experience with its use for sealing hospital floors in Indonesia. He successfully demonstrated the efficacy of the spray method in trials that led to further tests in early 1903 and to the Corps des ponts et chaussées' formulating construction specifications later in the same year (Earle 1974). Guglielminetti became known in France as le docteur Goudron (the Tar Doctor) as his efforts directly led to the creation of a dust-free road network in France.

Improved bitumens were available from 1904 on and greatly aided the development of spray treatments. When sprayed or brushed onto the pavement, they were certainly stickier and stronger than their predecessors. Initially they produced a slick and slippery finish, even with the application of sand to the surface. This problem was overcome by replacing the sand with small pieces of broken stone, called chippings. The better bitumens also permitted the use of thinner films, which decreased drying times and thus minimized damage to the public.

The next development was the technique of priming the surface of the underlying pavement structure by initially spraying it with a temporarily more fluid bitumen. This was obtained by either diluting (cutting) bitumen with lighter petroleum products or by emulsifying it with water. Although sprayed emulsified bitumen was first used as a dust suppressant in Philadelphia in 1903, its wider use had to await the development in the 1920s of soapy emulsifiers produced by abattoirs looking for markets for excess animal fats (Halstead and Welborn 1974). In 1910 the more fluid bitumen was the key ingredient in successful American spraying trials, probably on Riverside Drive in New York (Simon 1970). The resulting product was commonly called a *bituminous carpet* and, occasionally, an *oil mat* (Hveem 1971). Fluid bitumens also aided successful spraying in cold conditions.

There were still problems with the adhesion of the pieces of stone to the bitumenized surface, with obtaining a satisfactory distribution of the stones over the surface, and with the surface not responding well to solid tires. One major commentator of the time remarked, "Bituminous carpets on gravel roads have never been very satisfactory with lives of only six months being typical" (Agg 1916). Nevertheless, they were extensively used in the United States in the 1920s for roads carrying fewer than fifteen hundred vehicles per day (Sherrard 1930).

Between 1920 and 1930 the outstanding problems with spraying were solved by process improvements introduced in New Zealand and Australia. The motivation came from the realization that the flexible but waterproof sprayed bituminous coating could be applied to unsurfaced roads that had developed through gradual upgrading and staged construction, and where the traffic levels required a surface but not a significant increase in structural strength. Spraying offered this option, and its introduction was aided by the steady replacement of the damaging old solid tires with the gentler pneumatics. Thus, as vehicle technology changed, sprayed surfacing became the appropriate choice for more and more routes. However, the earlier failures under solid tires meant that the method's reputation had been permanently destroyed in many jurisdictions outside Australia and New Zealand.

Perhaps this diverted development partly reflects the delayed changeover from solid to pneumatic tires in these two countries. The move was accelerated in Australia, where the railways (e.g., in Victoria) increased their freight rates on the broken stone used in the asphalt process in an attempt to make roadmaking, and hence road freight, uncompetitive. Instead, the move forced further development by the road authorities of the much cheaper spray and chip seal method.

The method came to be called spray and chip seal when applied as a new surface on existing structural courses and as surface dressing when used to enhance existing bituminous surfaces. It is still widely and effectively used in Australia and New Zealand where, with modern practice, it can carry quite high volumes of heavily loaded traffic (Lay 1990).

Pavement Research

A number of instances of pavement research have already been discussed, beginning with the Oxford Street trials in 1838–1839, Adrian Mountain's woodblock trials in 1880, and Richardson's survey of asphalt experience at the end of the century. In the United States, national pavement-material testing began in 1900 in the Department of Agriculture's Chemistry Bureau. In 1909 Wayne County, Michigan, constructed a circular test track with the pavement subjected to load via steel-shod artificial hooves and a steel-rimmed wheel at opposite ends of a 6 m rotating arm. This device, called a paving determinator, was used to test brick, stone sett, and concrete pavements (Ray 1964).

The next major road-material testing machine was a 10 m diameter and 750 mm wide circular test track constructed at the British National Physical Laboratory (NPL) at Teddington in 1912. It was loaded by eight wheels, each applying a load of up to one tonne to the pavement (Boulnois 1919). The track was the initiative of Colonel Rookes Crompton, a pioneer motorist who had made his mark by using steamers to move freight on roads in India. He subsequently convinced the NPL to test the strength, stiffness, impact resistance, and wear resistance of a variety of road surfaces (Bird 1969). NPL continued conducting road research during the 1920s and opened a special experimental station in 1930. Within three years this station had evolved into the Road Research Laboratory, now known as the Transport and Road Research Laboratory, or TRRL (see Charlesworth 1987).

Chapter 4 described how the Office of Road Inquiry was established in the U.S. Department of Agriculture in 1893 and in 1918 became, after a sequence of changes, the Bureau of Public Roads. In 1912 Congress allocated half a million dollars to the office/bureau for an experimental program to study rural roads and thus improve rural postal services. This immediately led to the establishment of the Arlington Experimental Farm in Virginia near Washington. An early initial focus was on the sudden, impactive loading caused by truck wheels. By 1930 the facility had both large and small circular test tracks in operation (fig. 7.5).

Between 1920 and 1923 the bureau conducted a set of full-scale road tests at Bates, which is near Springfield in Illinois. It was the first in a famous American test series. World War I army trucks were used to load sixty-eight experimental pavements involving six different pavement types in a length of about 4 km. The tests were particularly directed at design methods for concrete roads. The results emphasized the importance of the strength of the natural formation and the critical nature of loads on the corners of a slab (HRB 1971). A similar set of tests on thirteen pavements was conducted on an elliptical test track at Pittsburgh in California.

The Western Association of State Highway Officials (WASHO) conducted the next series of full-scale pavement tests in Idaho between 1952 and 1954. One commentator cynically noted that the tests proved that 100 mm of asphalt was stronger than 50 mm (Hveem 1971). A major spinoff from the tests was the development by Albert Benkelman of the structural testing of pavements by measuring the vertical deflection of a pavement under load, using a device now called the Benkelman Beam and widely employed for assessing the remaining life of a pavement.

Probably the best known of the full-scale pavement test series was the AASHO Road Test initiated in association with the Bureau of Public Roads. The tests were conducted by the Highway Research Board between 1958 and 1960 to study the performance of pavement and bridge structures of known characteristics under measured moving loads with a known frequency of occurrence. The test site was at Ottawa, Illinois, and the vehicles were supplied by the U.S. Department of Defense. The test road consisted of 12 km of two-lane asphalt or concrete pavement and sixteen short-span bridges. There were 836 pavement test sections. The tests gave rise to the fourth power law, which is now widely used by pavement engineers and transport economists to predict the effect of axle load increases on pavement damage.

A deterioration in the longitudinal roughness of a road gives travelers an in-

Figure 7.5. Paving test at the Arlington circular test tracks in about 1930. *From FHA (1976) and with the permission of FHA.*

creasingly uncomfortable ride and also indicates changes in the structural sufficiency of the road. It is therefore a useful property to measure. Road roughness measurement was introduced in Belfast at the turn of the century by J. Brown, who developed a sled-based straightedge that both drew the road profile and produced a numerical roughness index. Brown called his device a Viagraph (Aitken 1907). The next roughness meter developments occurred in the United States. An elaborate, thirty-two-wheel profilometer was developed for the Bates Road Test in 1920 and a bicycle-mounted straightedge, also called a profilometer, was produced a few years later (fig. 7.6).

Vehicle-based roughness systems reflect the needs of a more mobile traveling public and began with the 1923 Lockwood Roughness Integrator, which measured the vertical movement of a weight hung by a spring from the body of a traveling vehicle. A variation of the device measured accelerations of the rear axle of the vehicle (Hogentogler 1923). The common current system of measuring relative movement between a vehicle body and its axle began with the Via Log, developed in New York and put into general service by the Bureau of Public Roads in 1926 under the name *roughometer* (Ilveem 1960; Public Roads 1926). Both the name and the technology remain unchanged today as a key method for determining the condition of modern roads.

Figure 7.6. Bicycle-mounted rolling straight edge, christened a profilometer. *From Hogentogler (1923).*

Summary

This chapter has shown that great technical advances occurred in the technology of pavement structures and surfacings during the nineteenth century. Almost in their entirety, these advances predated the development of the motorcar. The motivation came, not from the demands of the car, but from the rising aspirations of communities and the increased technical competence and confidence of individuals following the Industrial Revolution. Not only was there more travel, but there were also demands that travel be more efficient, effective, and comfortable. A new set of expanding aspirations had arisen, and there were widespread expectations that these could and would be met by technological development. Communities at last saw an alternative to a life full of mud, stench, dust, and noise.

Not coincidentally, the new bound surfacings and structural courses arrived just in time to satisfy these demands. Necessity was once again required to mother invention. The new cars and trucks, with all their promise, were almost useless without good roads. And when they did operate, they created major societal impacts; for example, speeding cars were creating enormous problems with dust and abrasion. A 1906 report commented that the need had become "still greater in the face of the growing prevalence of the motorcar, and especially of that much abused vehicle, the motor omnibus, the nuisances produced by these machines depending largely on the surface over which they travel" (Civil Engineering 1981).

Action was demanded on all fronts, and not only was the burgeoning car-owning population creating loud demands for better and longer roads, but, as one American technical correspondent noted in 1918, there were also the trucks!:

> From horse-drawn vehicles with concentrated loads of probably 3 ton at the most, travelling at the rate of four miles an hour, sprung almost overnight the

heavy motor truck with a concentrated load of from eight to twelve tons, thundering along at a speed of 20 miles an hour. The result? The worn and broken threads that bind our communities together. (Cron 1974)

Fortunately, the new bound surfaces were able to meet the demands of motorized vehicles and the social expectations of the new urban communities, plus they reduced travel costs and at least doubled the lives of pavements (Cron 1975b). Technological change had again spawned further technological change. The rate of road surfacing in the United States changed from 1 Mm per year between 1860 and 1890 to 15 Mm per year between 1890 and 1920 (Rose 1953).[13] To see this remarkable change as due to pressures from the car or to the availability of better surfacing technology is to see merely different reflections of the same truth.

Indeed, pre-nineteenth-century wheel loads rarely if ever exceeded 2 tonnes. Modern truck wheel loads can exceed 5 tonnes and are vastly more frequent, and yet pavements had become thinner rather than thicker. Roman pavements could be up to a meter thick. With the revival in roadmaking by the French, pavements dropped from an initial 450 mm down to 250 mm with McAdam's method. The same thickness of modern road can carry the much greater demands of modern traffic (Lay 1990), although the materials used—soil, stone, and a natural glue such as bitumen or cement—have not changed since pre-Roman days. What has changed?

First, the importance of drainage is now universally recognized. Second, the importance of assessing the quality of both the natural formation and the imported material being used is now recognized, and there are methods for assessing that quality and taking steps to account for any unavoidable quality deficiencies. Finally, how and why a pavement works is now understood; with that knowledge has come the power to determine and influence performance. If the understanding has at times come slowly, it can fairly be said that pavement technology did not always attract the best minds of a generation.

Pavement technology has not altered dramatically in the last fifty years. Engineers have become better at building the pavements, a little better at analyzing the stresses and strains in them, more conscious of the factors causing them to deteriorate, better at quality control, more aware of the properties of the natural formation, and better able to make skid-resistant and well-lit surfaces. Nevertheless, a pavement engineer from 1930 would have little difficulty in recognizing the way roads are built, the products used, and the results obtained. He would be impressed by the product, but not surprised by it.

Bridges

Chapter 8

Those men whose lot in life it is to build bridges—
are indeed singled out and blessed of God.
WILLIAM BLISS, LATE NINETEENTH CENTURY

Bridges are an essential part of the road system, and their history adds greatly to our overall story. However, they do not fit neatly into the sequence of this book. A discussion of individual bridges could possibly have been interleaved with many of the preceding chapters, but much of their contribution would have been lost. Furthermore, the author's view as an ex-bridge designer is that they are a separate and unique technology and require a chapter of their own. Indeed, they deserve a whole book, and a number of such excellent bridge histories will be referenced in the subsequent text.[1] This chapter is therefore only an essential abbreviation of the whole story. Some small mention of the history of tunnels will also be made, although many tunnels predominantly served the needs of canals and railways rather than roads.

One difference from the preceding chapters is a much greater emphasis on the theoretical studies associated with bridges. Over the centuries bridge technology has attracted some of the world's great minds, and there is value in exploring the reasons for their input, the timing of that input, and the consequences that arose therefrom.

In the Beginning

The first river crossings would have been by ford rather than by bridge. Many town names—Oxford, Stratford, Frankfurt—are indications of the importance of fords to our early ways. Some fords would have occurred as natural geological features, but many were deliberately enhanced using logs or stones. The associated technological development may have prompted the later construction of bridge piers.

Early Bridges

The first bridges would have either obviated the need to ford a shallow stream or provided a crossing over a narrow but otherwise impassable ravine. Of necessity they represented a relatively simple technological step forward. Crossing deep water or wide spaces was beyond the capabilities of natural materials and had to await major advances in technology.

The first simple bridges were typically horizontal stone slabs straddling stepping-stones in a ford, sometimes called clapper bridges; placed or fallen logs, sometimes called clam bridges; hanging vines across narrow streams; and natural stone arches.

Technically, a *beam* is a structural component that carries its payload by bending. The word *girder* usually now refers to a steel beam. The use of stone slabs and logs as bridge beams provided a ready way for inventive early societies to cross narrow openings. The oldest extant bridge is the Caravan Bridge, which is a 13 m stone slab over the Meles River near Izmir in Turkey.[2] Built about 850 B.C., it was used by Homer and Saint Paul (Steinman and Watson 1941). A three-span bridge built in 4000 B.C. of finished oak beams and spanning about 10 m was found buried in silt at Birkenhead in England in 1850. It is thus almost the same age as the Sweet Path (Edge 1990).

The use of stone slabs and logs as bridge beams rarely permitted spans in excess of 20 m to be crossed. Indeed, beams rely for their spanning capacity on both the inherent strength and stiffness of the material employed and the depth of the cross section. There is little that can be done in a simple way to improve on nature or to maneuver large logs across wide openings. Thus, bridge spans were not to improve much over those available from maneuverable natural timber until the coming of the truss and the arch. These new forms were to require considerable innovation. Therefore, it is not surprising that manufactured bridges were a millennia or so later in occurring than were the manufactured ways discussed in earlier chapters. The first records of significant manufactured bridges are from Egypt in about 2600 B.C. (Shirley Smith 1953; Wittfoht 1984).

The first practical extension of the beam bridge would have taken advantage of natural stone islands within a river to place a series of beams from one island to the next and hence cross the entire river with a multispan bridge. A bridge of this type over the Karkheh River existed at ancient Susa (see fig. 3.1).

Once this technique was seen to be successful, the next step would have been to build manufactured islands or columns of stone or wood within the waterway of a river. These columns are called *piers*. Their use increased the total distance bridged without requiring heavier beams or greater spanning capacity than had been used for the narrower crossings. Although pier building can still be quite demanding, it requires a lesser advance in technology than does the production of beams of twice the existing spanning capacity.

The bridge structure above a pier is called the *superstructure*. The *span* is the distance a bridge crosses between adjacent piers. The size of the span indicates the degree of technical skill needed in producing the superstructure. Of course, if long spans can be constructed, one can avoid the very costly and time-consuming task of building and maintaining strong and durable piers in the course of a free-flowing river.

In about 1100 B.C. the Chinese built the world's first major bridge, the Ju Qiao Bridge at Quzhou (Tang 1987).[3] The first major pier and beam bridge outside of China was probably the bridge Nabopolassar and his son, Nebuchadnezzar, built over the Euphrates in Babylon in about 605 B.C.[4] The bridge was 10 m wide with an overall length of close to 900 m, although the main structure comprised eight 14 m spans built of beams cut from the trunks of palm trees. The piers were built of stone, and remnants of seven still exist (Casson 1974). The bridge was much admired by Herodotus, who particularly commented on how the spans could be lifted up like a drawbridge. The bridge and the associated processional road described in chapter 3 both served the temple of the god Marduk.

An alternative approach to using piers was employed in a pontoon bridge in which the beams are supported on anchored boats. This method was first used in China over the Wei He River in 1200 B.C. and in Babylon around 800 B.C. The Chinese bridge was built for the wedding of the founder of the Zhou dynasty. The idea is said to have arisen from observations of ferry operations (Tang 1987). China has a long history of floating bridges over its wide rivers. Further west, before 600 B.C. the warlike Assyrians were using inflated animal skins to construct pontoon bridges to aid their military conquests, and the first recorded bridge builder in the West, Harpales of Samos, built a bridge using over 650 pontoons to permit Darius I and his son, Xerxes I, to cross the 900 m wide Bosphorus to attack the Greeks in about 500 B.C. (Wittfoht 1984).[5] Indeed, the pontoon bridge was the backbone of military bridge building for three millennia.

Deep, narrow valleys were sometimes crossed using cribwork, in which the whole valley was filled with layers of logs, each layer placed at right angles to the layer below. Each log was notched to sit securely on the one below, and water flowed between the logs. This was commonly called the American bridge.

Suspension Bridges

The common structural alternatives to beam action were the rope, which works in tension, and the arch, which works in compression. Rope was used for bridge building at an early stage because rope technology was important for many reasons other than bridge building and so was widely available to early societies.

Vines, rattan, oxhide, cane, and bamboo provided the first ropes with the tensile potential to span much larger distances than the existing natural beams. Vines and rattan could be used directly without any further processing because they grow up to 250 m in length. However, manufactured ropes allowed consistent quality to be achieved and commonly available material to be used. The Chinese used bamboo extensively for rope making. The bamboo was allowed to rot until only the cellulose fibers remained. These were then woven into ropes, which had about half the tensile strength of mild steel. The first recorded bamboo suspension bridge was built in 285 B.C. by the prince of Qin in Sichuan.

Thus, vegetable fibers were undoubtedly the basis for the first suspension bridges, with the rope simply suspended or draped from one support to the next. The traveler would cross by perilously hanging, sliding, being pulled, or climbing along the rope (see Deloche 1973). One writer included a picture of a bridge formed from a chain of monkeys and noted with a straight face that in 1916 such bridges could still be observed in South American jungles (Waddell 1916).

The structural attractiveness of the suspension bridge stems from the fact that ductile material is more effectively used in tension than in bending (Lay 1982). Hence, much lighter structures are possible with the suspension technique. Not surprisingly, suspension bridges have occurred in many parts of the world, particularly in South America, India, Southeast Asia, equatorial Africa, and China. Indeed, a suspension crossing was often a sine qua non for travel in precipitous regions.

The combination of new material and new structural form led to spans of up to 75 m, compared with the 20 m or so achieved with wooden beams. Major suspension bridges existed in China from before the time of Christ, and in about A.D. 700 there are stories of 200 mm diameter ropes being used to produce spans of up to 180 m at Tao Guan in Sichuan. However, Needham (1971) suggests that the bridge referred to may well have been a multispan crossing and casts doubts about reports of spans over 60 m. The extant An Lan Bridge at Guan Xian in Sichuan Province used bamboo ropes with a maximum span of 52 m before recent refurbishment with steel-wire rope (Tang 1987). Up to A.D. 1420, Chinese suspension bridges were of uncomplicated design using either a trolley or sling suspended from the rope and pulled along by a dragline, or a deck laid directly on the catenary (curve) of the suspension rope. Such bridges could not take wheeled vehicles and petrified most animals. They also swayed violently under any load and were rarely crossed in a strong wind.

The Inca's famous San Luis Rey rope bridge spanned about 50 m over a 100 m deep gorge on the wild Apurimac River at Aecha in Peru. Built about 1450, it survived until 1877 and was made famous in Thornton Wilder's novel of the same name. Inca suspension bridges were usually of hand-spun rope about 300 mm in diameter, with four ropes for the walkway and two for handrails. The Inca administration had a specific official, the Chaca Suyoyoc, whose job was to supervise all suspension bridges throughout the empire. Constant supervision was essential as many of the ropes had a life of only one to two years (Hyslop 1984). In Europe, it was common for major bridges to each be under the supervision of a bridge master (Home 1931).

Gradually, the East developed structural suspension systems that were more elaborate than the hanging rope (Steinman and Watson 1941). There are some questionable records of iron chain cables being used prior to the thirteenth century.[6] Iron ore was very common in Yunnan, and iron making advanced more rapidly in China (and in India) than it did in the West. Iron suspension bridges were in use in northern India by A.D. 400, and the Chinese technology was undoubtedly derived from there.

The change from fiber rope to iron chains had two major advantages. First, and most important, maintenance was dramatically reduced, and, second, the spans that could be crossed were somewhat increased.

The first recorded Tibetan builder of iron chain bridges was the monk Thangtong rGypal-Po.[7] In 1420 he significantly moved the technology forward from the difficult-to-use curved catenary deck of the Chinese system, when he built a pedestrian bridge with a flat deck suspended from the catenary by vertical hangers. Towers were needed at each end to raise the deck to an accessible level. The bridge crossed the Brahmaputra River near the Chak-sam-cho-ri lamasery at Phuntsogling in Tibet.[8] It was still in use in 1878 and was photographed in 1900, but was subsequently put out of service by a flood (Needham 1971). Thangtong built fifty-eight of these bridges to aid the movement of pilgrims (Tang 1987). He became a great lama and legendary figure in Tibet, where he also initiated literary reform and created Zang Opera.

Between 1465 and 1486 Liao Ran built an iron-chain suspension bridge over the Lan Cang River near Jinghong in Yunnan, China. Called the Ji Hong Bridge, it was repaired in 1545, described by traveler Xu Xiake in 1646, and finally replaced in 1846.

Iron chains were first used in Europe in 1237 to suspend galleries used as roads from the cliff face and for segments of the Stiebende Bridge. The work was part of a new route built through the Saint Gotthard Pass. Many of the so-called iron chains used in these early European bridges were actually formed from links made of pairs of timber slats joined together at each end by iron pins.

A further Chinese contribution was the provision of roads in precipitous regions by the use of cliff-face gallery roads. The four most famous were in Shaanxi Province, and work on the most notorious of these—the Shaanxi to Sichuan cliff road—began in the Warring States period in 475–221 B.C. (Needham 1971; Jiang 1986). None remain today. Gallery roads were either carved into the vertical cliff face or used timber ledges carried on poles cantilevered from holes drilled in the cliff face and additionally supported from the cliff face by struts or chains. In the most difficult areas the cantilever poles merely carried ladders or, at the extreme, only the holes were supplied, and intrepid travelers—with some dexterity—moved their own set of four poles from one hole to the next (Tang 1987).

Tunnels

The first traffic tunnel was a 900 m footway built under the Euphrates River in 2150 B.C. It ran from the royal palace of Queen Semiramis in Babylon to the Temple of Jupiter (Gies 1962).

The best known Roman road tunnel is the Grotta, which passes through the ridge of the Posilipo Peninsula between Naples and its suburbs of Bagnoli and the old Roman resort town of Baia. Built in about 30 B.C. to relieve pedestrian congestion, the kilometer-long tunnel has been widened over the millennia until it is now 6.5 m in diameter (Gies 1962). It was built, not by hammer and chisel, but by lighting a fire to heat the rock face and then splashing the face with cold water to cause the rock to fracture. The same technique was used by Chinese road builders of the time (Needham 1971).

In 1480 the ruler of Saluzzo in Italy built the first major solid-rock road tunnel since Roman times when he initiated the use of explosives to blast a tunnel 100 m long below a high pass in order to facilitate the export of salt. Two years later, the Italians were to rue their work when the French used the tunnel to invade Italy.

Between 1825 and 1843 the Brunels, with many trials and tribulations, built the first major soft-ground transport tunnel since Queen Semiramis and Babylon.[9] The 460 m tunnel passed under the Thames in London at Wapping, a kilometer downstream from the Tower Bridge. For eighty years it had the largest cross section of any soft-bore tunnel and was the first to use a protective shield to aid progress of construction. Although intended as a road tunnel, no approach roads were built and so it mainly served pedestrians until taken over by the London Underground in 1865 (Gies 1962).

After a hiatus of many centuries, a number of short road tunnels were built in mountainous regions around the world in the nineteenth century. The first in the United States was constructed in the 1870s, 10 km south of San Francisco. An urban road tunnel was completed in Los Angeles in 1901. The first major urban road tunnel was the Holland Tunnel under the Hudson River between New York City and New Jersey. Built between 1920 and 1927, it was named after its principal engineer, who died in 1924 before the tunnel was completed (Gies 1962).

The world's longest road tunnel is now the 16.9 km Saint Gotthard Tunnel on the N2 Highway in Switzerland. It was built in 1980.

The Arch

Natural arch bridges are formed when relatively soft material erodes, leaving in place a rocky span. A famous natural arch bridge is the Pont d'Arc in Ardèche in southern France, which has a span of 58 m and a rise of 35 m. In the United States, the Arches National Park in Utah contains some two hundred natural arches; the largest is Landscape Arch, which has an enormous span of 89 m and a rise of 32 m. Other famous natural arches are in Lexington, Virginia, and the now collapsed London Bridge in Victoria, Australia.

The first manufactured arches were created when inventive communities extended the steppingstone by carefully stacking a larger horizontal stone on top of a smaller one. The stone piers so produced thus became progressively wider, each layer of stone slabs cantilevering a little further out than the previous layer. Finally, adjacent piers were connected by a horizontal stone slab or wooden beam known as a *corbel,* and the resultant bridge was called a *corbel arch.* Such bridges could span up to 10 m, with unsubstantiated claims for 30 m spans (fig. 8.1a; Waddell 1916). A brick masonry corbel arch was used to cross the Euphrates at Babylon in about 700 B.C. (Davidson 1961).

The Semicircular Arch

Greater spans than could be achieved with the corbel arch required the use of the more subtle keystone arch. The idea for this structural form possibly arose from the use of two stone slabs leaning against each other to produce the pointed arch (fig. 8.1b). An advanced variant is the ribbed arch formed of two long ribs leaning on each other.

Keystone arches are composed of a sequence of stones depending on each other for mutual support. The stones, which are known as *voussoirs,* are each shaped into a taper to fit into the arch geometry, as shown in figure 8.1c. *Voussoir* means "element of an arch" (*voute*) in French. The word comes from the Latin *volvere,* which relates to the voussoir's wedged shape. The arch finally depends on the keystone at its top, which is—traditionally and practically—the last stone to be inserted. Indeed, during construction the arch must be supported on falsework until the keystone is in place. It thus requires its builders to possess administrative and conceptual foresight of a significant order. A Roman practice was to require the builders to stand under the bridge when the falsework was removed and the arch—hopefully—stood proud and unsupported.

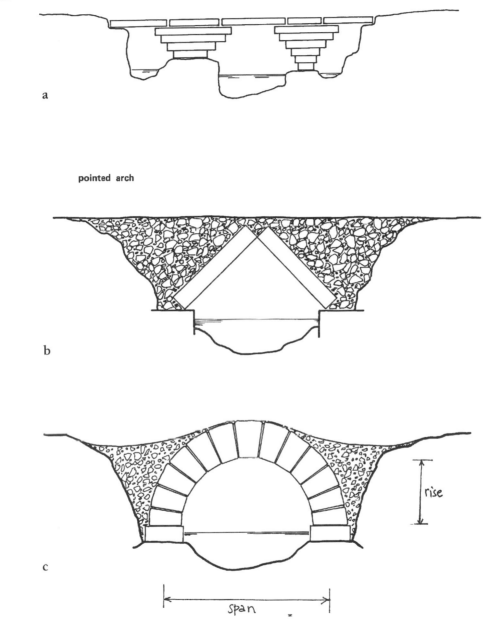

corbell arch

a

pointed arch

b

c

span

rise

Figure 8.1. Arch types: *a,* corbel arch; *b,* leaning slab, or pointed, arch; *c,* keystone or voussoir arch. The form shown is known as the muleback because the roadway follows the upper profile of the ring of the arch. A well-known partial muleback bridge is shown in fig. 8.2.

Figure 8.2. The Old Bridge (or Suleyman, Great, Twisted, or Roman bridge) over the Neretva River at Mostar in Yugoslavia. *Mostar* means "old bridge" in Serbo-Croatian. The 29 m span is an example of a partial muleback bridge. It was built in 1566 by the Turk Mimar Hayruddin (or Hajrudin or Khair-ad-din), a student of the famous Turkish bridge builder Kodza Sinan. *1969 photo by and with the permission of M. Verona.*

The arch works in service as long as it can maintain its compact, shoulder-to-shoulder geometry and so requires good, unmoving foundations. It is a natural form of construction wherever there is also good stone and deep, fast-flowing rivers. The first keystone arches were of simple semicircular shape, a form that is sometimes known as a radial arch or, in architecture, as a barrel vault.

The first keystone arch may have been built in a drain under a building in the Middle East in about 4000 B.C., perhaps initially with "permanent" timber falsework. However, the voussoir arch was relatively rare, and its common use dates from some Assyrian buildings built in 700 B.C. (Van Beek 1987). It was a valuable technique in a land where timber was in short supply. Stone was also unavailable in many parts of the Middle East, and construction was often of relatively low strength burnt brick.

The technology traveled from the Middle East to the Greeks and Etruscans and thence to the Romans. As with roads, the Romans took and expanded that knowledge to raise bridge building to a noble art (see Hopkins 1970 for a fine overview). Guilds of engineers and craftsmen, based partly on the army, ensured that the technology was disseminated and standards maintained. The work was also partly recorded by Vitruvius in his *De Architectura,* published in about A.D. 27 (Vitruvius 1960) but written in the first century B.C., before the work of such masters as Agrippa. Vitruvius recorded that the joints between the voussoirs should be on the radii of the semicircle, "their joints directed to a center."

The first recorded keystone arch bridge was the Pons Solares (or Solarius), which was built by the Romans across the Trevarone River in Italy in the seventh century B.C. The Romans soon made the semicircular keystone arch a major bridge-building form, although their technique was massive rather than adventurous. Their strength was as much in construction management, organization, and foresight as in engineering subtlety.

The Romans more publicly enter the bridge story with their Pons Sublicius—or Sublicus—(bridge of piles), a timber-beam bridge over the Tiber at Rome. It was built by Ancus Martius late in the seventh century B.C. Horatius held the bridge against the Etruscans in 598 B.C., to his everlasting fame, as recorded by Macaulay in his *Lays of Ancient Rome:*

> In yon straight path a thousand
> May well be stopped by three.
> Now who will stand on either hand,
> And keep the bridge with me?

The bridge was replaced with a stone arch in A.D. 350.

The two keys to building an arch bridge are its foundations and construction falsework. In difficult foundation conditions the Romans used both piles and coffer-dams to provide the necessary sound foundation for the arch. The piles were usually wooden logs driven vertically into and through soft ground to provide a firm, non-settling foundation. A cofferdam is a self-contained dam built in the river, from which the water is then emptied to permit workers to prepare a sound foundation in the riverbed. Building foundations for arch bridges is therefore a difficult technique that requires considerable skill and organization. Indeed, many of the foundations the Romans constructed in free-flowing rivers represented major civil engineering achievements. For building the falsework, the Romans often supported the timbers on stones protruding from the piers; these can still be seen in many of the extant bridges.

The Romans gave the position of chief bridge builder to a high priest known as Pontifex. The tradition began with the priest placed in charge of Horatius's Pons Sublicius and continues to the present day. One of the pope's titles is Pontifex Maximus or Pontiff.

For prestigious sites the upper surface of the arch was infilled with stone to allow the roadway surface to be brought to road level. Elsewhere the Romans used the muleback arch, with the roadway gradient following the semicircular curve of the arch, as illustrated in figure 8.1c.

A surprising number of the Roman bridges are still standing, although they have been continuously repaired over the centuries. Notable among these is the 37 m span Martorelli pointed-arch bridge in Spain, which was built in 219 B.C. and has the largest span of all the extant Roman arches. It is commonly called the Devil's Bridge.

"Devil's Bridge" was a common name for old bridges. Legend was that the devil had assisted in the construction of the bridge in return for the soul of the bridge builder—or of the first living creature to cross the bridge. In the latter version, the devil was often thwarted by the cunning locals, who ensured that an old dog was the first creature to cross the new bridge. A related belief was that a bridge would not be successful unless a living person was entombed within it during construction.

In Chinese lore, the devils were always in such a hurry to catch their prey that they could not turn round corners. Hence, zigzag bridges were considered devilproof. Perhaps the most famous Devil's Bridge was a perilous masonry arch across the Schollenen Gorge in the Saint Gotthard Pass. The bridge was destroyed during the 1799 campaign of the French revolutionary wars.

The following Roman arch bridges were built over the Tiber in classical Rome:

1. Pons Aenilianus (or Aemilius, Senatorius, Lapidius, Lepidi, or Palatinus), which was completed in 142 B.C. Aemilius Lepidus was the censor of the time. The remains are known as the Ponte Rotto (or Broken Bridge).
2. Pons Mulvius (or Ponte Milvio, or Molle), which was completed in 109 B.C., rebuilt in the fifteenth century, and restored in 1808 using four of the original arches. The bridge then successfully carried the traffic of three armies during World War II.
3. Pons Fabricius (or Ponte Fabricio), which was completed in 62 B.C., with two 24 m spans over an arm of the Tiber to a midstream island called Isola Tiburnia. It is still in service today as the Ponte a Quattro Capi and is the most complete of all the old Roman bridges. There is some irony in its longevity: the Romans paid the original builder over a forty-year period to ensure that he built a bridge that would last. Lucius Fabricius was the roads commissioner of the time, whereas the current name refers to the bridge's four-headed statue of Janus, the god of entrances.
4. Pons Cestius (or Ponte San Bartolommeo), which was completed in 43 B.C. to span the other arm of the Tiber to Pons Fabricius. Cestius was the Roman governor of the time. The bridge had a single 23 m span.
5. Pons Sublicius, discussed above.
6. Pons Aelius (or Ponte Sant'Angelo or Hadrian's Bridge), which was built by Hadrian in A.D. 134 and leads to his mausoleum, which is now the Castel Sant'Angelo. This bridge was named after Emperor Aelius. The bridge is the author's favorite structure and was described by Leonhardt—arguably this century's greatest bridge designer—in 1982 as the finest example of the Roman arch bridge. It has seven arches, the greatest of which has a span of 20 m. The waterway piers sit on the original foundations, which were built using the cofferdam technique. It was probably preceded at the site by, or was very close to, the Pons Neronianus. It was renamed Ponte Sant'Angelo in A.D. 600 when Pope Gregory I had a vision on the bridge that signaled the end of a plague then afflicting Rome. The statues of pairs of angels on top of the parapet over each pier were added in 1688 at the suggestion of the famous baroque architect and sculptor Giovanni Bernini.
7. Pons Probi, which was the southernmost of all the bridges of Rome.
8. Pons Aurelius (or Agrippae or the Janiculine Bridge), which was replaced by the Ponte Sisto in the fifteenth century, although the original Roman foundations were retained in the new bridge.

Note that there is some confusion over the names and locations of some of these bridges.

One of the most famous extant Roman bridges is the Puente Alcántara in Spain, built in A.D. 98 with 30 m arches towering 58 m over the Tagus River. The name is something of a mixup: *puente* is "bridge" in Spanish and *alcántara* is "bridge" in Arabic. The bridge is also called Puente Trajan-Alcántara. Trajan was the Roman emperor during its construction. The Alcántara was built by Caius Lacer, whose words are carved in a chapel beside the bridge: "Lacer, famous for his great skill, built this mighty bridge to last forever." Another famous Roman bridge is Agrippa's imposing Pont du Gard aqueduct at Nîmes in southeastern France, built in 19 B.C. and crossing a total of 265 m at a peak height of 46 m. These two bridges used fitted voussoir stones weighing up to 8 tonne each for the lower tiers and mortared joints for the higher tiers.

The greatest span achieved with a semicircular arch was possibly close to 50 m, although confirmed spans do not exceed 37 m (O'Connor 1975). Another Roman contribution was the timber truss-arch in which a timber truss forms the ring of the arch. Although it does not use the shoulder-to-shoulder concept of the masonry arch, it does use the idea of the arch ring as a compression member firmly held between two unyielding foundations. This structural form is discussed further when trusses are examined later in the chapter.

After the Romans

Chinese arch bridge construction began around A.D. 100 and probably borrowed from the West. The first Chinese traveler to the West was Zhang Qian, who penetrated as far as Balkh in about 100 B.C. (see fig. 3.2; Fugl-Meyer 1937). Chinese arches reached a pinnacle of elegance and structural form in the Sui dynasty (ca. A.D. 600); Li Chun produced major bridges that were about a millennia in advance of Western practice. A bridge design school he founded was active for some centuries. Li Chun's Zhaozhou Bridge, with its 37 m span over the Jiaohe River at Zhaozian in Hebei, was built in A.D. 590 and remains in use today as remarkable evidence of Chinese aesthetic and structural skills (fig. 8.3). The shape of the Zhaozhou Bridge is a less-than-semicircular segment of a circle and its 5:1 span-to-rise ratio was not to be bettered in Europe for 750 years. The name Zhaozhou relates to the location

Figure 8.3. The Zhaozhou or An Ji Bridge, a Chinese bridge of the Sui dynasty (ca. 600 A.D.) in Hebei Province. *Photo by the author in 1989.*

of the bridge at Zhaozian. It is also known as the An Ji Bridge and locally as the Dashiqiao—or Great Stone—Bridge. Li Chun also built a smaller but geometrically identical bridge, which is still standing, a few kilometers away.

The voussoir stones in the arch of the Zhaozhou Bridge are about a meter square and are surmounted by a second ring of 1,000 mm by 300 mm stones. A feature of the design is the use of pairs of bow-shaped iron connectors to key adjacent stones together and thus make the flat arch reassuringly possible. The eight bows attached to each stone are about 50 mm deep. The earlier Roman bridge builders had believed that the use of iron would offend the river gods, but later Roman bridge builders did use this technique to keep critical voussoir blocks together. Twenty-eight arch ribs are used to support each span of the 9.6 m wide bridge. In a second innovative and farsighted use of iron in this remarkable bridge, the arches are held together by five transverse iron ties.

Open arches at each end of the bridge both reduce its mass and permit easier passage of floodwaters. Another remarkable feature of the bridge is that it is founded on potentially deformable sandy clay. The foundation is apparently in the form of a raft 5 m long and 8 m deep. The potential for the iron bows to resist tension in the arch may have made possible the longevity of an arch bridge on such less than perfect foundations, as any appreciable movement over the last fourteen hundred years would have caused its collapse.

The Zhaozhou Bridge was well restored in 1955–1958 and has only carried light traffic since restoration, due more to a downstream bypass and the incorporation of the bridge into a park dedicated to Li Chun, than to any inadequacy of the structure. An admirer wrote: "The bridge is like a rainbow in the sky, or lightning flashing through a cloud. . . . It seems to fight with the banks and struggle against heaven" (Tang 1987). Others have described it as like a crescent moon rising from the clouds.

The legend of the bridge is that it was built in one night by Lu Ban, the founder of Chinese stonemasons. The feat alarmed two of the immortals, who tried to destroy it with an overloaded donkey and wheelbarrow. Lu Ban successfully held the bridge up with his hands, but the gouges of the donkey and wheelbarrow are still visible in the stone deck of the bridge.

In A.D. 1154 the Chinese also took the record for the longest length of bridge with the completion of the Su Di Bridge at Tian Teng, near Quanzhou. Of stone pier and timber beam construction, it used 23 longer spans and 140 smaller ones to produce a total length of some 6 km (Tang 1987).

In the Western World, the art of bridge building died with the Roman Empire. The semicircular arch virtually disappeared and was replaced by the less demanding *ogive,* or pointed, arch typical of Gothic architecture. Technically, the pointed arch is a reversion to the earlier leaning slabs (fig. 8.1b), which preceded the development of the semicircular arch. Because it exaggerated the muleback tendency, it was less appropriate than the semicircular arch for bridge building. The arch skills that were maintained were in the hands of the Moors and the Turks at the periphery of the old empire.

As bridges were often more essential to travel than were roads, some priority was given to their maintenance in the post-Roman era. To aid this task a group of churchmen at the Hospice of St. James near Lucca in Italy began an informal brotherhood of bridge builders (*fratres pontis*) to build and maintain bridges for pilgrims.

Legend has it that their emblem was the Greek letter tau made to look like an auger and an ax (Watson and Watson 1937). The brotherhood, which attempted to copy the Roman style of bridge building without the support of technical understanding, a slave labor force, or a knowledge of cement, was not successful (Rose 1952a).

The equivalent French groups were sometimes known as the Frères Pontifes or Frères du Pont (brothers of the bridge). They did not belong to an established order, but operated as independent fund-raising groups around Avignon, Lyon, and Pont-St.-Esprit. More legend than fact surrounds their activities, and their impact was probably both minor and localized (Grattesat 1982). One of the legends is that the brothers were distinguished by a white costume with a red badge representing two bridge arches under a cross.

In twelfth-century England, the clerical bridge-building role was frequently linked to the Benedictines. Chapter 3 discussed a guild devoted to road maintenance, and chapter 4 described the Scandinavian commemorative practice incorporated in the Salna Stone. Chinese Taoists and Buddhists had a similar tradition, stressing the spiritual benefits of bridge building and roadmaking (Needham 1971). This spiritual tradition is thought to have stemmed from the Zoroastrians, who believed that bridge building was one way for sinners to expiate an evil deed.

London was located at the lowest ford across the Thames, at the current site of London Bridge. Evidence of Roman bridgework has been found at the site, and in about A.D. 980 the Saxons built a wooden bridge there. In 1010 a Viking fleet, commanded by Olaf Haraldsson, pulled the bridge down to prevent the fleet's progress up the Thames from being hampered by English soldiers located on the deck of the bridge (Gramsborg 1986).[10] This act gave rise to the nursery rhyme, the common version of which is:

> London Bridge is falling down, falling down, falling down,
> London Bridge is falling down, My Fair Lady.

An earlier (1760) version of the rhyme, which could also have referred to subsequent fire, water, and ice damage, reads:

> London Bridge is broken down
> Dance o'er my Lady Lee.

Or, from Otto Svarte's poem "Heimskringla" in the Olaf Sagas (Watson and Watson 1937);

> London Bridge is broken down
> Gold is won and bright renown.

The bridge was rebuilt but, in 1016, again proved an obstacle, this time to King Canute of Denmark. Rather than pull it down, he dug a canal around one end and simply sailed past it. A subsequent bridge was destroyed by flood in 1091.

In 1176 the Normans started building a masonry bridge of 360 m total length across the Thames, under the guidance of Peter, the chaplain of St. Mary Colechurch, a parish church in Grocery Alley (Home 1931). Peter was familiar with the conditions because from 1163 he had supervised the construction of a timber bridge on the site. He received technical help from Isembert, a successful French bridge builder from the town of Saintes, where there was a famous extant Roman arch bridge—England, on the other hand, had no extant Roman masonry arches. King John had brought Isembert to England because he was impressed by French bridges. The bridge was 6 m wide and was supported by nineteen pointed masonry arches and a drawbridge, each spanning between 5 and 10 m. The piers were built by the cofferdam method and reduced the waterway by 75 percent. The work was partly funded by a tax on wool and by major contributions by the archbishop of Canterbury and the papal legate. Work on the bridge was completed in 1209, three years after Peter's death and with the loss of 150 lives during construction.

It was Britain's first major masonry bridge. The superstructure carried many timber buildings as well as providing a roadway. Indeed, the 5 m wide bridge carriageway operated as a narrow and cluttered street of wooden shops, inns, and houses, which were not removed until 1761 when the bridge was also considerably widened. Many lives were lost due to the old narrow carriageway and its lack of sidewalks (Home 1931).

London Bridge was severely damaged by fire in 1212 and by ice in 1282 and was partly rebuilt in 1396. Indeed, four great fires swept the bridge, which nevertheless lasted six hundred years until it was demolished in 1832. A prime reason for the demolition was that the wide and frequent piers created a major restriction in the waterway, causing fast river currents, making boat passage hazardous, and creating high maintenance costs as the water scoured the foundations. Boat passage was called "shooting the bridge" and claimed thousands of lives.

The bridge was the only London crossing over the Thames until 1750, when Westminster Bridge was opened. Tolls were collected from 1282 on (its financial success as a toll bridge was discussed in chapter 4; see Syme 1952). The bridge was so important to Londoners that it became common to leave money to "God and the bridge." When London Bridge was demolished in 1832, the greater river flows and tidal movements scoured the foundations of Westminster Bridge, which soon had itself to be demolished despite the novel and ingenious work that designer Charles Labelye had put into those foundations.

Bridges of the time were structures of great esteem and importance, and it was common for them to be dedicated to saints and to contain bridgehead shrines. On a midriver pier, London Bridge had a beautiful chapel, which Peter of Colechurch dedicated to Thomas à Becket. Peter was buried in the chapel, which survived until 1553 when a poor posthumous view of Thomas (who had been made a martyr saint in 1172) and his actions led to his chapel being demolished by the new Protestant powers (Home 1931; Watson and Watson 1937). Between 1304 and 1678, the southern half of the bridge included the Traitors, or Drawbridge, Gate, above which were placed the staked heads of citizens executed for treason. In 1598 some fifty heads were on simultaneous display. The first such "traitor" was the Scottish resistance leader William Wallace and the last was a Catholic banker, William Staley. Sir Thomas More and Guy Fawkes also graced the gate in this manner.

One of the few even partially extant bridges built in the millennium after the Romans is the Pont St. Bénézet over the Rhône River at Avignon in southern France. It was built between 1177 and 1187 by the former shepherd boy Bénézet, who was said to be under divine inspiration (Davison 1961). He had an association with the Frères du Pont and was probably the only bridge builder to be canonized, although for ethereal rather than masonic miracles. The work gained Bénézet his sainthood posthumously; he died before it was completed and was buried in its chapel (Upton 1975). He is now the patron saint of bridge builders.

The initial structure of Avignon consisted of stone piers and a timber deck. The three centered stone arches were constructed progressively from the second quarter of the thirteenth century (Grattesat 1982). They were approximately semi-circular and were possibly inspired by the Pont du Gard some 30 km away. At its peak the Pont St. Bénézet had twenty arches, each spanning about 30 m and with a maximum span of 34 m.

The bridge was neither well designed or well constructed. It lost its first arch in 1680, and only four, built by Pope Clement VI between 1348 and 1352, subsequently survived the ravages of the Rhône. For much of the bridge's working life, its administration was strained—the king of France controlled the right bank of the river, and the pope controlled the left bank. The bridge's total length of about 900 m makes it the longest masonry arch bridge ever built.

The bridge is the famous Pont d'Avignon of folk song:

On the bridge at Avignon,	Sur le pont d'Avignon,
Here one dances, here one dances.	l'on y danse: l'on y danse.

The song originally began "sous le pont" and referred to dancing on the riverside verges beneath the arches of the bridge. The chapel at Avignon is over one of the piers and is dedicated to Saint Nicholas. It also served as a poorhouse and hospital for travelers.

In 1189, completing a remarkable trio of long, masonry arch bridges built around the world over twenty-five short years, the Chinese constructed the Lu Gou (or Marco Polo) Bridge over the Yongding River in Beijing. It has eleven semicircular masonry arches with clear spans of between 12.3 and 13.4 m and crosses a total of 266 m. An awestruck Marco Polo brought news of the bridge to the West. The restored bridge is still standing and gained new fame in 1937 as the site of a key early incident in Japan's war of aggression against the Chinese.

Span-to-Rise Improvements

Given the difficulty of bridge building and the relative ease with which fords could be constructed, the major bridge construction challenges naturally arose on unfordable rivers. Such bridges were usually enormously expensive, and the semi-circular arch, with its need for frequent piers, was often unable to meet the challenges posed by major crossings. Once such a "great" bridge was established at one of these sites, it became a foci for all the roads of the region (Belloc 1923). It was thus a major military and administrative feature and was usually built with fortified towers and a narrow carriageway to ease the task of defenders.

The first European arch that was flatter, and thus more able to provide clear crossings, than the semicircular Roman structures, was the Ponte Vecchio (or Old Bridge) designed by Taddeo Gaddi and built over the Arno in Florence in 1345.[11] Located on the site of earlier Roman and medieval bridges, it remains in use today (fig. 8.4). Its three arches of approximately 29 m span are each made up of circular segments, which together form much less than a semicircle. The resulting

Figure 8.4. The Ponte Vecchio. *Photo with the permission of the Science Museum, London.*

"flat" shape avoided the need to build additional piers in the fast-flowing Arno River but placed more reliance on the piers and abutments to resist outward-acting horizontal forces. The bridge therefore required greater constructional skills than the semicircular arch. The task was made simpler at the Ponte Vecchio site by the presence of solid foundations on either bank of the river. The design concept was closely linked to the like development of the flying buttress in medieval cathedral construction.

The span-to-rise ratio (see fig. 8.1c) is a measure of the technical difficulty of an arch bridge design. For the semicircular arch the ratio is, of course, two. The Ponte Vecchio's span-to-rise ratio of 7.5 surpassed the record that had been held by Zhaozhou Bridge for 750 years and was not to be bettered for a further 400 years.

The largest masonry arch built in this period was a 72 m span bridge over the Adda River at Trezzo in Italy, which was described as "the boldest bridge of the Middle Ages" (Straub 1952). It had a span-to-rise ratio of just over three. The bridge was built by Bernabe Visconti in 1377 and was destroyed in 1416 during a local war. Its span was not surpassed until the 85 m Pont Adolphe was built by Paul Séjourné over the Petrusse River in Luxemburg in 1903. Arch spans are limited by the need to support the arch by independent falsework until the keystone is dropped into place. The record-breaking Pont Adolphe span was made possible by using newly developed self-supporting falsework to carry the incomplete arch. The Adolphe Bridge held the record for only two years, then the record passed to the current holder, a 90 m span at Plauen in Germany.

After the Adda Bridge, the next major step in Western bridge design occurred when Bartolommeo Ammanati—a well-known Florentine sculptor and architect—designed the three-span Ponte della Santa Trinita over the Arno in Florence, just downstream from Ponte Vecchio. He introduced the elliptical, or basket-handle, arch, in which the arch profile approximates half an ellipse (fig. 8.5). This new geometry permitted relatively high span-to-rise ratios. When completed in 1569 with a 29 m maximum span and a span-to-rise ratio of seven, it was a radical departure from the then-current high, pointed Gothic arches with their relatively small spans, large piers, and steeply sloping roadways.

Apart from An Ji and Ponte Vecchio, previous span-to-rise ratios had not bettered four to one. Indeed, the high span-to-rise ratios of these two bridges over the Arno were considered excessive by subsequent builders and constricted by the Counter-Reformation movement and so were rarely copied. Local story is that vehicle drivers feared to use the bridge, given its extreme flatness. They were only convinced of its safety when Napoleon took heavy artillery over it during his invasion of Italy some two hundred years after its construction.

Span-to-rise ratios of eighteen to one are feasible, and so the completed works of the time with their seven to one maximum had not pushed the technology to its limits. The Pont d'Alexandre III over the Seine in Paris has a span-to-rise ratio of seventeen.

The actual arch shapes used by Ammanati were not his own design but were copied from Michelangelo's sarcophagi on the nearby Medici tombs constructed some thirty years earlier (Hopkins 1970). A kink where the two halves of the arch meet is hidden by a stone escutcheon. The bridge was described by essayist E. V. Lucas in 1914 as "the most perfect union of two river banks imaginable," and a leading contemporary bridge designer, Fritz Leonhardt, called the Ponte della Santa Trinita "possibly the most beautiful bridge of the medieval period." It was destroyed by the Germans in 1944, but has subsequently been restored.

The most critical attributes of the elliptical-arch bridge were to be able to resist the horizontal forces at the piers and abutments with little or no movement, as any spreading would cause the voussoirs to drop out of position, and to avoid having the underside (intrados) of the voussoirs come apart under tension induced by bending of the arch. This tension occurs if the line of thrust of the force from the abutments and the weight of the traffic rises too far above the centerline of the voussoirs. Its occurrence is not difficult to calculate mathematically, but must have required great skill in those pre-nineteenth-century years when no appropriate analytical tools existed.

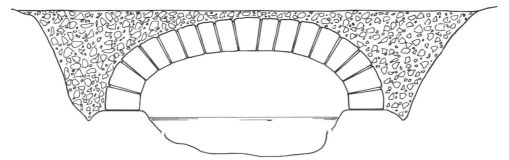

Figure 8.5. Elliptical arch bridge.

In 1673 the Japanese built the four-span Kintai Bridge over the Nishiki River at Iwakuni. It is a muleback bridge composed of timber arches that are less than semicircles. Excepting that one arch has been replaced every five years since 1673, it is the oldest extant timber arch bridge.

A famous story concerns a bridge built by William Edwards over the Taff River at Pontypridd in Wales in 1746. His contract required his triple arch to stand for seven years. It was destroyed in a flood after two years. His replacement bridge failed when the falsework collapsed during construction. The third try failed when the arch was being backfilled. His fourth attempt—a slender 42 m span—remains standing today.

The father of modern bridge building was the Frenchman Jean Perronet (1708–1794), the first director of l'Ecole des ponts et chaussées. His major technical contribution was the understanding that, in multiple-arch bridges, the horizontal thrusts on a pier from adjacent arches canceled each other out, thus permitting a 60 percent or so reduction in the thickness of such piers. It appears from Vitruvius that the Romans were aware of the concept, but there is no evidence of its application before Perronet. As a consequence, Perronet's bridge superstructures were much less massive and permitted much easier passage of water and boats than had earlier bridges. Of course, this balancing technique between the thrusts meant that the arches had to be built symmetrically out from each pier. David Steinman, himself a great bridge builder, wrote of Perronet, "In his hands the masonry arch reached perfection" (Steinman and Watson 1941). Perronet certainly deserved his reputation as the "master of the stone arch bridge" and his work is beautifully recorded (Perronet 1782).

Technically, his best bridge was probably the Pont St. Maxence over the Oise River some 60 km north of Paris on the road to Flanders. It had a twelve to one span-to-rise ratio, thus almost doubling the previous limit. The bridge was destroyed by Napoleon in 1844, repaired, and then destroyed twice by the Germans, once in 1870 and most recently during their retreat from the Marne in 1914 (Upton 1975). It has been described as "the most slender and daring stone arch" (Gies 1962).

Perronet's most beautiful bridge was probably the Neuilly Bridge with its five 37 m span elliptical arches over the Seine in Paris and on the alignment of the Champs-Elysées. Built in 1770–1772 and described as "the most beautiful and graceful stone bridge ever built," it was demolished in 1932 to aid navigation (Gies 1962; Grattesat 1982). Perronet's last and greatest existing bridge is the Pont de la Concorde, also over the Seine in Paris. Its original design was too revolutionary for the French authorities, and Perronet was required to thicken the piers. He died in 1794 at age eighty-seven in a shed at the end of the bridge, from where he was supervising its construction.

Improved falsework to overcome the vexing problem of supporting the arch during construction was developed by Robert Mylne and used on the 1769 Black-friars Bridge in London. The next major breakthrough came from ex-millwright John Rennie, who was influenced by the Parisian bridges to further develop the semi-elliptical arch for London's Waterloo Bridge, opened in 1817 and replaced in 1942. Rennie also used the elliptical arch for the five-span replacement of Colechurch's London Bridge. This bridge was opened in 1831 and demolished in 1968 when its foundations proved inadequate. It was sold to developers in the United States and now spans part of Lake Havasu in Arizona. The new, third London bridge is a three-span concrete structure of unobtrusive proportions.

Chapters 3 and 4 discussed the major contributions that Telford made to pavements and roads. Following his appointment as county surveyor for Shropshire, he also designed a number of splendid masonry arch bridges. His first bridge, a triple arch over the Severn at Montford, was built in 1787. One of his best was the 1828 Over Bridge, a 46 m elliptical arch that crossed the Severn at Gloucester and was based on Perronet's Neuilly Bridge (Heyman and Threlfall 1972). The Over Bridge today carries the A40 towards Wales. However, Telford's major contribution to bridges was probably in iron bridge construction.

In A.D. 1048 the Chinese built an interesting 18 m span scissor-braced timber arch bridge known as the Rainbow at the then capital city of Kaifeng (Tang 1987). French engineers constructed a number of timber arch bridges in the eighteenth century, using layers of timber planks clamped together to achieve the arch. A span of 45 m was achieved when the method was used in the construction of the Pont Louis over the Isar near Freising in Germany.

The next major advance in bridge technology required materials with either some tensile strength or a much higher strength-to-weight ratio than either stone or timber. Concrete and steel met these requirements. The largest concrete arch bridge is currently the 204 m span Usa-Gawa Bridge in Yamaguchi, Japan.

The Metal Arch

The Chinese had used iron for making suspension bridge chains and for ties between voussoirs in their masonry arches. The idea of using iron in bridge building slowly percolated to the West and was used in England early in the eighteenth century. In 1755 the French tried to use iron in a bridge at Lyon, but it proved too expensive and the project was abandoned. In addition to its relatively high cost, iron was only cast in relatively small amounts of uncertain quality, contained various impurities—particularly carbon and slag—and was relatively brittle.

Coalbrookdale, in Shropshire, had a long history of ironworking and was blessed with local supplies of iron ore, low-sulphur coal, limestone, water power, and water transport. Not surprisingly, it became arguably the birthplace of the Industrial Revolution. Abraham Darby arrived in 1708 and subsequently introduced high-volume iron production by substituting coal-based coke for wood-based charcoal in the iron-making process. The first in a sequence of iron-making changes that occurred at Coalbrookdale between 1720 and 1750, the coke permitted cast iron with very low (3 percent) carbon content to be produced. This result was important because, the lower the carbon content, the more malleable and ductile is the iron. The development of iron is pursued later in this chapter.

In 1781 Coalbrookdale, which is on the Severn River, was also the site of the first iron bridge. The 30 m span Ironbridge used five semicircular cast-iron arch ribs with a 300 × 150 mm rectangular cross section. It was designed by Thomas Pritchard, who died before it was completed, and was initially conceived as merely providing falsework for a masonry arch. Described as Britain's best-known industrial monument, Ironbridge is still standing today, although it has been restricted to pedestrian traffic since 1931. The tolls levied on the bridge are shown in figure 4.1.

Although the bridge broke new ground in material use, its design is a copy of a masonry arch with joints made in accordance with mortice procedures based on timber carpentry. No bolts or rivets were used. The two uppermost incomplete semi-circular arches, which have no historical antecedents, also have little structural merit (Ruddock 1979).

The next iron bridge was initially designed by political writer and activist Thomas Paine for erection over the Schuylkill River in Philadelphia. Paine is best known for his book *The Rights of Man,* but his initial training as a corset maker probably better prepared him for structural engineering. His bridge design in 1788 was based on block-shaped cast-iron voussoirs, which were used to form an arch bridge with a twenty to one span-to-rise ratio (fig. 8.6). Whereas Ironbridge copied the jointing methods used for timber bridges, Paine copied masonry jointing practice. A portion of Paine's design was manufactured in England, but then his attention was caught by the French Revolution and his financial backing evaporated. The redesigned 72 m span bridge was erected in Sunderland, England, in 1796 and was known as the Wear Bridge (Smiles 1874; Hamilton 1958). It lasted until 1858.

Also in 1796, Telford erected his first, and England's second, iron bridge at Buildwas, also on the Severn and about 3 km upstream from Ironbridge. Although its 39 m span was 30 percent greater than Ironbridge, its superior design using iron for its own sake meant that, instead of needing more iron, it required only half the amount of iron (Billington 1981). The Buildwas Bridge survived until 1906. Another of Telford's famous bridges was a 51 m cast-iron arch over the Severn at Gloucester. These Telford bridges were flat arches formed from latticed trusses rather than from iron voussoirs. They were thus similar in form to the Roman truss-arches. The central third of the arch directly carried the roadway. It has been said that Telford's use of iron in bridges made him the first to practice the art of structural engineering (Billington 1983). As Billington commented, "His iron arches are more visually attractive than those of his contemporaries, and they were also technically superior."

An incentive for the thoughtful use of iron was that its high price made it essential that it be used as effectively as possible. Billington (1983) notes that, as a consequence, cast iron "literally founded the modern engineering profession by

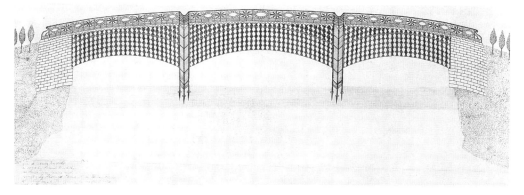

Figure 8.6. Thomas Paine's design for an iron arch bridge. *Drawing from and with the permission of the Science Museum, London.*

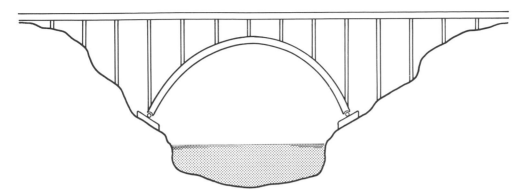

Figure 8.7. Modern two-hinged arch bridge.

forcing a group of designers to think deeply about structures at a new scale." For example, the first cast-iron arch using the structurally efficient tubular cross section for its load-carrying components, rather than voussoir blocks, trusses, or solid bars, was built in 1824 in Braunschweig in Germany. The largest cast-iron bridge was Rennie's 73 m Southwark Bridge in London, which operated from 1819 to 1922 (O'Connor 1975). Rennie was aided in his design task by the brilliant theoretician Thomas Young.

To date all the metal arches discussed copied the keystone approach in which the ends of the arches were geometrically integral parts of the abutments, producing so-called fixed arches. Technically, they were fixed because the ends of the arch could not rotate relative to the abutment. The fixed arch reached its zenith with James Eads's 1874 St. Louis Bridge (to be discussed below), after which two-hinged arches began to take over (fig. 8.7). The two-hinged arch is pinned at each end, so that no rotational effects can be transmitted between structure and foundation. This technique, which required some insight to invent, permits lighter members to be used for the arch and eliminates some of the dependence on keeping abutment movement under strict control. The two-hinged method was epitomized in Gustave Eiffel's iron bridges of the 1870s and 1880s. While relatively simple to employ in iron and steel arches, two-hinged construction was difficult, if not impossible, with masonry arches.

In many bridges (such as the through bridge) it is possible to carry the considerable horizontal abutment forces produced by an arch by using an internal tie spanning between abutments, to produce a tied arch. The abutments no longer need to be able to resist such forces because the tie braces the arch. This efficient form was first used in timber roof trusses in 1600 and was advocated by Faustus Verantius of Venice in his 1617 book.

The longest arch bridge for many years was Othmar Ammann's 511 m steel span Bayonne Bridge across the Kill van Kull in New Jersey, completed in 1931 and a meter or so longer than the Sydney Harbour Bridge in Australia, completed in 1932. Bayonne Bridge held its span record until the 518 m span New River Gorge Bridge was built in West Virginia in 1977. Not coincidentally, the construction of the Bayonne Bridge began after construction had commenced at Sydney Harbour.

Arch Theory

As in many other cases in this book, the theory of arch action lagged behind, rather than led, construction practice. Once the span of a semicircular arch is selected—usually on the basis of local conditions—the only important design parameter to be determined is the depth of the voussoir blocks (i.e., the thickness of the arch ring). The Romans developed simple empirical relationships that permitted the depth to be determined once the span was known (Vitruvius 1960). The relationships were retained through the Dark Ages, "built into the secret books of the masonic lodges," and used frequently until "swept away by the Renaissance" (Heyman 1988).

An arch can become unstable and fail when it somewhat dramatically snaps through, or buckles, from its intended semicircular shape to some new, inappropriate one. Leon Alberti, an archetype of the Renaissance man, had carefully studied ancient engineering, Roman practice, and the work of Vitruvius. His treatise on building called *De re Aedificatoria* (Ten books on architecture) was written in 1452 and published in 1485 (Mesqui 1986). It was the standard Renaissance text on architecture and engineering, and he was known as the Florentine Vitruvius. His book gave the first empirical, geometric rules for preventing arch buckling, requiring the thickness of the voussoir ring to be at least one-tenth of the span (Straub 1952). A number of such rules were to develop over the centuries, long before a theoretical solution becoming available.

In about 1500 Leonardo da Vinci looked at the failure modes of arches and at different arch shapes and came close to producing a correct analysis of arch behavior. He also correctly recognized that the key to the continued functioning of the arch lay in its foundations. Referring to potential foundation movements, da Vinci said, "An arch comprises two weaknesses which both work for its collapse and can be transformed into a strength" (Parsons 1939).

After some two millennia of arch construction, the theoretical analysis of the structural behavior of arch bridges began in 1675 with the work of Robert Hooke, who saw the reciprocal analogy between the arch and the loaded cable; that is, he saw that the same load calculation methods could be applied to the arch as had been developed for loaded cables. In order to avoid losing precedence for his discovery to Isaac Newton or Christopher Wren, his perceived competitors, he published it in 1676 hidden as a Latin anagram at the end of a paper on helioscopes. Translated, the anagram read: "The riddle of the arch. As hangs a flexible cable, so, inverted, stand the touching pieces of an arch" (Heyman 1972). Nevertheless, Hooke cooperated closely with Wren in the design of the dome of St. Paul's Cathedral—a dome is a three-dimensional arch and so provided Hooke with a new challenge.

Hooke had adopted the same ploy in the same article to protect his careful experimental establishment of the principle of elasticity. He hid it in another Latin anagram, which when solved reads: "The extension of a loaded element is proportional to the force applied to it" (*ut tensio sic vis*). In his anagram he simply put all the letters in alphabetical order; thus, "ut tensio sic vis" became "ceiiinosssttuv." He announced its solution some two years after the anagram was first published. Hooke claimed that the anagram was introduced merely "to fill the vacancy of the ensuing pages." Jakob Bernoulli used a similar trick in 1691 when he used a logogram to secretly announce his important analogous discovery that there is a linear relationship between the moment (torque) applied to a beam and the resulting curvature (Heyman 1972).

Thomas Young, a multitalented physician, explained and expanded on Hooke's work in a series of lectures published as a book in 1807. The book, *A Course of Lectures on Natural Philosophy and the Mechanical Arts,* was to have a major impact on structural analysis. With the heightened publicity, Hooke's principle of elasticity subsequently became widely known as Hooke's law.

Needless to say, in the intervening 130 years Hooke's discoveries were little known or applied. Philippe de la Hire reintroduced Hooke's analogy in 1695 in his book *Traité de Mécanique,* assuming that there was no friction between adjoining voussoirs. Two years later, David Gregory expressed the concept in terms similar to the lower-bound theorem, which forms the basis for modern methods of structural analysis (Lay 1982). The understanding was first applied practically by Benard Bélidor and Pierre Couplet in France in 1729 (during Perronet's time). In 1730 Couplet was able to overcome Hire's understandable but unrealistic assumption of zero friction.

In 1739 Charles Coulomb used the arch failure modes established in 1730 by Couplet and in 1732 by Augustin Danyzy.[12] Louis Navier published a comprehensive theory of arches in 1826 in his definitive *Leçons sur l'application de la mécanique.* In 1846 William Barlow used models to help develop a method based on analyzing the shape of the collapsing arch, and in 1854 Jacques Bresse introduced the concepts of bending and stability to arch analysis, in order to allow significant changes in length and curvature to be considered (Kuzmanovic 1977).[13] In the same year, Yvon Villarceau produced a theoretically valid design procedure. Finally, the key method permitting the elastic analysis of all arches was introduced by Karl Culmann in Switzerland in the 1860s.

Work on arch theory has returned to prominence in recent years as it has become necessary to accurately assess the strength of many old masonry arches under modern traffic.

The Truss

Simple timber beams could span up to 7 m, or 10 m with the addition of a little technology and 20 m if large tree trunks could be maneuvered into place. To cross greater distances, it was necessary to either use additional piers or develop a more effective structural form. One such form was the semicircular arch discussed above, and early versions permitted spans of up to 40 m.

Early Timber Trusses

An alternative structural form to the arch and the beam is the truss, which uses the geometric rigidity of the triangle to produce a manufactured member by greatly enhancing the bending and buckling strength available from a given amount of material. The truss relies heavily on the interconnection and geometric separation of its relatively small components. The need for easy connection between their components meant that early trusses were constructed solely of timber. For a long time, truss bridges were known as artificial bridges.

The structural form that probably led to the development of the truss was the strutted beam or raked frame. This technique uses a raked strut to prop a beam from underneath and thus increase the distance that it can span (Wittfoht 1984). The method was widely developed in early China. By cantilevering the strutted beams from each pier and spanning the distance between two facing cantilevers by a conventional beam—called a suspended span—the Chinese were able to cross spans of up to 35 m (Tang 1987). There are claims that, with elaborate propping, spans of up to 45 m were achieved (Needham 1971). Telford occasionally employed the technique, and it is further demonstrated in figures 8.16 and 8.17. The method suggested the further strengthening and stiffening value of adding diagonal components.

A frieze on Trajan's Column depicts his A.D. 104 bridge over the Danube at Iron Gate near Orsova in Romania (see fig. 3.3). The bridge had twenty timber arches with spans of about 35 m. Designed by Appolodorus of Damascus, it was built to aid the defeat of the barbarians and was destroyed by Emperor Hadrian as the barbarians began to get the upper hand. Seventeen centuries were to pass before the lower Danube was bridged once again, this time at nearby Turnu Severin.

The frieze suggests that the bridge had a number of trusslike features. Referring forward to figure 8.15, the chords of a truss are the two members which run the length of the truss and provide the bending strength. They are separated by the diagonals, and sometimes by vertical components, which carry the vertical shear forces. In this context, Trajan's Column shows horizontal trusses with parallel chords and crossed diagonals, partly supported by four ribbed timber arches, which, in turn, are carried on substantial stone piers. An alternative, lower technology, explanation for the horizontal trusses is that they were merely handrails braced in their plane. Perhaps handrail bracing led to a realization of the strength provided by trussing. However Hopkins (1970) is confident that the column shows genuine trussing. He argues that truss technology would have been needed to provide falsework for the masonry arches that the Romans were building at the time.

Timber trusses did not have a long life, so their early history is obscure. Nevertheless, there was a strong demand for trusses in mountainous regions with clear waterways, because the rapidly flowing streams often washed away the obstructing piers of the masonry arches. The oldest extant truss bridge is the zigzag Kapellbrucke (Chapel Bridge) over the Reuss River at Lucerne in Switzerland. Built in 1333, it consists of strutted and crudely triangulated trusses of moderate span, simply supported on piled trestles, suggesting an evolution from the strutted bridge.

The modern truss bridge probably developed from the example set by the truss roof construction techniques used in medieval churches and similar major structures. The available space between roof and ceiling provides an ideal opportunity for truss construction. The skill of medieval masons in timber splicing also overcame the old limit on truss spans imposed by the need to use a single piece of timber for the lower chord of the truss, which is in tension.

Andrea Palladio was a famous Italian architect/engineer who built many palaces and revived the classical style of building in Europe. In 1570 he used trusses to form the arches of a four-span bridge, with each span about 14 m. The bridge crossed the Cismon-Brenta River at Bassano and lasted for three hundred years (Davison 1961; Mesqui 1986). In the same year he published a treatise called *Four Books on Architecture,* which included sections on truss design.

Between 1756 and 1759 a major strutted truss-arch with spans of 60 m and

51 m was built over the Rhine at Schaffhausen, Switzerland, by two Swiss carpenters, the brothers Hans and Johannes Grubenmann of Teufen (Kirby et al. 1956). The Grubenmanns designed the bridge to span 110 m and thus cross the entire river. However, they were required by their commission, against their wishes, to use a midstream pier remaining from a previous bridge. To demonstrate the merit of their design, Hans Grubenmann is said to have removed the bearing between the bridge and the center river pier to demonstrate, albeit temporarily, that his design could indeed span 110 m.

The Grubenmanns built intuitively, testing their ideas by constructing wooden models, one of which is still on display in the Allerheiligen Museum in Schaffhausen. Their bridges were not structurally sophisticated and have been described a little harshly as "ponderous . . . with a needless amount of timber" (Fletcher and Snow 1932). Their largest working span was a 70 m crossing built in 1757 over the Rhine at Reichenau. There are claims that the brothers built a 119 m timber arch bridge over the Limmat River at Wettingen near Baden in 1758, which would have been the largest such bridge ever built (O'Connor 1975). However, the bridge probably never passed the model stage as it lacked the structural strength for its intended span and appears to have been replaced by a similar 60 m span structure (Wittfoht 1984; Fletcher and Snow 1932; Hopkins 1970). All three bridges—Schaffhausen, Reichenau, and Wettingen—were destroyed by Nicolas Oudinot's French army while retreating from Austria during the 1799 campaign of the French Revolutionary War. This war also destroyed the miraculous Devil's Bridge in Saint Gotthard Pass.

The American Influence

In the New World there are records of small bridges built in both Virginia and New England between 1611 and 1620 (Rose 1950a; Newlon et al. 1985). The first bridge structure was probably a wharf built in 1611 at Jamestown, Virginia. However, major American bridge building began in 1662 with a crossing of the Charles River in Boston between Old Cambridge and Brighton called the Great Bridge. The bridge had thirteen piers and a total length of 80 m. It was not a pier and beam bridge but consisted of "cribs of logs filled with stone and sunk into the river—hewn timber being laid across it" (Plowden 1974). The initial structure was destroyed by a flood in 1685, but the bridge survived in various guises for over a century. By 1793 the Charles River had been crossed by the West Boston Bridge, which had a total length of 1.05 km achieved by using seventy-five piers with a 14 m average timber-beam span.

The Charles River bridges represented the limit of simple beam technology. The large crossings in the United States demanded something better, and yet there were often neither the materials nor the foundations for the arch. Instead, it was necessary to adopt the earlier European initiatives in the use of trusses. The adoption was so successful that it effectively created a new, cleaner structural form. Without doubt, the main thrust for further timber truss development was to occur in the United States. There were three main reasons the United States went in a direction contrary to that of Telford and his colleagues in the Old World. First, iron and masonry costs in the United States were impossibly high toward the end of the eighteenth century, whereas structural timber, particularly white pine, was cheap and plentiful. Second, the country was in an expansionary mood following the

Revolutionary War. Third, carpentry skills had been developed in the numerous shipyards along the New England coast.[14]

There are reports of a timber truss bridge over the Connecticut River at Bellows Falls in Vermont in 1785. However, the initial lead in truss development came from Timothy Palmer, a New England shipyard carpenter who constructed a 48 m span truss-arch in 1792. The site of this bridge over the Merrimack River, 3 km above Newburyport, Massachusetts, was later to be used for James Finley's famous Essex-Merrimac suspension bridge. In 1794 Palmer used his growing understanding of truss technology to build a 73 m span truss-arch over the Piscataqua River near Portsmouth in New Hampshire. It was a bridge of some public wonder and was known as the Great Arch. Eleven years later he completed the Permanent Bridge over the Schuylkill River at Market Street in Philadelphia, using three spans of 46, 59, and 46, m. The Permanent was a structural breakthrough in that it used no supporting raked strut on the underside of the truss-arch and was the first of a famous series of American covered bridges. Called Permanent because it replaced a series of pontoon bridges, it lasted until gutted by fire in 1875.

There is a record of a covered bridge having been built in Babylon in 780 B.C., the fourteenth-century Kapellbrucke was a covered bridge, and notable builders such as Palladio and the Grubenmanns built covered bridges. Covering a bridge kept water off the wooden structural members and hence increased their lives beyond the ten years common for unprotected timber bridges (Shank 1980). At their peak there were some twelve thousand covered bridges in the United States. The last new covered bridge was built in 1917, and some eight hundred are still in existence. American covered bridges were also called "kissing bridges"—they provided a convenient place for unobserved lovemaking (Barth 1980).

German immigrant Louis Wernwag built the Colossus Bridge over the Schuylkill River at Upper Ferry in Philadelphia in 1812 (Nelson 1990). It was a local showpiece and, at 104 m, it was probably the first bridge permanently to span over 100 m (Peters et al. 1981). The structure was a timber truss-arch 1.07 m deep and, as a significant innovation, used iron rods for the lighter truss tension diagonals. The main compression chord was formed from three timbers laminated together to produce a 1070 × 300 mm member. The bridge was destroyed by fire in 1838.

The next step in truss development was to use the truss to create a beam rather than as the main member in an arch. The beam has the great advantage that it just sits on its supports, whereas the arch is jammed between them. Thus, the beam exerts much lower horizontal thrusts on the piers than does the arch. Its disadvantage is that it carries its load by bending and is thus inherently more flexible and more demanding on its components than is the arch, which works in direct compression. The breakthrough occurred in two stages, beginning in 1804 with Theodore Burr's bridge over the Hudson at Waterford, Connecticut, which lasted for 105 years before being destroyed by fire. The structure used constant-depth beam-trusses outside the plane of, but connected to, a timber arch. By adding the arch to the truss, Burr was able to reduce the longitudinal sag in the road surface caused by the beam deflecting under load. Burr patented the arrangement, which was to be used for many American covered bridges, commonly called a Burr truss. His design is said to have been inspired by a drawing by Palladio.

Another significant spanning of the time was Burr's 110 m timber truss-arch across the Susquehanna River at McCall's Ferry in Pennsylvania, on a route between

York and Lancaster. The bridge was built in 1815, and, although destroyed by an ice jam in the river two years later, it was the longest timber arch ever built, if the temporary record at Schaffhausen is ignored. Hundreds of Burr's bridges remain in use today. He was a designer of great ingenuity, and his work attracted wide national and international attention.

The final stage in the conversion from arch-truss to beam-truss occurred between 1813 and 1850. The first pure beam-truss in the new style was built in 1813, anonymously and with little recognition, over Otter Creek in Vermont. It was of a lattice construction.

Technically, the lattice is a highly indeterminant structure and is very difficult to analyze for stresses, due to the many components that share the same load. However, the early lattice trusses were sufficiently strong to obviate any need for such academic analyses. The practical advantages of the form were that it was easy to fabricate because it only required simple internal connections, and it was much easier to construct than were the older truss-arch shapes. Lattice trusses were initially built of timber, but the arrangement subsequently formed the basis for many iron truss bridges. In 1820 Ithiel Town constructed the first publicized lattice truss bridge over the Mill River near New Haven, Connecticut. Although an architect, Town was more a promoter and a salesman than a builder. He boasted of the lattice truss that he could "make it by the mile and cut it off by the foot."

The structurally complex lattice beam was followed in 1829 by a return to a simpler, but structurally more sophisticated, internal framing system using crossed diagonals. The system was due to Stephen Long, who had been a colonel in the mapping section of the U.S. Army. His technological breakthrough was that he calculated the forces in the members and so was able to size them according to the load that they would carry. Long's first bridge carried the Baltimore–Washington road over a railway line and was thus also one of the world's first overpasses. The main motivation for the simplification and codification of truss bridge design was a growing lack of trained carpenters to meet the increasing demand for timber trusses. Structural systems therefore had to be as simple as possible, standardized, and capable of prefabrication. An alternative was to look for other solutions—after all, iron was now becoming a little more commonplace.

Indeed, iron began to dominate bridge building on major routes from 1850, and timber trusses were reduced to serving the low end of the market. They experienced a resurgence in 1904 when Hetzer introduced glued, laminated members in Germany, but few were produced after the First World War.

The Metal Truss

As a consequence of high levels of carbon and various impurities, seventeenth-century European cast iron was very brittle. It slowly became more malleable, beginning with the work of René Réaumur in France in 1722. By midcentury, the common way in which ductility and tensile strength were enhanced was to remove impurities by the slow process of repeatedly heating and hammering the material to produce wrought iron.

One of the first effects of the new industrial muscle supplied by steam was the capacity to manufacture iron in large quantities. The initial stages in this sequence occurred in 1776, when the new steam engine was used to increase the air supply to the blast furnace, and in 1782, when its power was used to allow hammer forging of the wrought iron. In 1783 Henry Cort cast and rolled lengths of iron beams at Coalbrookdale. Although of simple cross section, the beams provided the first opportunity to use iron efficiently in bending. From 1784 limited quantities of simple structural shapes were sold as welded iron. The first wrought-iron bridge was built in 1791 in Wörlitz, Germany; however, the practice did not become widespread for another seventy years.

One intermediate solution, developed by Henry Cort in 1784, was to stir a molten puddle of cast iron prior to casting to drive off as much carbon as possible and thus increase malleability (Lay 1982). The resulting material was called puddled iron. Cast iron and wrought iron have the same compressive strength, but the purer wrought iron has superior tensile strength and ductility. The process was further enhanced in 1828 when oxygen was introduced into the puddle. This method provided malleable iron suitable for use as feedstock in the mechanized wrought-iron processes. As the ultimate step in this process, Cort replaced hammering and forging with rolling in 1783. The use of malleable irons and continuous rolling permitted much higher production rates to be achieved.

The new material soon had its impact on the existing truss technology, providing a totally new way of making tension members and of connecting them together (Peters et al. 1981). For instance, Long's truss was improved by millwright William Howe's 1840 use of turnbuckled iron rods for the vertical tension members (fig. 8.8) and Caleb and Thomas Pratt's 1844 use of these new components for the diagonals (fig. 8.10).[15] The turnbuckles allowed changes in the shape of the timber over time to be accommodated. These innovations produced the characteristic truss configurations that still carry their originator's names, although Palladio had preempted the Howe configuration.

Howe's first truss was erected over the Congaree River in South Carolina in 1827, but it was not publicized or patented until 1840. Apart from the use of metal rods, his truss was otherwise a combination of the simple supports of the Town truss and the simple framing of the Long truss.

The Howe and Pratt trusses, using compression and tension diagonals respectively (figs. 8.9 and 8.11), were the first modern beam-trusses. They used iron for major members and exerted no horizontal thrusts on the abutments. They could span up to about 50 m and, with their simple framing, soon became the popular favorites both in the United States and throughout Europe (Kuzmanovic 1977). To reduce the danger of the diagonals buckling, it was initially common to use crossed diagonals connected at midlength, rather than the modern layout shown in figure 8.8, which is made possible by the use of more efficient structural shapes for the diagonals.

The first all-iron Pratt truss was produced in 1852. Variations on the Pratt truss used curved chords to increase the midspan depth and hence the bending resistance of the truss. They were known as Parker or camelback trusses when the top chord was curved and bowstring trusses when the bottom chord was so treated. The technique permitted spans of up to about 80 m. The first bowstring truss was built in 1833 over the Timis River at Lugoj in Romania by Hoffman and Madersbach. Squire Whipple, an instrument maker by training, built a bowstring truss across the Erie Canal at Utica in 1840, using cast iron for compression members and wrought iron for tension members. To Whipple also belongs the credit for the first truss bridge with the iron members fabricated with large, accurately machined holes at

Figure 8.8. Photograph of a Howe truss bridge, the 1895 Hampden Bridge over the Murrumbidgee River at Wagga Wagga in New South Wales. Each truss spans 34 m. *Photo by permission of the Roads and Traffic Authority of New South Wales.*

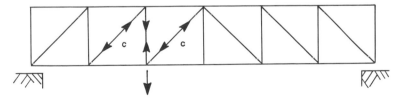

Figure 8.9. Diagram of a Howe truss. The *c* indicates compression and the *t* tension. Compression members must be stockier than tension members.

each end (eye-bars) and connected together by iron pins inserted through these holes. This took advantage of the simple metal-fabrication methods available in the days before riveting, bolting, and welding. The 44 m span bridge in Troy, New York, was completed in 1853, and the eye-bar soon became the trademark of the iron truss bridge (FHA 1976).

The Bolmann truss (fig. 8.12) made effective use of iron in tension and was more like a raked or strutted beam than a conventional truss. Bolmann trusses frequently suffered from vibration problems and none were built after 1875, although one survives today in Lynchburg, Virginia.

The main European contribution to this burgeoning truss technology was the

Figure 8.10. Photograph of a Pratt truss bridge, the 1962 Glenelg River Bridge at Nelson in Victoria. *Photo by permission of Vic Roads.*

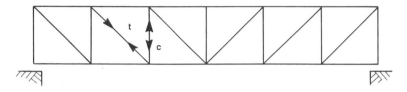

Figure 8.11. Diagram of a Pratt truss.

Warren truss (fig. 8.13 and 8.14), which omitted the vertical members and relied solely on the diagonals for carrying the vertical shear forces. It was developed in 1846 by the Belgian engineer Neuville for some small canal bridges and is named after James Warren, a British engineer who built a similar bridge over the Trent at Newark in 1848.

A later development (ca. 1880), which increased the spanning ability of the Pratt and Warren trusses, was the use of subdivided panels (as in fig. 8.16) to carry the deck loads between the truss connections. This technique was used in the largest truss bridge ever built, the 549 m span Quebec Bridge, erected over the St. Lawrence River near Quebec City in 1917 after a couple of unsuccessful earlier attempts (to be discussed below). It was a suspended span bridge in the manner of figures 8.16 and 8.17 below and held the world record until 1929, when the 564 m span Ambassador suspension bridge was completed over the Detroit River between Detroit and Windsor, Canada.

Another truss form was the Fink, or Belgian, truss, which became popular in building construction because its overall triangular shape fit many roof lines and roof loads could be applied to its sloping upper chord.

The advent of the new steam railways, with their heavier loads, need for flatter grades, and expansionary advocates, brought great demands for more bridges—all

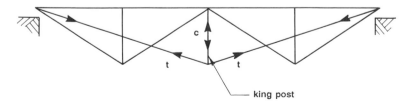

Figure 8.12. A Bolmann truss.

Figure 8.13. Photograph of a Warren truss bridge, the multispan McKillops Bridge crossing the flooded Snowy River in Victoria. Built in the 1930s, it was one of the first welded trusses. *Photo by permission of Vic Roads.*

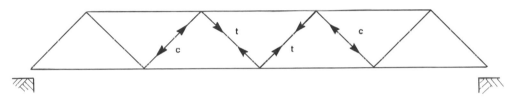

Figure 8.14. Diagram of a Warren truss.

to be built in minimum time and for minimum cost. As a result of these and other incentives arising from the Industrial Revolution, the nineteenth century was a period of major development in bridge form and capacity. A whole new range of bridge types was invented, and the arch lost its structural dominance. At a very measurable level, the maximum span achieved increased from 120 m in 1800 to 420 m in 1900. These large new structures began to have a major impact on the built environment.

A great deal of work was required to form a structural member from wrought iron. Wrought-iron plates were made by continually rolling a series of hot iron bars together until they formed a plate. This required much mechanical effort and made the members very expensive. Rolled wrought-iron beams up to 300 mm deep became available in about 1850 (Humber 1864).

The wrought iron of the time had other problems apart from its high structural cost. Mechanically, it was far from reliable, suffering seriously from impurities and internal discontinuities. Some 502 iron bridges—or a peak of 25 percent of all metal bridges—failed in the United States between 1870 and 1890, forcing many railways to revert back to masonry arches (Kuzmanovic 1977). Not all the failures were due to the iron; part of the blame was due to "fly-by-night bridge salesmen and promoters, who, sometimes involved in political and business corruption, provided cheap and inadequately designed structures" (FHA 1976). The most spectacular of the failures was the Ashtabula rail bridge in Ohio—a wrought-iron Howe truss—which collapsed due to design faults in 1876, killing ninety-two passengers on the Pacific Express.

With the invention of the Bessemer converter in England in 1856 and the open-hearth furnace in 1861, steel became a potential structural material, whereas its previous use had been confined to weapons and cutlery. Although small quantities of steel had been used in 1828 for a 102 m span suspension bridge built by Ignatius von Mitis over the Danube Canal near Vienna (Plowden 1974), its first major bridge use was in three truss bridges built in Holland in 1862 (two years after the demise of the von Mitis bridge) with spans of about 30 m (Hopkins 1970). They were not very successful. Success came with Eads's great St. Louis steel bridge of 1874. The first all-steel bridge was built in 1876 over the Missouri River in Montana, employing five 150 m truss spans. Basic, or mild, steel was introduced in 1878 and remains the workhorse of today's bridge steels. When the British firm Dorman Long began rolling steel beams in 1885, bridge construction entered a new era, particularly with respect to short spans.

Riveted joints came into use in about 1908, and all-riveted trusses appeared in 1910. Welding was not used for bridgework until the 1930s.

Metal fatigue began to arise in iron railway bridges in the 1840s and forced some railways to revert to timber trusses. In 1847 Queen Victoria established a royal commission to "ascertain such principles and form such rules as may enable the Engineer and Mechanic, in their respective spheres, to apply the Metal with confidence, and shall illustrate by theory and experiment the action which takes place under varying circumstances." Like many such commissions, their 1849 report had its basic principles wrong, incorrectly blaming fatigue on "recrystallization" of the metal. The report did, however, formalize the concept of the safety factor. A safety factor of N means that a member should be capable of carrying a load N times greater than the expected load. Engineer witnesses suggested that N should lie between three and seven, and the commission selected six. The number fell gradually toward two over the next century as materials became more reliable and methods of design and analysis became more refined.

Truss and General Structural Theory

There was little theory around to help the medieval masons with their new-found truss technology. Apart from the work of Jordanus Nemorarius on statics late in the twelfth century, the study of forces had progressed little since the time of Archimedes in the third century B.C. It took the genius of da Vinci to open the doors. The structural design of bridges is about resisting the forces of traffic and the bridge's own mass. The useful concept of force and the first analysis of the forces in a beam are both due to da Vinci, the latter in 1500. He also began modern truss theory when he recognized the value of the triangle as the only inherently rigid geometrical shape, as in the simple form of the Warren truss in figure 8.12.[16]

In 1586 the Dutchman Simon Stevin introduced two graphical techniques, called the parallelogram of forces and the funicular polygon, which permitted straightforward force analyses and checks on whether all forces had been considered and were in equilibrium. Stevin began his career as a merchant, became a professor of mathematics and later state treasurer, army quartermaster, inspector of dikes, and superintendent of waterways. Stevin's work was put into modern form by Pierre Varignon in 1722.

Galileo was slightly younger than Stevin and became a professor of mathematics at the age of twenty-five; however, his impact on structures came later. While under sentence by the Inquisition from 1633 to 1638 and prohibited from his heliocentric celestial studies, he reverted to the safety of structural theory and explored the strength of rectangular beams in bending in his 1638 book *Dialogues Concerning Two New Sciences*. The work showed that the strength of a rectangular beam was proportional to its width and to the square of its depth. It also extended da Vinci's force theory by adding the concept of the force acting at a distance, which is called a *moment* and determines how much a member bends. Gottfried von Leibniz made the next major advance in 1675 when he examined the effect of bending on a beam and suggested that the resulting internal stresses were linearly distributed within the cross section of the beam.[17] However, he mistakenly thought that the line of zero stress was fixed at the bottom of the beam. Jakob Bernoulli corrected this error and was then able to calculate the deflection of a loaded beam in 1694.

The first book on bridges was Gautier's *Traite des Ponts* published in 1714, although it was entirely descriptive and contained no theory. The famous Dutch scientist Pieter van Musschenbroek empirically assessed the strength of many types of structural members. As part of this work, he accurately observed and recorded the buckling of a column in 1742. The great mathematician Leonhard Euler produced the associated theory some fifteen years later. In 1773 Coulomb provided the first unified theoretical basis for the structural behavior of materials, including the functioning of beams, using a theory that is still taught today (Coulomb 1773). Thomas Young took the work of Euler, Coulomb, and Hooke and developed the analysis of bent and eccentrically loaded columns in his 1807 book. In 1831 Eaton Hodgkinson produced a practical method for column design and—in conjunction with William Fairbairn—extended the work to the new iron members.

Navier took Coulomb's work further and in 1826 published a standard textbook for beams, including the first completely correct solution for their bending under working loads and a new theory for indeterminant beams. The text drew on the theoretical work of Jakob Bernoulli, Euler, and Coulomb, bringing it together in a brilliant and coherent way (Heyman 1972). The book was also partly inspired by visits to England to examine suspension bridge construction. These visits are discussed below.

In a determinant structure (e.g., fig. 8.12) the forces in the components do not depend on the composition of the structure, only on its geometry. However, the indeterminant structure is of sufficient framing complexity that the forces in it cannot be determined by Newton's laws alone. The stiffnesses of the individual components must also be considered if the forces are to be known. Navier solved the case of the indeterminant continuous beam, which is a beam resting on more than two supports—then a somewhat exotic situation. The market demand was for a method for analyzing oversafe indeterminant trusses—such as the lattice truss. Although the lattice truss might have been easy to build, it was hell to analyze.

The important theoretical concept of reducing indeterminant trusses to determinacy by considering only key diagonals in their analysis was introduced by Dmitrii

Zhurawski in Russia in 1847 (Kuzmanovic 1977). A century was to pass before this brilliant intuitive concept was theoretically justified by the lower-bound theorem of structural analysis (see Lay 1982). The theory of indeterminant structures was put on a sound general basis with B. P. Clapeyron's three-moment theorem in 1857 and Luigi Menabrea's least-work theorem in 1858.

Whipple created a firm basis for practical truss analysis in 1847—some two millennia after their first use—with the issue of his book *A Work on Bridge-Building.* He showed which members were in compression and which in tension and so allowed much more efficient member selection. Unfortunately, his work did not become widely available until Whipple republished it in a larger book in 1874.

Truss buckling was first analyzed by Herman Zimmerman in 1851. In the same year, a Bavarian railroad engineer named Karl Culmann was sent by his employers to the United States to study America's new bridge technology. He was inspired by the practical truss developments he saw there and subsequently promoted them in lectures and reports. In 1859 he was appointed the first professor of engineering in the Swiss Federal Polytechnic, and in 1865 he published his key work on graphical statics, *Die Graphische Statik,* which was particularly useful for lattice truss design. His successor, Ritter, continued to develop the graphical methods that were particularly important in those precalculator and precomputer times.

In 1864 James Clerk Maxwell published his theory of determinant structures and his graphical method for structural analysis in the *Philosophical Magazine.* The work was not known in continental Europe, where the theory was generally named after L. Cremona, who developed it independently eight years later, drawing from the earlier work of Culmann and Ritter. Even in Britain the use of Maxwell's method only became widespread with the introduction of the convenient Bow's supplementary notation in 1873. Robert Bow was a Scotsman whose design services were rejected for the ill-fated Tay Bridge. In 1850 he had published a major book on truss analysis, only three years after Whipple. In 1875 Williot introduced his graphical method for estimating truss deflections, which Heinrich Müller-Breslau expanded to a widely used format.

Influence lines indicate the loads in a member as a vehicle crosses a bridge. In 1868 Emilio Winkler published his work on the subject, which was to be the basis for most subsequent practical bridge design (Lay 1964a). At about this time William Rankine codified the concepts of stress and strain. His textbooks provided the first English-language equivalent of material that the French had had for almost a century. In 1874 Otto Mohr produced the widely used virtual displacement method of theoretical truss analysis, and eighteen years later showed how the secondary stresses in a truss could be calculated. Alberto Castigliano in 1876 published the strain energy and virtual work theorems that are still used today and thus put most existing methods on a sound theoretical basis.

Now that the overall structural member could be designed, the more detailed analysis of local or secondary stresses in the members required better stress theories. Augustin Louis Cauchy had introduced the concepts of stress analysis for members behaving elastically in 1827, and Mohr subsequently developed theories for combined elastic stresses (the Mohr circle). Stephen Timoshenko began his many contributions to stress analysis in 1910.

In 1882 H. Manderla introduced slope deflection equations, permitting algebraic rather than graphical solutions for indeterminant beam bending. In 1932

Hardy Cross used these equations to introduce the moment distribution method for the easy analysis of indeterminant bending structures by an iterative method known as "relaxation." This was important in precomputer times for structures not amenable to graphical analysis.

The successful analysis of the truss therefore occurred well after the analysis of the beam and the arch had been resolved, even though they are much more complex structures than the truss. Since modern trusses were not invented until the 1840s, their analytical solutions lagged well behind.

This new scientific approach to bridge analysis could not flower until the widespread availability of iron and then steel in the second half of the nineteenth century provided a well-behaved material that could be readily subjected to linear mathematical analysis. The first step was to know the mechanical properties of the material. Although Musschenbroek had produced the first test results for the strength of iron in 1742, a century passed before Fairbairn provided practically useful test data in response to the needs of his own major designs.

A major paradigmatic jump was necessary before designers could bring themselves to use the much more ductile steel beyond its elastic limit and capitalize on its postelastic, plastic properties. *Plastic* refers to deformation under constant load and is a characteristic of structural steel once it has reached its elastic limit. Although Mohr had examined the plastic strength of materials a quarter of a century earlier, the first useful tests on the plastic range of steel were conducted by E. Meyer in Germany in 1908. Further extensive testing was conducted in Germany between 1928 and 1938 (Lay 1964b). The theoretical basis for the application of plasticity to structures finally occurred in 1940 when John van den Broek showed how to design for a specified failure condition by using his *limit state design* method.

Modern Bridges

It is now appropriate to discuss recent developments in bridges. The earlier taxonomy based on structural form is discarded in order to illustrate the rapid changes that have taken place in the last century—without the invention of any new structural forms. Instead, the advances have been in materials, techniques, and structural understanding.

Suspension Bridges

Modern suspension bridges became possible as the demands of oceangoing ships led to the availability of better and better iron chains. The first modern suspension bridge was constructed in 1801 at Jacobs Creek in Uniontown, Pennsylvania (see fig. 3.10), by Finley, an Irish-born politician, lawyer, and judge and good friend of Gallatin (Kirby et al. 1956). His legal background is not too surprising because "in the absence of an organized engineering profession [in the United States] at the time, it was the lawyers, who were largely self-taught in land surveying and the law, who became the first American engineers" (Kemp 1979). Finley's bridge spanned 21 m, using wrought-iron chains made from 25 mm sq bars formed into links about 2 m long. Although severely damaged by a horse and oxen team in 1821 (or 1827), it lasted for fifty years (Shank 1980). It was clearly the precursor of modern suspension bridge construction.

Finley went on to construct forty such bridges before his death in 1828 (Kemp 1979). His Essex-Merrimac Bridge, built in 1810 with a 73 m span over the Merrimack River at Newburyport, Massachusetts, was the first suspension bridge whose deck was sufficiently stiff and strong to permit use by heavy vehicles. Finley achieved this with the Tibetan technique of using freestanding piers rather than merely suspending the chains from abutments (Shank 1980). The bridge was rebuilt after being damaged by an ox wagon in 1827 and remains standing today, as the world's oldest major suspension bridge. Using experimental techniques, Finley was the first to determine the forces in the suspension chains. In 1808 he introduced a further innovation by using trusses for the deck structure. Fifty-three years passed before Rankine showed how to analyze this system theoretically. Finley's bridges have been described as "astonishing structures" and bold for their time (Kemp 1979).

In 1806 Captain Samuel Brown introduced anchor chains into the British navy. In private service he developed chains made from eye-bars and used them on a 135 m span suspension bridge he built in 1820 over the Tweed River at Kelso. A number of his suspension bridges were subsequently destroyed by high winds. They were not alone in suffering this fate; a 79 m span suspension bridge designed by John and William Smith over the Tweed River at Dryburgh Abbey near Melrose in Scotland—less than 20 km from Kelso—failed in strong winds just six months after completion.

The prototype of the modern bridge was Telford's Menai Straits suspension bridge (fig. 8.15; see also fig. 1.1), which was an essential part of his London–Holyhead Road. It spanned 174 m and had a shipping clearance of 30 m. Building

Figure 8.15. Menai Straits Bridge. *Photo with the permission of the National Museum of Wales (Welsh Folk Museum).*

on Samuel Brown's work, Telford used 2.8 m long eye-bar links formed from 75 ×
25 mm wrought iron to create the suspension catenary, which was hung from masonry towers. Telford designed by intuition rather than calculation, and his work
possessed acknowledged similarities to Finley's, whose techniques had been disseminated by Thomas Pope in a book published in 1811 (Kemp 1979). However,
Telford claimed that the Menai Bridge had taken a new approach (Penfold 1980).
Completed in 1826, it was the first bridge to span a reach of ocean and the first
British bridge to hold a world span record. The Menai Bridge was rebuilt in 1940
after being troubled by wind effects over much of its life. In the rebuilding, its eye-
bar chains were replaced by steel cables. Of the original bridge only the towers and
abutments still remain in service.

Jakob Bernoulli had solved the mathematics of the suspended (i.e., caternary)
cable in the seventeenth century, but Davies Gilbert did the first structural analysis
of a suspension bridge in 1821 as part of the design work for the Menai Bridge
(Kemp 1979). This was followed by the better-known approach introduced in 1823
in Navier's *Memoire sur les ponts suspendu*. The book was written after the author had
been sent on missions to England in 1821 and 1823 to study the new techniques
being used on the Kelso and Menai bridges. Navier provided a theoretical explanation for the British methods, particularly acknowledging the prior work of Finley
and Gilbert. He also designed one suspension bridge, the 167 m span Pont des
Invalide over the Seine, which collapsed in 1826, only a few years after completion.
The collapse resulted from a variety of technical and administrative factors, many of
which were outside Navier's control (Straub 1952).

Wire had previously been avoided in bridge construction out of fear of cor-
rosion. In 1816 a local Philadelphia wire business, White and Hazard, was the first
to use wire cables in place of chains and eye-bars when they refurbished a 122 m
span pedestrian bridge over the Schuylkill River in Philadelphia. The next major use
of cables was by the Seguin family in France, whom Louis Vicat had convinced of
their superiority over eye-bars (Kemp 1979).[18] The Seguins subsequently used wire
cables for an 18 m span experimental footbridge at Annonay in 1823 and, two years
later, for the Tain to Tournon pedestrian suspension bridge, which they built in two
85 m spans over the Rhône. This latter bridge was so successful that it was later
permitted to also carry heavy vehicles. The Seguins went on to build some 180
suspension bridges (Grattesat 1982). Marc Seguin had initially been inspired by
reading Thomas Pope's 1821 book on the earlier American suspension bridges.

Vicat's advocacy of wire carried considerable weight as he was well regarded.
He had first made a name for himself in 1818 with the development of hydraulic
cement for bridge foundations, using lime-clay mixtures in the Smeaton-Parker manner. He had also invented a test for concrete and mortar viscosity.

Two particular advantages of wire cable were that it provided increased tensile
strength and that it was not prone to the hidden cracks that had plagued iron bars.
Nevertheless, eye-bars remained in some use until the advent of rust-resistant, gal-
vanized high-strength wire in 1870. Eye-bar chains were particularly popular in
Britain, where empiricism rather than theoretical analysis prevailed.

Vicat conducted revolutionary work on spun cables of iron wire for a suspension bridge he built in 1829 in Argentat over the Dordogne River on the road to
Aurillac (Hopkins 1970; Reverdy 1980). *Spinning* is the in situ forming of the suspension cables from bundles of smaller wires. It made possible the construction of

large suspension spans and was the death knell for the use of chains in bridgework. The Argentat bridge, which was also the first to use cement in the construction of its suspension piers, survived until 1903.

The Menai Bridge kept its span record for eight years until 1834, when Joseph Chaley built the spectacular Grand Pont Suspendu over the Sarine River valley near Fribourg in Switzerland. Using cables made from a thousand wrought-iron wires, it spanned 273 m with its deck 50 m above the water. The bridge was demolished in 1923. Chaley was a French army medical officer who had been trained in bridge engineering by one of the Seguin brothers. His main technical innovations were the introduction of bundled cables and aerial spinning.

The Fribourg span record was exceeded in 1849 by a 308 m span designed by Charles Ellet to take the National Road over the Ohio River at Wheeling, West Virginia. He had won the right to design the bridge in competition with John Roebling. The record was held for just one year, after which it passed to a New York State suspension bridge. Ellet, a self-taught canal surveyor, had studied bridge engineering during a year at l'Ecole des ponts et chaussées (Sayenga 1983). He subsequently brought the French wire cable technology to the United States. The Wheeling bridge had six cables, each containing 550 wires, supporting a wooden deck truss (Kuzmanovic 1977). After five years in service, it was damaged when a high wind caused torsional resonance. Ellet supervised the repairs to the bridge, which reopened in 1859 and is still in service, although further repairs were made on five separate occasions, including work by Roebling in 1886 (Kemp 1979; Sayenga 1983). Ellet also designed the ram ships used by the Union in the American Civil War. He died as a result of a wound suffered in a naval battle during that war.

John Roebling was a civil engineer and German immigrant who arrived in the United States in 1831. Together with Ellet, he introduced the United States to the use of the spun wire cable, although the two differed strongly on the method of manufacture. Roebling learned of the use of wire from his undergraduate engineering lectures in Germany, whereas Ellet attended a lecture Vicat gave on the subject in France in 1830 (Sayenga 1983). Roebling studied Ellet's Wheeling bridge carefully and learned from its problems.

Roebling's tenth and greatest major suspension bridge was the 478 m span Brooklyn Bridge, which crosses the East River and provides a land route between the large commercial centers of New York City and Brooklyn. Roebling had been working on the concept since the 1850s, but his plans had been severely set back by the Civil War. A brutal winter in 1866–1867 renewed the demand for the bridge and led in May 1887 to the formation of a company to build the bridge. Roebling was appointed chief engineer. His design introduced special provisions for stiffening the deck and the large-scale use of steel wire in place of wrought-iron wire for suspension cables. Roebling died in a construction accident during the early stages of construction.

Kuzmanovic (1977) says of Roebling, "His grasp of the principles of the suspension bridge was far ahead of the understanding of other engineers of his days." Rated as one of the world's great bridge designers, Roebling built his first suspension bridge in 1846 on Smithfield Street in Pittsburgh (Billington 1981). He was the first to design a suspension bridge that was stiff enough both to resist wind and to carry the loads imposed by steam trains, and he greatly improved cable technology.

Roebling's son, Washington, took charge at Brooklyn following his father's accident, but he was partially paralyzed soon after from the bends while working in a compressed-air caisson producing foundations for the bridge. Compressed air allowed work to occur in difficult foundation conditions. The technique had been introduced on the Medway Bridge in England in 1851. Roebling directed the remainder of the work from his bed. The bridge was completed in 1883. The span extended the world record by 26 percent and held the title for seven years, until it was surpassed by the cantilever suspended span bridge across the Firth of Forth in Scotland.

The first bridge to span 1,000 meters was Ammann's 1931 George Washington suspension bridge over the Hudson in New York, which had a span of 1,067 m and almost doubled the existing record held by the Ambassador Bridge. Amann was a Swiss-born immigrant to the United States who has been rated as one of the world's major bridge designers (Billington 1981). His Kill van Kull arch bridge has already been discussed. The George Washington Bridge holds the record as the world's most used bridge, due to its wide double deck with fourteen lanes of traffic and its high traffic flows. It lost its span record in 1937 to Joseph Strauss's 1,280 m span Golden Gate Bridge in San Francisco. The record then passed to the 1,298 m span Verrazano Narrows Bridge in New York in 1964, to the 1,410 m Humber suspension bridge at Hull in England in 1981, and to the 1,780 m span Akashi Straits Bridge in Japan in 1991.

An alternative to the catenary suspension bridge is the cable-stayed bridge, where the "suspension" cables remain practically straight. The deck therefore moves far less under load than it does in a suspension bridge. This form began with Faustus Verantius, who used it in conjunction with catenary cables for a bridge in Venice in the early seventeenth century.

The cable-stayed method was used spasmodically, and often unsuccessfully, between then and 1950, when German engineers reintroduced the form to cope with the reconstruction of bridges over the Rhine after the Second World War. The development of this bridge form was intimately linked with the introduction by German ex−aircraft designers of the orthotropic, stiffened-steel deck plate, which had strong military antecedents and which permitted much shallower beam sections to be used in the deck structure. The method was first used in the Kurpfalz Bridge at Mannheim and in the German designer Franz Dischinger's 1955 Strömsund Bridge in Sweden. The cable-stayed era began in earnest with three harp-shaped bridges built in Dusseldorf: the 1958 Heuss with a maximum span of 260 m; the Kniebrucke with a main span of 330 m; and the Oberkassel with two 260 m spans.

In the hierarchy of modern bridge types, the span lengths achieved are typically as follows: beam and truss, up to 200 m; cable-stayed, 150−350 m, although the maximum span achieved is currently the 465 m Alex Fraser Bridge in Canada; suspension, 300 m and over (Podolny 1976).

Other Great Bridges

Three other great bridges built in the nineteenth century are still in service and will now be described, although as rail bridges, neither the Britannia nor the Firth of Forth Bridges should feature in this history. However, both were revolutionary in form and had major long-term influences on bridge building.

William Fairbairn and Robert Stephenson's 1850 Britannia Rail Bridge is located only a few kilometers north of Telford's Menai Straits Bridge. The Britannia is a remarkable bridge. The four spans are formed of pairs of massive hollow, rectangular iron tubes through which trains pass. The bridge crosses a 450 m opening with a maximum span of 138 m. The bridge was designed by empirical testing, rather than by using the theories of Navier and others. Indeed, it was originally planned to incorporate suspension chains, but these were omitted at a late stage when a test showed that the bridge could carry a locomotive twelve times heavier than the design load. Stephenson, who also introduced the wrought-iron plate girder in 1841, was the son of George Stephenson, the inventor of the Rocket steam engine.

James Eads's 1874 St. Louis Bridge was the first bridge to use structural steel and high-strength steel in its three—151 m span, 156 m span, and 151 m span—tubular truss-arch ribs. The bridge was initially a railway bridge and was subsequently converted to carry road traffic. Eads began his career in marine salvage in the Mississippi River and had been honored as a Civil War hero for his design, construction, and operation of armored ships known as ironclads. The St. Louis was his only bridge.

John Fowler, William Arrol, and Benjamin Baker's Firth of Forth Bridge has two 520 m cantilever spans. The bridge highlighted the effectiveness of a mix of both cantilevers and suspended spans (see figs. 8.16 and 8.17), a technique that had been reintroduced by Heinrich Gerber in Germany in 1866 with a 128 m clear span over the Main River at Hassfurt. The Firth of Forth Bridge contained the world's longest span for twenty-seven years, until the Quebec Bridge—which also employed cantilever and suspended spans—opened in 1917 (Steinman and Watson 1941). Built between 1883 and 1890, the Firth of Forth Bridge has been described as the pinnacle of Victorian engineering.

Concrete

The first use of cement in bridges since the Romans occurred in 1796 on the Isle of Sheppey Bridge in England. Concrete was first used for bridges in France in 1840 for the construction of a 14 m arch bridge over the Garonne Canal at Grisolles. Unreinforced concrete has no significant tensile strength and thus presents little advance over masonry blocks. The lack of tensile strength is overcome by adding steel reinforcing bars to carry any tensile loads. Reinforced mortar has been found in Roman roofs and in the Great Wall of China. The modern introduction of the idea possibly began with Jacques Soufflot, who used iron bars to prevent cracking in the masonry dome of the Church of St. Genevieve in Paris in the late eighteenth century (Hamilton 1958). Telford placed iron bars in the concrete masonry of his Menai Bridge towers in 1825. An alternative approach used on the Homersfield Bridge near Bungay in England in 1870 was to embed a cast-iron arch frame in concrete (Stanley 1979).

In 1849 reinforced concrete was being used in France for minor domestic items. It was first applied structurally in bridges in 1875 using a method invented by a French gardener, Joseph Monier, who had begun by making wire-reinforced tubs and tanks in association with his work at Versailles. The first Monier bridge, a pedestrian bridge on the grounds of a French castle at Chazelet, was an arch with a 16.5 m span and a high span-to-rise ratio. Matthias Koenen in Germany initiated the theoretical analysis of reinforced concrete in the 1880s.

Probably the first modern concrete arch bridge was the triple-arch Pont Neuf,

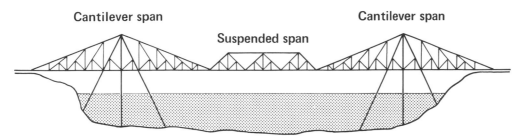

Figure 8.16. Diagrammatic illustration of a suspended span bridge.

Figure 8.17. "Living model" example produced by Firth of Forth engineer Baker to demonstrate suspended spans. The rods and chair legs are in compression, and the men's arms are in tension. On the matter of scale, Baker commented, "The chairs a third of a mile apart, the men's heads 360 feet above the ground." *From Peters (1981), with the kind permission of the Institution of Civil Engineers.*

designed by Belgian stonemason François Hennebique and built in 1899 at Chatellerault in France. The three arches spanned 40 m, 52 m, and 40 m. Hennebique was the first to realize the commercial advantages of reinforced concrete and, between 1899 and 1909, over a thousand bridges were built using his system. He also introduced beams with a practically useful T cross section to reinforced concrete, although they had previously been studied in the 1850s by Thaddeus Hyatt, an American experimentalist.

Reinforced concrete was brought to perfection by Ritter and Hennebique's Swiss student Robert Maillart and the Frenchman Eugène Freyssinet. Maillart developed completely new and very elegant structural forms. In 1901 he used a hollow rectangular member made of reinforced concrete to provide the 38 m span of his Inn Bridge at Zuoz in Switzerland. His most spectacular bridge was the Salginatobel, built in 1929–1930 near Schiers in Switzerland. It is a 90 m span, three-hinged arch employing hollow rectangular sections.

Prestressing means applying a permanent load to a bridge to enhance its load-carrying capacity. The first deliberately prestressed bridges were the ancient Chinese

Bow Bow bridges, which had their bow-shaped main timber-beam span bent upward by forcing it to fit into too small a gap between two masonry cantilever spans. Spans of up to 20 m were achieved (Tang 1987).

Prestressed concrete is a relatively new invention. It overcomes the low tensile strength of conventional concrete and the different stiffnesses of concrete and steel by using tensioned cables to place the whole concrete element into permanent compression. In reinforced concrete the steel is passive, waiting for a tensile stress to be applied. In prestressed concrete the steel is active, applying compressive stress to the structural member before any tensile loading stresses arise.

Prestressed concrete came into its own following the cement shortages after World War II, but owes its development to the experiences Freyssinet had in 1910–1911 on his 72 m span flat concrete arch bridge over the Allier River at le Veurdre. The bridge was destroyed in 1940 during the Second World War, but it very nearly did not even make the First World War; it sagged some 130 mm soon after construction when the concrete deformed under load and over time, a phenomenon known as *creep*. Freyssinet saved the bridge by jacking the arch apart horizontally at the keystone. This prestressed the bridge and placed most of the concrete in compression. Freyssinet realized the broader significance of his repair technique and then conducted an extensive and pioneering study of concrete creep and shrinkage. In 1917 he also introduced mechanical vibration to produce a much higher quality concrete. His eventual appreciation of all the phenomenon involved in stressed concrete led him to patent the prestressing process in 1928.

Dischinger built the first intentionally prestressed concrete bridge in 1928, an arch bridge over the Saale River at Alsleben in southeastern Germany, tensioning the concrete after it was set by using cables placed outside the structural element (Wittfoht 1984). To handle ongoing creep of the concrete under load, the Dischinger system allowed the cable tension to be continually adjusted. Unfortunately, he used low-strength steel—success only came when high-strength steel was used.

In about 1930 Freyssinet realized that the problem of adjusting for creep could be circumvented by using higher-strength steel in the cables to provide greater initial prestressing tension, thus covering any long-term stress relaxation due to concrete creep. Ironically, Freyssinet had some problems having his method accepted in his native France. He parted company with his successful Limousin firm in 1929 and went through a number of difficult years, although his 1930 Pougastel Bridge over the Elorn River was the most significant bridge of the time. In 1936 he built the first modern prestressed-concrete-beam bridge, a 19 m span across the Oued Fodda in Algeria. His next bridge, built in 1938 in conjunction with the Wayss firm, was an autobahn overpass in Olde, in Westfalen, Germany. In 1939 he developed his still-famous loading jack and conical anchorage system for prestressing wires, dramatically improving the usefulness of the prestressing system. In 1942 he built a 42 m prestressed bridge to extend an autobahn over the Neisse River in Germany (Stanley 1979). The first major postwar prestressed bridge was a 42 m span built in Hamburg in 1950.

Movable Bridges

The first movable bridges were drawbridges over defensive moats. This defensive role was a supplementary function of many early bridges. A second reason for using movable bridges was to avoid building much higher bridges to permit the passage of shipping. The drawbridge, or *bascule bridge,* operates by a relatively simple technology compared with swing, traversing, and lift bridges. The lifted span hinges about a transverse horizontal axis; counterbalancing the span on the other side of that axis eases the lifting process. In swing bridges the span rotates around a vertical axis. Traversing, or retractile, bridges move the span longitudinally, usually on rollers, and lift bridges raise it vertically.

Nevertheless, swing bridges permitted much larger spans. The swing bridge record was a 105 m span built in 1861 over the Penfeld River at Brest. In 1867 a 17 m drawbridge in Copenhagen created a record for movable spans. A new era of movable bridges began in 1894 when four new types were introduced: the double-leafed drawbridge (the 60 m bridge at the Tower of London); a rolling drawbridge-cum-traversing rail bridge over the Chicago River; a novel lift span designed by John Waddell to span the southern leg of the Chicago River; and a drawbridge with an underfloor counterweight, also in Chicago.

Bridge Failures

Bridge failures have had a particular significance in bridge development, providing very visible reminders of a technology operating beyond its bounds. The primary purpose of a bridge is to carry traffic, and one would expect excess traffic loading to be a major cause of bridge failure. However, major bridges appear to fail for every reason except routine traffic loading. Perhaps this is because maximum loads are well understood and usually well regulated. However, recall how the advent of the steam train caused a number of railway bridge failures as a result of designers considering only static loads and failing to account adequately for the very high impact loads produced by the new trains.

Initially, traffic loadings were a problem. In 1824 a 78 m span suspension bridge over the Saale in Nienburg in southeastern Germany collapsed when carrying a crowd of people. Seven years later the Broughton suspension bridge in Britain failed when troops were marching in step across it. This introduced the widely held belief and subsequent official regulation that marching squads should break step when crossing a bridge. In fact, a later investigation showed that the bridge would have fallen down anyway and that the troops in step were merely the straw that broke the camel's back. Humans—whether in step or not—produce a very heavy bridge loading.

Bridge traffic loading studies began in 1873 when Bindon Stoney investigated crowd loading for the design of some Thames bridges. The work had its precedent in earlier live loading studies, which used three hundred workmen to test panels for London's Crystal Palace in 1851. Stoney's experiment involved packing Irish laborers onto a weighbridge. Other studies found that, for a given deck area, the loads produced by wagons and by herds of cattle were only 70 percent and 60 percent, respectively, as heavy as a crowd of people (Cron 1976c). There are reports of at least four bridges collapsing under the weight of a herd of animals and one—Broughton—under human loading.

In addition, people are very sensitive to bridge oscillations, which usually result when the loading frequency matches the natural frequency of the bridge and are unrelated to structural failure. Many light footbridges have "failed" their clients in this respect.

Wind loads became significant as bridges became larger. Despite the prior warnings from the wind problems with Samuel Brown's suspension bridges and the well-publicized collapse of the Chain Pier Bridge at Brighton, England, in the wind, it took the Tay Bridge disaster to finally show how little designers knew of wind loadings. In 1879 gale force winds in Scotland blew down the Tay Bridge at Perth while it was carrying the loaded Edinburgh mail train (Prebble 1956). The train and thirteen 60 m span, high-level, wrought-iron beam-trusses fell into the Tay River. Seventy-five passengers died in the icy waters. The bridge had been in service for just eighteen months.

Thomas Bouch, the designer of the bridge, had been knighted for his work. After the collapse, he said that he had made no special provisions for wind pressure, although high winds were commonplace in the area. He died in the year following the disaster. The tragedy had a major influence on the design of the Firth of Forth Bridge in the 1880s; the designer, Benjamin Baker, conducted extensive wind-pressure measurements.

The Tay disaster also led the infamous Scottish poet William McGonagall to pen the following inimitable lines in 1890:

> Beautiful Railway Bridge of the Silv'ry Tay!
> Alas! I am very sorry to say
> That ninety lives have been taken away
> On the last Sabbath day of 1879,
> Which will be remembere'd for a very long time.[19]

The Wheeling suspension bridge failed in 1854 due to torsional resonance in the wind. In 1940 Leon Moisseiff produced the 840 m span Tacoma Narrows suspension bridge in Washington State, with its daring span-to-depth ratio of 350, which more than doubled the Golden Gate record. The bridge self-excited in torsion and shook itself to pieces in a moderate 70 km/h wind after four short months of operation. This self-destruction emphasized a further aerodynamic factor. Up until then, suspension bridge decks had been designed on the basis of their deflection under vertical load from self-weight and traffic. It came as a shock to realize that the increasingly thinner decks could act as aerofoils in the wind. The Wheeling example was ever present, and both Telford and Roebling had warned of the possibility a century earlier.

Indeed, suspension bridges were particularly prone to failure. Alfred Pugsley, a well-known expert, estimated that 120 suspension bridges were built between 1800 and 1900, with an average life of forty years and a one-in-seven failure rate, which was about ten times the rate for ordinary bridges. The current rate is about one in four thousand.

The 1907 collapse into the St. Lawrence River of a 150 m cantilever anchor arm of Theodore Cooper's Quebec Bridge highlighted how little was known about steel members in compression, despite the extensive prior theoretical work described earlier. There is often a large gap between theory and practice. The bridge failed for a second time due to an erection mishap in 1916 and was completed, on the third attempt, in 1917.

There was a plethora of iron bridge failures in the United States in the 1870s and 1880s. In particular, the 1876 collapse of the wrought-iron Howe truss in Ashtabula, Ohio, emphasized the need to better understand materials and structural analysis. We also saw how some bridges were destroyed by flood and water scour, some burned down or rotted away, and some were damaged by passing traffic.

Inadequacies in the temporary falsework supporting a bridge during construction have led to many collapses. Two famous failures of this type were the Second Narrows Bridge in Vancouver in 1958 and the Barton High Level Bridge in England in 1959. Foundation failures have also contributed to many bridge collapses, such as that of the Peace River suspension bridge in British Columbia in 1957.

The collapse of the Duplessis Bridge at Trois-Rivières in Quebec in 1961, followed by the Kings Bridge collapse in Melbourne, Australia, in 1962, demonstrated the critical nature of the brittle fracture of steel and the need for the careful specification of steel intended to be welded.

The 1967 collapse of the thirty-nine-year-old, fifty-three meter span Silver chain suspension bridge over the Ohio River in West Virginia took forty-six lives.[20] The tragedy underlined the need for careful bridge inventories and regular and sophisticated bridge inspection techniques. Nevertheless, the collapse of the Mianus Bridge in Connecticut in 1983 suggested that further underlining might be required.

The destruction in 1975 of the Tasman Bridge in Hobart, Tasmania, as a consequence of ship impact was perhaps the most spectacular of a series of collapses from ships striking bridges. Bridge engineers were slow to accept the high probability of such collisions.

The failures of the Fourth Danube Bridge in Vienna in 1969 and of Milford Haven in Wales and Westgate Bridge in Melbourne in 1970 highlighted the quite different dual needs for a better understanding of plate buckling in box girders and of construction management. Steel box or tubular girders—usually of rectangular cross section—were introduced by Robert Stephenson on the Britannia Bridge. They enabled relatively thin steel plate to be used on large structural members. The resolution a century later of a number of problems experienced with box girders, such as those used at Milford Haven and Westgate, significantly advanced the state of the art of the design and construction of those components.

Summary

Bridges have always played a key transport role, but their major development awaited the availability of new materials following the Industrial Revolution. In comparing bridge types over history, the masonry arch was the only significant permanent structural form from Roman times until 1780. The timber arch had a period of dominance from 1750 to 1820, but the masonry arch stayed in common use until replaced by the concrete arch in 1900 (O'Connor 1975). Since about 1820 the suspension bridge has been the major structural form for the largest spans and was the first structural type to cause bridge spans to increase significantly over Roman achievements.

Bridges have also illustrated the role that genius and persistence play in the advance of a technology. The history of bridges has demonstrated that it has not been enough for a technology to be ripe for the plucking; it has also been essential for there to be some individual willing to learn from the lessons of the past, to have the perception to select appropriate but radical concepts and technologies, and to possess the courage and stubbornness necessary to put the new ideas into practice.

From the Past
into
the Future

━━━━━━━━━━━━━━━━━━━━━━━━━━━━━━━━━━━● *Chapter 9*

Make no little plans; they have no magic to stir men's blood.
DANIEL BURNHAM, AMERICAN TOWN PLANNER, 1907

The past provides the platform from which the world moves into the future. With an understanding of the past now established, this chapter begins by reviewing current aspects of that inheritance and then examines emerging new directions. We now look at how the past gave rise to the present and, using that knowledge, suggest extrapolations that can be made from the present into the foreseeable future. One feature dominates the discussion: the twentieth century has been the century of the car, a technology that has seen both overadoption and overreaction.

The Dominance of the Motor Vehicle

One of the great virtues of the motor vehicle traveling on the modern road has been the variety of travel purposes it has served with such effectiveness and flexibility while operating within a wide range of urban and rural forms. The major travel purposes have been journeys to and from work; social visits; domestic trips, e.g., shopping; recreational trips; industrial, wholesale, and retail freight distribution; and farm to market trips.

The road and the motor vehicle have been able to fulfill these needs well because the road developed in law as a way providing free passage "to all the King's subjects." This right has proved to be fundamental to many of our key twentieth-century needs, leaving the road well placed to serve rising community demands for unrestricted mobility and extensive door-to-door travel, demands fueled by the effectiveness and low cost of the motor vehicle.

The motivations for roadmaking in this century have thus changed dramatically from those explored in Chapter 4. Administrative and militaristic motives have passed the road by, but new suitors have taken it by the hand. Today, the main driving forces for road development are industrial and community needs—a disparate association that has sometimes made for uneasy bedfellows. The four main reasons for this new association have been the wide availability of the car, the efficiency of the truck, the ease with which motor vehicle purchase and fuel usage can be taxed, and the effectiveness of road construction in creating employment and development.

There has thus been a continuing community pressure for investment in new roads. The construction of these new roads has usually arisen as a result of either (a) a benefit-cost analysis showing the road to be a good investment; (b) land development for homes, industries, and farms being concomitant upon the construction of road access; or (c) citizens paying for good roads as part of an improved standard of living package.

In developed economies transportation currently takes about 10 to 15 percent of most national expenditures and road transport commonly consumes about 45 percent of transportation expenditures—i.e., about 10 percent of national funds are spent on road travel (Lay 1990). Thus, roads and road transport are in the same spending league as defense, health, and education. And—unlike many of the other categories—all citizens participate daily and effectively in the road system and hence directly enjoy the benefits of the investment. Virtually every person in today's developed world relies on or is affected by road transport. Domestically, road transport is often our second largest financial commitment.

For reasons such as these, roadmaking has been a vote winner for most of this century, particularly as many road projects can be completed within the term of a political office (Foster 1981). Popular signs beside road construction projects use some variant of Your Tax Dollars at Work. In 1960 Daniel Patrick Moynihan, a prominent American political figure, commented, "Highway construction is especially important to professional politicians, since it provides the largest single supply of money available these days to support their activities."

The all-pervasive dependence on the car and the pace-setting American way of life were perhaps best summarized in the famous words of U.S. Secretary of Defense and ex–General Motors employee Charles Wilson, when he expressed the opinion to a Senate Armed Services Committee in 1953 that "what was good for our country was good for General Motors, and vice versa." Indeed, for many years the seven largest companies in the United States—General Motors, Standard Oil of New Jersey, Ford, Chrysler, Mobil, Texaco, and Gulf—were all directly linked to roads and road use.

During the twentieth century there has been occasional reference to the roads lobby, road gang, or road establishment, suggesting an ongoing amalgamation of vehicle builders, oil companies, motoring clubs, and road constructors aimed at achieving their common ends at all costs.[1] Such organized lobbies have existed, most noticeably in the form of the American Road Builders Association (ARBA). The ARBA's official history, when seen in an American context, indicates an active, determined lobby but does not reveal any suggestions of overaction (ARBA 1977). Other strong American lobby groups have included the American Association of State Highway Officials (AASHO, now AASHTO, with the new T for transport), the American Automobile Association, the Highway Users' Federation for Safety and Mobility, the Automobile Manufacturers Association (AMA), the American Petroleum Institute, and the American Trucking Association. The Highway Users' Federation began as the National Highway Users' Conference in 1932 and changed its name in 1970. It has a strong insurance industry input, but nevertheless has a wide membership spectrum. The International Road Federation has also been an active lobbyist from its Washington headquarters.

Naturally, lobby groups representing those concerned with increasing the levels of road construction and/or usage should not be expected to adopt a completely rational stance. Indeed, a May 1969 *Reader's Digest* article described the highway lobby as "a pressure packed alliance of all who promote highways for profit—from truckers and construction unions to billboard firms—and it's riding roughshod over development of a sane transportation system in the United States" (Dennison and Tomlinson 1969).

Although organized lobbies have asserted considerable influence, particularly in the United States, their independent effectiveness has often been more in the minds of the antiroads protagonists than in reality. The underlying truth is that the road system has offered so many attractive features to the general public that the most widespread and effective lobby has always been the population at large, particularly those with a license to drive and a home in the suburbs. As will be discussed below, the combination of organized and natural lobbies resulted in some marked roadmaking excesses, and the activities of the antiroads groups were frequently natural and important reactions to those excesses (Kelley 1971).

Road and Town Planning

It is valuable to now concentrate on those aspects of road planning and town planning where dramatic changes have occurred this century. Many, but far from all, of these changes have been in response to the car and the truck. Changes in rural planning have not been as noteworthy, and those that have arisen are discussed in the next section.

Urban Development

Urban development is very much an issue of our times and one that is inextricably linked to roads and streets. The recent nature of the issue is seen by considering the proportion of the world's population living in towns and cities: 1885, 3 percent; 1965, 30 percent; 1985, 50 percent. Even if the definition of towns and cities is extended to also cover villages, it is still found that less than 20 percent of pre-nineteenth-century populations lived in any sort of urban area. It is important to realize in considering these events, that we recently witnessed the first major change to have occurred in the structure of cities over the five millennia of their existence (Blumenfeld 1965).

The urbanization of England and Wales in the nineteenth century can be seen from the percentage of people living in cities with over twenty thousand people: from 17 percent in 1801, this figure grew to 35 percent in 1851, to 54 percent in 1891, and to 69 percent in 1951. Likewise, the percentage of people living in all urban areas increased from 50 percent in 1851, to 77 percent in 1901, and to 81 percent in 1951 (Weber 1899; Hall 1973). By any measure, there has been a recent dramatic change in our collective lifestyles.

Historically, most towns were dominated by the needs and capabilities of their rural hinterlands, so the pace of change remained measured and deliberate. The Industrial Revolution signaled the beginning of a dramatic change in that situation. The workers of the revolution had, perforce, to live in cities, wherein were found the necessary sources of supplies, steam-produced mechanical power, plentiful labor, and markets for the products made. The demand for city living was further fueled by the advent of the cross-country train and the long-distance telegraph. At the same time, increased farm productivity much diminished the need for rural labor (Blumenfeld 1965).

As demonstrated in chapter 1, city growth had traditionally occurred within the long-standing city limits. Indeed, Elizabeth I and Oliver Cromwell had both tried to limit the slow, natural expansion of London by surrounding it with great estates. Elizabeth in 1580 also banned any building outside the walls of the city and within 5 km of a city gate.

In a portent of the twentieth century, London's geographical expansion between 1650 and 1850 was largely through the conversion of about thirty-five great estates and noble estates. These conversions were planned, controlled, low-density developments that permitted the wealthy to move westerly and out of the crowded city. They gave rise to such features in London as Covent Garden, St. James's Square, and Grosvenor Square. Covent Garden was the first, with work beginning in about 1630 under Inigo Jones as architect. St. James's Square was planned in 1662, but the major work was undertaken by Christopher Wren in about 1680, following the Great Fire of 1666. Likewise, the first deliberate expansion of Paris was to permit aristocrats to use the parklands of the Bois de Boulogne, then located outside the city walls.

Early in the eighteenth century, residential villages servicing, but outside of London proper, existed at Clapham, Streatham, Peckham, Lewisham, Greenwich, Edmonton, Tottenham, Highgate, Hampstead, and Kentish Town. The first planned suburbs were probably "new towns" built by estate owners to meet the desires of the prosperous merchants in late eighteenth-century London for out-of-town residences. This resulted in the creation of towns such as Somers Town (1786), Camden Town (1791), and the two Paragons on the Old Kent Road and at Blackheath (Stern 1981). Suburban villas for the well-to-do could be found from Greenwich to Hammersmith, between 5 and 10 km out of London. Nevertheless, in 1801 rural land still existed within 7 km of the city. A similar pattern occurred in the United States, with spacious upper-class suburbs first advertised in 1719 at Barton's Point in Boston, followed by like developments in Philadelphia and New York.

Previously the land outside the city walls had been a place for the poor and destitute (Jackson 1985). With the Industrial Revolution, it became a desirable place to live and the demand slowly widened to embrace the middle classes. The working class stayed confined to decaying city cores and to unplanned gray areas on the city's downwind fringes.

Over the period of the Industrial Revolution, the population within London's walls declined steadily, as the following data illustrate: the population grew from seventy thousand in 1695 to eighty-seven thousand in 1750, then dropped to seventy-eight thousand in 1801 and to fifty-eight thousand in 1811 (Mitchell 1972). The decline has continued to the current day. However, with a total population over this period of about a million, most people were already living outside the original city walls. By 1820 in the United States and by 1850 in Europe, using any measure, the urban fringes of cities were growing more rapidly than their cores.

An empirical inevitability with such a spread was that the population density of suburban developments declined exponentially with distance from the city center (Clark 1977). Despite overall population increases, the population densities of greater New York and of nearby Philadelphia reached their peak in 1860 and continued to drop thereafter (Blumenfeld 1965). Inner-city areas continued to attract business and administrative facilities, further lessening the areas available for housing.

Thus, the increasing population pressures on cities were being met by continued inner-city crowding (fig. 9.1) and by a gentle movement of the upper classes to the new suburbs. Not surprisingly, the residential areas within the cities in the latter half of the nineteenth century were crowded and congested, and significant parts were decidedly unpleasant urban slums. In 1880 land reformer Henry George described urban tenement slums as "breeding the barbarians who will destroy the new Rome." As a solution to the crowding and associated "moral decay," city planners and social reformers further encouraged the flight of the population to the suburbs (Foster 1981). Others tried to reform cities via the City Beautiful movement and its righteous belief that the physical appearance of a city could be used to enhance "moral purity and harmonious moral order."

An alternative to cities expanding radially via peripheral, circumferential suburbs was a revival of the self-contained *new town*. Conceived by nineteenth-century industrial visionaries and social reformers such as Robert Owen, Joseph Rowntree, George Cadbury (Birmingham), and William Lever (Liverpool), the idea led to Ebenezer Howard's 1898 garden city concept—although the term *garden city* was first applied to an estate on Long Island, New York, in 1869.[2] The key elements were compactness (walking was the major internal travel mode); location in the countryside and physical separation from other towns; and self-contained employment and social facilities. The town center was served by both commuter rail and arterial highway and offered insulated travel by foot and bicycle as well as by car.

Examples of these new towns are Howard's Hertfordshire towns of Letchworth (1903) and Welwyn (1919). Howard played little part in their actual development, and he was closely aided by Lever and Cadbury. The garden cities were to have populations of about thirty thousand, a diameter of about 2 km, a rural setting, tree-lined ring and radial roads, and direct park access for all houses. Unfortunately, the grand visions were savagely cut back by funding realities. As Brindle noted in a review in 1984, "These early trials proved to be abortive, achieving at the best only partial development before major revision or being swamped by suburban growth."

Nevertheless, garden cities showed the way forward and led to such later versions as Clarence Stein and Henry Wright's 1929 Radburn in New Jersey (Brindle 1983). Radburn in the 1930s was called the "town for the Motor Age," and Lewis Mumford in 1964 called it "the first major departure in town planning since Venice." The key advance was that in Radburn, cars are separated from local residents. In order to achieve residential privacy and quiet, cars come to the house—not by a freeway, but by a cul-de-sac.[3] Separate walking paths are an integral part of the design. In 1980 Birch wrote that Radburn "has acted as a permanent reference for generations of planners, and it persists as a respectable icon in the field's literature."

Figure 9.1. A London traffic jam in the late 1860s, as seen by Gustav Dore. The street is in the Temple-Tower area of old London. *From Dore and Jerrold (1872).*

A related development was Clarence Perry's 1923 concept of the residential precinct self-contained within a larger suburb and protected from any through traffic. It was first applied at Fairlawn, New Jersey, in the 1930s (Hall 1988).

Urban Transport

Until well past 1850 the bulk of the working population could not afford to use public transport. Daily travel was only feasible for most citizens if undertaken on foot. There was as yet no improvement in urban transport to match the land use and population changes which were occurring. But the demand was there and it was soon to be satisfied.

The burgeoning new railways took advantage of the situation in two main ways. First, nineteenth-century rail development was usually based on private companies being given wide powers of compulsory land acquisition in urban and rural areas (Kellett 1969). The companies naturally acquired the land for their routes in the working-class housing areas where land prices were lowest, as can be seen from the location of London's arterial rail routes. It is estimated that half a million Londoners were dispossessed of their homes in this process. Thus, the urban railways usually represented a completely new and dominant grid superimposed over the old street systems. This pattern, which is most obvious from an aerial view, was often repeated throughout the world during the urban freeway construction periods in the 1960s and 1970s (James 1984).

The urban renewal through urban destruction idea was often well regarded, from George Haussmann to the present day. Indeed, the process of urban renewal by demolition was sometimes referred to as "Haussmannizing," in recognition of Haussmann's Parisian methods. As an example of the attitudes at work, an 1838 select committee in London wrote:

> There were districts in London through which no great thoroughfares passed, and which were wholly occupied by a dense population of the lowest class of persons who being entirely secluded from observation and influence of better educated neighbors, exhibited a state of moral degradation greatly to be deplored. It was suggested that this lamentable state of affairs would be remedied whenever the great streams of public intercourse could be made to pass through the districts in question. It was also justly contended that the moral condition of these poorer occupants would necessarily be improved by communication with more respectable inhabitants, and that the introduction at the same time of improved habits and a freer circulation of air would tend materially to extirpate those prevalent diseases which not only ravaged the poorer districts in question, but were also dangerous to the adjacent localities. (Edwards 1898)

What a fine charter for road constructors! And there must have been some perceived benefits, as sixty years later Charles Robinson continued the theme in his 1901 book *Modern Civic Art:*

> It has been found that often there is no better way to redeem a slum district than by cutting into it a great highway that will be filled with the through

traffic of a city's industry. Like a stream of pure water cleansing what it touches, this tide of traffic pulsing with the joyousness of the city's life and purpose, when flowing through an idle or suffering district wakes it to larger interests and higher purpose.

The second way in which the railways used poor urban conditions to their advantage related to suburban issues. Although the early railways were designed for long-distance travel, they naturally located intermediate stations wherever their lines passed through nearby rural villages or the perimeters of large cities (Jackson 1985). In addition, as railway lines extended, the station was usually located in the center of any new development. A typical example is Chestnut Hill, Philadelphia, developed from 1884. This meant that commuters in those precar days had to walk to and from the station every day, thus leading to a compact urban layout (Stern 1981).

In cities such as London the railways also contributed to suburban developments, particularly over distances of 30 km or so, where previously self-contained towns like Croydon, Redhill, and Reigate were able to provide dormitory living for city workers. Continuing the earlier trend, these workers were very much from the middle and upper classes. The towns the railways created in the period between 1860 and 1880 were thus towns with railways, rather than railway suburbs, and catered to people "who want to buy 20 or 30 acres of land and build [them]selves a good house." The railway only began meeting the needs of the lower-class traveler as the nineteenth century waned and passed and as the competitive pressures of the electric tram came to be felt (Kellett 1969).

Thus, it cannot be said that the train or the tram or the car or the freeway caused the urban sprawl of both housing and industry, for they were all easily predated by that sprawl. Clearly, factors far larger than transport have been at work, and it would be wrong to see cities as merely the passive consequences of transport developments (Lampard 1955). Transportation is merely a means to some other end, a fact observed with some perception by the commentator who in 1899 noted of the sprawl, "The American penchant for dwelling in cottage homes instead of business blocks after the fashion of Europe is the cause, and the trolley car the effect" (Jackson 1985).

Nevertheless, the role of transport in aiding and abetting the process was significant; three of the real transport villains have already been discussed. The horse bus and then the horse tram had had minor influences on city size, but the prime catalyst from 1880 to 1914 was the electric tram (Jackson 1985). The city of Boston provides a quite quantitative and chronological example of what occurred:

> From 1852 to 1873 the horse railroads of Boston merely stretched out the existing city along established paths. The outer boundary of dense settlement moved perhaps half a mile, so that at the time of the great Depression of 1873 it stood two and a half miles from Boston's City Hall. During the next fourteen years, from 1873 to 1887, horsecar service reached out a mile and a half further. . . . In the late 1880's and 1890's the electrification of street railways brought convenient transportation to at least the range of six miles from City Hall. (Warner 1962)

During this expansionary age it was common for tram advocates to take up the theme encountered earlier and to speak of the "moral influence" of their services on the life of the city, allowing citizens to flee the crowded tenements for the spacious suburbs.

As another, somewhat later, example of the influence of the tram, Los Angeles—the epitome of the suburban city—began to expand under the influence of its electric tram system, well before the car had its opportunity to exert an influence (Lay 1983). It is sometimes forgotten that in its heyday in the 1920s the Los Angeles electric tram operated some 2 Mm of track, often in its own right of way. The same pattern applies in the author's hometown of Melbourne.

The availability of suitable transport is certainly a necessary condition for the urban sprawl to occur, as the following discussion will illustrate. Technically, one of the key parameters is the time that people have available for work-related travel. There are only twenty-four hours in a day, and most travelers to work are prepared, it would seem, to devote about an hour of their day to travel, with about 15 percent prepared to spend an hour and a half (Lay 1990). When travel was on foot, this time budget constraint limited the home to work separation to a few kilometers. With the coming of the affordable horse bus, it was about 5 km. The electric tram raised the possible separation to 10 km, and the suburban electric train pushed the limit to 20 km.

The car and its attendant freeway now permits a 50 km separation. Modern cities have expanded accordingly, and urban population densities in persons per hectare have dropped from the original six hundred discussed earlier to one hundred in urban areas and to fifty or less in suburbia.

Land speculators frequently encouraged this pattern (Cannon 1967; Foster 1981). For example, between 1900 and 1920 Los Angeles grew in population from one hundred thousand to six hundred thousand. During this time the owners of the Pacific Electric Railway Company developed thirteen new towns in the area.[4] As one consequence, the famous Beverly Hills started in 1906 from a subdivision created near Pacific Electric's transit track along Santa Monica Boulevard. Although Los Angeles is now known for its urban freeways, they were relatively late arrivals in that city's transport history.

The general situation in the United States has been described as follows:

> The original [urban transit] franchise holders had set the fares too low. . . .
> They did this because their profits resulted from the increased land value
> transit generated. . . . As long as real estate profits could be realized by build-
> ing and subsidizing transit lines, there was an impetus for the private sector
> to provide the needed subsidies and build, operate and even refurbish transit
> lines. (Schaeffer and Sclar 1975)

A similar pattern occurred in many cities. When economic reality hit the urban transit operations established by the land speculators and the fare box proved unable to cover operating costs, the car was ready to take up the slack and sustain the suburban sprawl promoted by the land speculators. As will be seen later, the car lobby was also willing to help, most noticeably by buying up teetering urban transit companies.

In the private enterprise environment of Los Angeles the transit network did not survive. In many other cities, such as Boston and Melbourne, similar networks survived only by moving to public (i.e., government) ownership and then to operating subsidies. The role of urban transport is pursued further later in this chapter.

This century has thus seen the galloping urbanization and suburbanization of much of our society, placing great demands on the cities and on their urban transport systems, many of which were found wanting. Most public transport systems are now unable to function without major operating subsidies. Nevertheless, such systems are widely recognized as desirable and worthy of subsidy, as cities can only perform their intended function if they are well served by public transport. Only in very small or low-density cities is the road system alone adequate for travel needs.

The Suburbs

The desire for suburban living is indeed very ancient. The following example was found on a clay tablet dated 539 B.C.: "Our property seems to me to be the most beautiful in the world: It is so close to Babylon that we enjoy all the advantages of the city, and yet when we come home we are away from all the noise and the dust" (Jackson 1985).

Strong pressures for suburban living existed throughout the nineteenth century, particularly for the wealthier classes. The role of the new transport facilities was to help satisfy the preexisting demands; they did not cause those demands.

The advocacy for the development of new American suburbs after the Civil War (1861–1865) was undertaken by Andrew Downing and his disciple Frederick Olmsted. Their advocacy was based on the concept of class-segregated suburbs "for the emerging elite and middle income groups of American cities. The new suburbanites sought country-like surroundings, part of Victorian romanticism's ideal life. Country living and home ownership also suited traditional American values emphasizing rural life styles" (McShane 1979).

The provision of new transport facilities merely accelerated a process that had been in operation throughout the nineteenth century. By the turn of the century, the process had resulted in new cities "segregated by class and economic functions and encompassing an area triple the territory of the older walking city [which] had clearly emerged as the center of American urban society. The electric streetcar was the key to the shift" (Jackson 1985). Indeed, as discussed in chapter 5, for some decades the streetcar played the major immediate role in urban expansion.

In Britain the desire for suburban living was also prompted by a reaction to the side-by-side, back-to-back low-income housing developed to meet the central city, walk-to-work accommodation demands of the Industrial Revolution. This desire was legalized by the British Public Health Act of 1875, which specified such items as house spacings and road widths and resulted in the spread of the individual suburban house.

A key element in the British pattern was the garden suburb, which adapted the village to the suburb—epitomized by Barry Parker and Raymond Unwin's development of Hampstead Gardens in London between 1905 and 1914. The two had cut their planning teeth working on Howard's Letchworth new town.

From a needs viewpoint, the sprawl of modern cities resulted partly from the demand of many citizens for a low-density suburban lifestyle and partly because the land on the urban fringe was usually considerably cheaper than land at the urban core. Citizens have commonly traded off poor accessibility for lower housing costs and then, with innate shrewdness, demanded that governments compensate them for their accessibility loss by building a transport system to service their relatively inaccessible properties.

At the turn of the century, the move to the suburbs was strong and dramatic, despite the fact that city centers received a strong fillip from the invention of the lift (elevator), whose vertical transport allowed much higher density development of valuable city core sites.

Many city centers were dependent for their existence on being their community's only shopping and work center. When the car offered a closer suburban alternative, such centers could no longer either compete or live up to their common name of central business district (CBD). The construction of radial freeways often aided the process by providing for outward as well as inward movement. The resulting decline in CBD land values was self-defeating, for the centers then deteriorated and became even less attractive. Consequently, the population of most CBDs declined steadily from the early 1930s. Frank Lloyd Wright saw this decline of the dominance of city cores and the growth of suburban centers as inevitable and largely desirable. There was more than a little accuracy in Henry Ford's prediction that "cities are doomed, we shall solve the city problem by leaving the city" (Ford 1924).

The move to suburban living with its individual houses, self-contained yards, and reliance on the private car had an important secondary effect. No longer was the street pavement the major place in which to play and socialize. People retreated to their yards, and the car took over the street.

Another important phase of suburban development was the early belief that the wide avenue was the best location both for prime residential development and for shopping facilities. Strip development became commonplace. An incentive for a change from residential strip (or ribbon) development was that the early trams, which were very noisy, frequently used these strips and turned them into residentially undesirable "trolley streets." The vacated sites became ideal for commercial development and the ugliest form of strip development (Armstrong 1976). Nevertheless, once travel speeds began to exceed walking speeds, one casualty was the centralizing role of the single marketplace in city development.

Many of these strip roads, with their gasoline stations, restaurants, fast-food stores, advertising signs, and various car-oriented operations, came to be major urban eyesores. They were described (Blake 1963) with only mild exaggeration as "poisoning . . . the townscape with festering sores along their edges." Nevertheless, strip development long preceded the car. It is a natural development for any community located along a major road and can be found in many older cities, such as Bristol. It took half a century for many communities to learn the need for access control and land-use management along such roads and to discard the long shopping strip for more compact, market-type arrangements of streets, squares, and malls.

The first car-based shopping center was the Country Club Plaza, opened in Kansas City in 1922. The first motel was probably a set of tourist cabins built at Lincoln Park, Los Angeles, in 1922 (Belasco 1979). However, the word was first used in 1925 by the Motel Inn on Highway 101 in San Luis Obispo, California. The first drive-in movie theater opened in Camden, New Jersey, in 1932. Fast-food outlets began with the White Castle chain in Kansas City in 1921.

The millennia-old tendency toward strip development was based on the view that the roadside was always ripe for commercial development. The problem was epitomized by the long fight in the United States between the billboards of the advertising sign lobby and those opposed to roadside advertising. As Ogden Nash wrote in his "Song of the Open Road":

> I think that I shall never see
> A billboard lovely as a tree.
> Indeed, unless the billboards fall
> I'll never see a tree at all.[5]

In 1965 the Highway Beautification Act banned new roadside advertising along federally funded roads in the United States. The legislation lasted just three years before the advertising industry caused it to be emasculated by congressional amendment. The battle ended in uneasy compromise. The truce is still shaky.

Perhaps the most famous roadside advertising sign was the Burma-Shave series, which appeared alongside American roads between 1927 and 1963. Burma-Shave was a shaving cream company that advertised with a series of five red signs with white writing, placed about 50 m apart and carrying lighthearted verse. Typically:

> Henry the Eighth
> Sure had trouble
> Short term wives
> Long term stubble
> Burma-Shave.

The Role of the Car

There were prophets of the car's impact. The president of the American Automobile Association accurately predicted in 1903 that, "with the coming of the low priced automobile you will find [suburbs] building up with homes, each on its ample plot of ground . . . [with] . . . the pleasure of existence enhanced immeasurably" (Foster 1981). In the same year another American commentator noted with equal perception:

> The automobile will hurry this tide [begun by the tram] countrywide with accelerated speed. It will begin its work with the well to do—those who are capable of keeping up a private establishment. This is the class which is now in the van of the movement to escape from the city. . . . [In the future] the inevitable slump in price [of the car will see it] become the vehicle, not only of the few, but of the many, bringing with it a relief from city life . . . opening up the country for modest homes, to which the city worker may daily escape. (Flink 1974)

Although the car did not initiate suburban developments, it certainly gave them great encouragement. The result left a lot to be desired. As Mumford (1961) described the situation: "As soon as the car became common, the pedestrian scale of the suburb disappeared, and with it, most of its individuality and charm. The suburb ceased to be a neighborhood unit: it became a diffuse low density mass. . . . The suburb ceased to be a refuge from the city and became part of [it]."

The car also took up the slack in another way. The precar, streetcar city had commonly developed in a star pattern, with each finger of the star corresponding to an electric tram (or other transport) route. People had to live within walking distance of these lines. With the coming of the car, no such restriction existed, so all the undeveloped areas between the fingers were soon occupied by new suburbs. Similarly, the availability of the urban truck meant that factories and warehouses no longer had to be located near a railway siding or at wharfside. Soon large tracts of the cheaper land on a city's outskirts became industrial zones.

In the United States, Charles Robinson's turn-of-the-century (twentieth) City Beautiful movement—with Haussmann in mind—began imposing grand radial and concentric boulevards on existing city layouts. The idea was aesthetically grandiose but largely ineffective in a democratic community. His views had strong righteous and moral overtones. Nevertheless, some of the plans formed the basis for the parkways to be discussed below. The garden city movement was in many ways a pragmatic outcome of the City Beautiful concept.

These high hopes for new routes occurred in the spirit of technological optimism that existed in the period between the two world wars. Planning philosophy generally held that building radial roads to met the growing demands of the car would manage rather than incite urban sprawl. Faith was placed in greenbelts, which were rings of open space planned to encircle a city. Nevertheless, the sprawl continued, and most greenbelts succumbed to land development pressures. Until the 1970s it was also widely thought, in the same optimistic spirit, that engineering refinements to the car and the street would eliminate most of the problems created by the car (Foster 1981).

Pre-nineteenth-century cities tended to develop on a grid pattern, with only the roads bringing services from the hinterland providing radial routes. As the city core expanded, the need for circumferential routes became more obvious, but increasingly difficult to establish because the cities continued to expand and no pre-existing transport right of way was available. Neither the ring road (beltway) nor the town bypass were natural consequences of transportation evolution. Both needed to be superimposed solutions; that is, they required either good luck or perception and persistence on the part of planners. In Europe good luck prevailed, and ring roads were frequently built on the sites of old city walls (Morris 1972). Beijing has similarly used its ancient walls.[6] Even when an old wall site was available, creation of the ring road often also needed a strong and authoritarian central administration; the first Parisian ring roads occurred in just such circumstances in the eighteenth century (Stave 1981).

Indeed, the common experience with planned rather than evolved ring roads was that they dramatically attracted new developments to their wayside. They thus quickly lost their enveloping role and became internal service rings. A famous example is Boston's Route 128. Within several years after Route 128 opened, it had attracted twelve thousand new employees to its corridor. A cynical view of its effect was that "no sooner was the circumferential Route 128—originally conceived to get traffic around the city—completed than the fields through which it passed became the sites of new factories that created urban traffic problems outside the limits of Boston" (Whitehill 1966). However, a retrospective review of eleven American cities with ring roads indicates that the ring road had been just one of many influences on the pattern of urban development (Mills 1981).

A major related road development has been the imposed planning and construction of bypasses around rural towns where the old highway alignment naturally led into and through the town center. As travel speeds increased, the effect of such through traffic became a growing nuisance. The bypass largely removed that nuisance. Nevertheless, common experience has been that local traders initially oppose a bypass but subsequently praise its benefits. In the United Kingdom the construction of bypasses began in 1920. Bypass construction in the United States in the 1930s was not successful, because it was found that most highway traffic was visiting rather than bypassing each town. This pattern was to change as improved transport permitted longer travel distances.

In this overall context, it is worth recalling the comments of Buchanan (1958) on the frustrations of designing cities for the car: "We always seem to have been a lap behind the motor car, by the time we had thought of 30 foot carriageways, it was duals that were really needed, by the time we had thought of duals it was motor roads that we wanted, by the time we had thought of roundabouts it was flyovers that were needed." Part of the problem was a very understandable lack of foresight as to the impact that the car would have. Writing in 1930 one official (Blamey 1930) explained, "Twenty years ago no one could foresee the growth of the motor car, and in the layout of cities this could not possibly have been taken into account, and we and those who come after us have the problem to solve."

There has also been a tendency to construct new roads along easy rights of way, which usually meant undeveloped river valleys and similar urban open space. This tactic slowly disappeared as the strength of the urban opposition to untrammeled roadmaking increased. On the other hand, as roadmaking technology improved, it became possible to place roads in previously impossible locations, particularly in swampy or marshy zones. As a result, many cities became separated from their waterfronts by new road developments.

Travel by private car has been the desired method of transportation in most cities throughout much of this century. Yet road systems and car parks soon became so congested that the level of service provided by the private car was lowered to such an extent as to make some alternative travel mode at least as attractive. The concept that congestion could be used to provide a balance between private and public transport was first proposed by the economist Arthur Pigou in 1920.

One hard lesson to learn from Pigou's hypothesis was that roadmaking could only partly alleviate congestion, because the construction of more road space usually attracted more road travelers. In many cases the major change effected by road construction was an increase in the number of people traveling by road, rather than improvement in the level of service offered to existing road travelers. In addition, cities usually reached a political and social equilibrium between the amount of road space needed for road transport and the effect of the consumption of that space on the city's urban fabric.

In this equilibrium, local improvements—particularly to local networks and to intersections—commonly produced locally desirable traffic benefits and were particularly useful in alleviating the problems stemming from road layouts inherited from the precar days. Outside peak hours, the effects of such works were particularly noticeable.

To date this review has concentrated on urban areas, where rapid changes have occurred throughout this century. In rural areas, the major planning influence of the car has been to remove the viability of many of the small towns established when a reasonable day's journey was 10 or 20 km. These towns often disappeared, with only their skeletons remaining.

Predicting the Pattern

Analytical attempts to predict and explain the amount of travel that people undertake, and hence provide forecasts for planning, have two bases. First, it is necessary to explain why people locate their activities and their consequential trip origins and destinations in particular areas. Such a study of land use is outside the scope of this book, although we have flirted with the issues involved. The second stage is to predict the quantity and quality of the travel people undertake between the established origins and destinations. Henry Carey gave the basis for this work in 1858 when he suggested by analogy with Newtonian physics that a "great law of molecular gravitation" applied. Some empirical confirmation was subsequently obtained from studies of migration patterns. The 1927 Boston Transportation Study used a gravity model in that it assumed that travel was inversely proportional to the square of the distance between origin and destination. However, the application of the gravity model was not commonplace until a seminal application by Alan Voorhees in 1955 (Heightchew 1979).

Numerate transport studies began with John Burgoyne in Ireland in 1837, but the approach was little used. However, as a result of the burgeoning car, in the 1920s a number of cities undertook formal transport studies, considering road capacities, traffic systems, and traffic forecasts. The leader of the American thrust was Herbert Fairbank at the Bureau of Public Roads (FHA 1976). In a farsighted move the United States Congress, beginning with the Federal Aid Highway Act of 1934, provided 1.5 percent of road funds for surveys, plans, and investigations (Weiner 1986). By 1940 all American states were participating in such activities. The surveys included road inventories, traffic counts, and vehicle use studies. Urban travel studies began in 1944.

Freeways

The first thing an approaching observer from outer space would notice about the developed world's land transportation system would be its freeway network. Over the last half-century these ways have been major consumers of space and effort and have been very obviously well used. They have also been major causes of community concern and dissension as they have attempted to cater to the seemingly insatiable demands of the car. Freeways clearly deserve careful review.

Parkways

The early military history of the freeway was discussed in chapter 4, with particular emphasis on the German autobahn and the Italian autostrade. However, they were preceded historically by the Long Island Motor Parkway in the United

States. The parkway had been built on private land in 1906 as a racecourse for the Vanderbilt Challenge Cup, first run in 1904 and sponsored by the American Automobile Association, with a trophy donated by William Vanderbilt.

Initially, the parkway was a single asphalt carriageway, 10 m wide and 10 km long. In 1908 it was opened to the public as a toll road for cars only. Construction continued, using reinforced concrete as well as asphalt, and by 1910 some 80 km were available for public toll use (FHA 1976). It was the first public road to use overpasses to eliminate intersections and the first to employ banking on curves, a frill that had not been needed in the low-speed precar days. The road was not successful economically, and its sharp curves led to its waning popularity. Nevertheless, it operated as a toll facility until 1937.

The term *parkway* predated the Long Island facility. It originated in Williamsburg, Virginia, in 1699 and was initially applied to roads with wide, grassy central medians. The concept was popularized by the development of Central Park, New York, in 1858 by the pioneer landscape architects Frederick Olmsted and Calvert Vaux. Olmsted's design sank the four main transverse traffic roads below ground level and used bridges to carry local surface traffic over these arteries. These surface roads then crossed over the path network on a separate set of bridges. Olmsted was a disciple of Downing, the author of America's first book on landscape gardening, and had also been influenced by Haussmann's activities in Paris. He saw the major power of roads in visually shaping the landscape. His son was also prominent in the landscaping of roads. Both were named Frederick Law Olmsted. They were not the first to consider the landscaping of roads, as the great fifteenth-century Italian Renaissance architect, Leon Alberti, had written in his 1485 book *De re Aedificatoria* that a road should be made "rich with pleasant scenery."

An 1868 Brooklyn park proposal by Olmsted and Vaux led to the construction of the Ocean and Eastern parkways (Patton 1986). Consequently, a number of fine landscaped parkways were built in New York and Washington in the period between the world wars. They were effectively linear parks containing a road built for noncommercial traffic. In many ways the parkways were similar to current-day freeways, but were usually designed for lower speeds and thus had less generous alignments and permitted side access to the roadway.

The first of this series of parkways was the 27 km Bronx River Parkway, proposed in 1907 by a commission established by New York State to clean up the banks of the Bronx River and protect animals in the Bronx Zoo from water pollution. The parkway ran from the Bronx Zoo to White Plains and was built between 1916 and 1923 as a joint venture between Westchester County and New York City. The original design consisted of four unseparated lanes with the alignment comprised largely of long curves. There were no intersections, and crossroads passed over the parkway. The design speed was 60 km/h and no trucks were permitted (FHA 1976). The parkway was widely acclaimed by the local community, and Westchester County real estate developers actively supported the extension of the concept. By 1932 the county boasted some 270 km of parkway.

Robert Moses, later builder of New York's urban freeways and said by many road protagonists to be this century's greatest road builder, began building parkways as commissioner of the Long Island Parkway Commission from 1924 to 1963. He deliberately kept the overpasses on his early parkways low to keep out the buses used by the lower-class non−car owners.

Further parkways were built by the federal government as job creation schemes during the Great Depression (Patton 1986). The first of the federal parkways was the initial 24 km Alexandria section of the George Washington Memorial Parkway along the palisades of the Potomac. It was designed and built by the New York Parkway Commission team and was opened in 1932. However, the main roadmaking emphasis in the United States between the wars was on applying a hard and wide running surface to the existing two-way road network. By the early 1930s most of the country was linked by a road system usable in all weather.

A freeway has safely separated carriageways for each travel direction, no access from abutting properties, and grade-separated intersections. None of the above roads had quite the geometry to qualify as freeways, although the distinction that excludes some is only marginal.

European Enhancements

The carriageway is the pavement available for travel, and a dual carriageway separates travel in each direction. Dual carriageways were built in Paris, Berlin, New York, and other American cities prior to the automobile age. However, none had limited access. The title of the world's first limited-access, dual-carriageway road belongs to a German road. Between 1913 and 1921 a private group called AVUS— Automobil Verkehrs und Uebungsstrasse—built a perfectly straight, dual-carriageway, controlled-access road of about 10 km between Nikolassee and Charlottenburg, running through Berlin's Grunewald Park. Had it not been for its narrow, unprotected median, it might have been the world's first freeway. An expanded AVUS remains in operation today as part of the E51's entry into Berlin. The AVUS experience led to the establishment in 1924 of a society devoted to the study of roads designed for the car—SUFA—which presented plans for a 22.5 Mm network in 1929.

The German autobahnen were originally planned in the late 1920s to alleviate massive unemployment—historically, a common subsidiary reason for road construction was to occupy large numbers of relatively unskilled people in construction and maintenance. The autobahn's chief advocate was Fritz Todt, engineer-manager of a Munich construction firm, Nazi party member, and close confidant of Hitler. He was to become director-general of German roads. The autobahn designers had inspected the existing American parkways, and they returned to Germany intent on building their new roads on a grander, straighter scale. Design speeds were set at 165 km/h, about double the values then in use on the parkways.

The first autobahn was a 20 km stretch from Bonn to Cologne, which was built in 1929–1932. It was built by the Rhineland Province, which was then the only province with road construction powers, under the control of the Cologne mayor, Konrad Adenauer, who was later to lead postwar Germany with distinction. The road had a divided carriageway with two lanes in each direction and was probably the world's first freeway. The German Third Reich took over all autobahn construction in 1933. Hitler's subsequent subversion of the program was discussed in chapter 4.

Another early contender for the title of first freeway was the Italian autostrada. A great deal of support for autostrada construction came from Mussolini when he rose to power in 1922, intent on restoring Italy to the glories of the Roman Empire (Hindley 1971). He quickly adopted Piero Puricelli's 1921 plan for a new Italian road network. The first autostrada was a two-way, single-carriageway, controlled-access road built 50 km from Milan to Varese and Lake Como (Cron 1976a; Earle 1974). The first 21 km stage was opened in 1924 and the remainder in the following year. The design speed was 60 km/h and the paving was predominantly concrete (PIARC 1969). It was built as a toll road by a private company originally called Strade e Cave, which later changed its name to Societa Anonima Autostrade, and then to Azienda Nazionale Autonoma della Strade (ANAS). It was given capital funding from government-raised fuel and general taxes.

The Interstate System

Up until the 1920s the roadmaking impetus had been to get the United States "out of the mud." With that task largely accomplished, economic and social imperatives began to take over.

Work on the first urban parkway, the Henry Hudson Parkway down the west side of Manhattan Island in New York City, began in 1934. Parkways soon became a significant part of American travel mores. Their success and that of the new German autobahnen whetted the American appetite for freeways. Whereas the parkway was designed for beauty and recreation and private travel, the freeway was designed for movement and commercial travel. These definitions were proposed by Edward Bassett while he was chairman of the American National Conference on City Planning. In 1930 he coined the word *freeway* in opposition to the growing view that major roads should all be toll roads, and therefore not free (Patton 1986). Today *free* usually implies freedom from intersections and uncontrolled access. A New Jersey alternative to *freeway* is *speedway*. Indeed, the epitome of the extremes of these concepts is the highly efficient and heavily trafficked—but awesomely ugly—New Jersey Turnpike, built in the early 1950s (Gillespie and Rockland 1989).

A marked general fillip for freeway construction in the United States came from the experience of Los Angeles with its Pasadena Freeway. The freeway was opened in 1940 over the 6 km between downtown and Pasadena, and it was called the Arroyo Seco Parkway for the first year. Its name change signaled a major change in public attitude, encouraged no doubt by the dramatic increase in Pasadena land values that followed the opening. The champion of the freeway had been Lloyd Aldrich, the Los Angeles city engineer since 1933. In 1939 he produced a report advocating 1,000 km of freeway for Los Angeles, beginning with the Pasadena Freeway. In the 1950s and 1960s many cities around the world were also to produce grandiose and unachievable road plans.

Similarly, the construction viability of interurban freeways was demonstrated in the United States by the completion of the 270 km Pennsylvania Turnpike through the Appalachian Mountains between Harrisburg and Pittsburgh in just two years between 1938 and 1940, thus meeting election commitments. What is often overlooked is that the turnpike did not require the acquisition of new land, but utilized an unused railway right of way known as Vanderbilt's Folly (officially the South Pennsylvania Railroad), which had been built to about 60 percent completion between 1883 and 1885 (Shank 1973). The project had the personal support of President Roosevelt and was funded by direct federal, reconstruction, and unemployment (WPA) grants.

The designers drew significantly on the recent German autobahn practice, so the turnpike had dual 7.2 m carriageways separated by a 3 m median, providing two lanes of travel in each direction. It halved travel time between the two cities and provided an alternative to the Lincoln Highway (U.S. 30). Moreover, it was soon carrying ten thousand vehicles a day, whereas a decidedly pessimistic 1939 Bureau of Public Roads (BPR) report had predicted about seven hundred. The turnpike was the very first section of the mammoth interstate system (to be discussed below) and operates today as part of Interstate 76 (I-76).

The parkway-freeway concept caught the imagination of the American people. The General Motors Highways and Horizons Futurama exhibit at the 1939 World's Fair in New York was the most popular show at the fair, attracting some five million visitors (Geddes 1940).

In a famous incident in 1937, President Roosevelt called BPR chief Thomas MacDonald into his office and handed him a map of the United States on which the president had drawn six crisscross lines representing a national road system. Mac-Donald was told to study the program. Returning to the BPR, he quickly passed the map and the task on to Fairbank. The BPR, using newly available planning data, reported back to the president in 1939, recommending that a national highway system be constructed, but questioning whether the traffic demand was adequate to justify basing the system on toll roads (U.S. Congress 1939). The pessimism of BPR estimates has already been encountered.

Roosevelt appointed the National Interregional Highway Committee in 1941 with MacDonald as chairman and Fairbank as secretary. With confident foresight, part of Roosevelt's charter to the committee was to study "the possibility of utilizing some of the manpower and industrial capacity expected to be available at the end of the war." In 1944 the committee's International Highways Report recommended establishing the National System of Interstate and Defense Highways (Weiner 1986). The concept first appeared formally in the 1944 Federal Aid Highway Act (FHA 1976; see Rose 1979 for a different view of this period). Its strong military basis was described in chapter 4.

The immediate postwar years created other demands for public and political attention, and the idea of an interstate system went into hiatus. Nevertheless, there was a strong feeling in the American community that better roads were essential. A paragraph in the March 1952 issue of the *Ladies Home Journal,* noted, "Some of our more frugal taxpayers at the club jump me when I argue that we need more four-lane highways in our State. 'Too expensive they say. . . .' They forget how our narrow roads frazzle every motorist's nerves and ulcerate his disposition" (25).

In 1952 and 1953 General Motors ran an expensive national essay competition on the theme "How to plan and pay for the safe and adequate highways we need." The two winners were General Lacey Murrow from the American Association of Railroads and Robert Moses. The modern era of the freeway began soon after this seminal essay competition, as lobby groups began to encourage a political vision of a nationwide road network.

In 1954 President Eisenhower appointed a committee under General Lucius Clay to study American highway needs. Clay was an ex–military engineer and hero of the Berlin airlift during the cold war, and then a board member of General Motors. Eisenhower's secretary of defense, Charles Wilson, was also from the GM stable. Eisenhower had appointed a personal confidant, Francis du Pont, to succeed Mac-Donald as head of the Bureau of Public Roads. Du Pont just happened to be a member of the family that owned GM. In retrospect, the influence of ex-military technocrats was possibly stronger than that of the auto industry, although the two groups had been intimately linked in the United States since World War I. The Clay Committee advised Eisenhower that an interstate system was needed.

Arguments over alternative funding methods (tolls, loans, and pay-as-you-go) delayed early passage of the legislation. Tolls were not favored by the federal bureaucrats. The loans would have to be repaid through tolls or fuel taxes, possibly levied on the roads built by the loan. The pay-as-you-go approach funds works from fuel taxes raised on other roads before the work starts. The loan versus pay-as-you-go debate was in many ways a surrogate for a larger debate between federal and states' rights. Senator Harry Byrd, a past governor of Virginia, was a strong advocate of the pay-as-you-go concept and coined its terminology. His states' rights side won, and the subsequent outcome was described by the Clay Committee secretary, Francis Turner, as a system of national highways rather than a national system of highways.

The system was initiated by two acts, a new Federal Aid Highway Act and a Highway Revenue Act, both passed in 1956 and signed into law by President Eisenhower on 29 June of that year. The acts required the system to be designed for traffic projected for 1975 and, in an exceptional move, the initial funding was authorized for twelve years, from 1957 to 1969. The federal government was to meet at least 90 percent of the cost from a trust fund maintained by motor vehicle taxes, predominantly a gasoline tax, and based on an existing California model.

A new act in 1968 added 2.5 Mm to the proposed network. Although initially scheduled for completion in 1972, the 70 Mm system was, though very close to completion, still under construction in 1990. It was by far the world's largest public works program.

The Urban Freeway

Between cities the freeway has made a major contribution to this century's transport development, providing an economical and socially attractive method of travel. The freeway happily accommodates the independently operated car, bus, and truck traveling at speeds of 100 km/h or more. This accommodation is achieved with increases in both travel efficiency and safety.

The high-speed urban freeway paints a different picture. Conceived by enthusiasts such as Moses and his Henry Hudson Parkway in the early 1930s, by the commencement of the interstate system in 1956 the United States already had some 800 km either completed or under construction. It has been suggested that the grid layout of the streets in the typical American city, extending to the horizon, provided an open invitation for the freeway to enter the town (Patton 1986).

Unfortunately, with transverse dimensions of over 120 m, the size of these urban freeways exceeds human scales, consumes significant urban space, and creates major urban barriers. Other side effects such as noise, air pollution, and poor aesthetics are often unacceptable. Landscaping—so much a feature of the early parkways—played only a minor role on most of the initial urban freeways. Understandably, inner suburban residents objected to the disruption of their living areas to provide facilities for commuters between outer suburbs and the city core. Perhaps all this is not surprising, considering a view expressed by Moses in 1964: "You can draw any kind of picture you like on a clean slate . . . in laying out a New Delhi,

Canberra or Brasilia, but when you operate in an overbuilt metropolis you have to hack your way with a meat axe" (St. Clair 1986).

The objections of commuters had a precedent. In the late nineteenth century the inner London boroughs had successfully resisted the operation in their streets of horse buses serving outer London suburbs. Subsequently, many cities retrospectively rued the devastation caused by metropolitan rail construction, particularly the hideous and noisesome "elevateds." Lewis Mumford articulated the new concerns when he said in 1957 that "the highway engineers have no excuse for invading the city with their regional and transcontinental trunk system" (Mumford 1964). The concerns were epitomized in 1959 by citizens halting the construction of the Embarcadero Freeway in San Francisco and then the Spadina Expressway in Toronto.

In addition to the complaints of the community at large, relative travel efficiency of the urban freeways is such that they often attract more usage than they were designed to accommodate. This excess demand reduces the service provided until road travel again reaches an inefficient equilibrium with other transport modes. Commentators (Meyer and Gomez-Ibanez 1981) have said that "the greatest disappointment with the interstate highway program . . . was that it did not seem to achieve its major objective of reducing traffic congestion. . . . In essence, an effective reduction in urban traffic congestion proved an elusive goal that did not yield easily to the simple expedient of building more and bigger highways." The urban freeway was cynically described as the shortest link between two traffic jams (Meyer and Gomez-Ibanez 1981).

Furthermore, the urban freeway produced other dramatic negative consequences, as one commentator has pointed out: "American urbanites made a decision to destroy the living environments of nineteenth century neighborhoods by converting their gathering places into traffic jams, their playgrounds into motorways, and their shopping centers into elongated parking lots. These paving decisions made obsolete many of urban America's older neighborhoods" (McShane 1979).

What the freeways missed was usually scavenged by the urban renewal programs, leaving little of the past remaining. The situation in one city is told strongly in the language of 1966 by Whitehill: "A monstrous overhead highway . . . cut ruthlessly through the center of Boston. The Storrow Drive, which separated the Back Bay from the Charles River Basin by six lanes of fast-moving traffic, hopelessly destroyed a great human amenity."

Mumford, writing in 1964, commented that five years earlier, "when the American people, through their Congress, voted" for the interstate program, "the most charitable thing to assume about this action is that they hadn't the faintest notion of what they were doing." One story suggests that even Eisenhower fell into that category. Upon seeing the construction of the first urban freeway entering Washington, he telephoned officials expressing disbelief that such roads could be part of his interstate system. He was soon to learn precisely what he had initiated some three years earlier (Patton 1986). Indeed, sometime between 1944 and 1962 a new urban direction was subtly added to the interstate philosophy. The first mention of urban routes was in 1947; by 1956 about 20 percent of the system length was in urban areas (Weiner 1986).

In retrospect, the urban freeway has only worked in suburban areas where it, or at least its planning, has occurred in advance of land development. Successful examples would include Atlanta, Georgia, and Perth, Western Australia.

The situation was recognized to the extent that the 1962 Federal Aid Highway Act created a federal mandate for urban transportation planning. The concept spread quickly in the United States, and the urban freeway movement went hand in hand with many grandiose urban transportation plans produced in the 1950s and 1960s. These plans were predicated on the false assumptions that providing extra road space would satisfy urban travel demands, that the diversion of urban space to roads was of first priority, and that the funds existed to put the full plans into practice. Many of the plans floundered in all three areas; they did not satisfy demand, community opposition grew, and adequate funds were not forthcoming. A key commentator (Orski 1979) observed the "unfulfilled dreams of grandiose projects of the 1960's and early 1970's."

This was not the first time that the optimism of transport and traffic engineers had exceeded practicalities. Half a century earlier there had been a superstreet and superhighway craze when double- and triple-deck urban streets were seen as the solution to car congestion (Heightchew 1979). It was widely believed that these superstreets would soon be the rule rather than the exception. When Arthur Tuttle proposed a three-level solution for an urban street in 1924, he was criticized by Harvey Corbett because he did "not go far enough." The superstreets reappeared in the General Motors Futurama exhibit at the 1939 World's Fair.

The seeds of the concept had come from the work of Eugène Henard in Paris at the turn of the century. Henard's vision also gave rise to the interchange and the roundabout. The visionaries, in turn, had perhaps built on the ideas of da Vinci, who in the fifteenth century had conceived of a two-level city—with the upper level for the convenience of the upper class and their carriages and the lower level for the working class and their wagons.

An important development dating from the 1970s was a greater emphasis on the people-moving capacity of urban and suburban freeways by giving priority to buses and high-occupancy cars. Such measures have had a noticeably beneficial effect on freeway effectiveness. Another development has been the realization that, at the urban scale, equivalent benefits to the freeway can be obtained with lower design speeds and less-generous alignments. Indeed, it is strange in retrospect to realize that most urban freeways have been built to the same design speeds and alignments as rural freeways. They were, in fact, rural freeways located in urban areas. But an urban traveler's primary need is to minimize total travel time; the need to be able to travel safely at 130 km/h for extended distances is relatively minor.

Miscellany

The grade-separated, multilevel cloverleaf intersection so characteristic of the freeway was invented by Henard in Paris in 1906, although its essential features had previously been used in railway practice. Cloverleaf intersections were introduced into the United States in the 1920s. A number of them—such as the ones in Woodbridge in 1928 (fig. 9.2) and in Rahway, both in New Jersey—were built on roads that were not freeways (HEB 1930). As the cloverleaf and its kin came to be applied to the new restricted-access roads, freeway intersections became particularly large land consumers (fig. 9.3). The more complex of these often came to be known as "spaghetti junctions" or "spaghetti bowls." The most notorious of these are the M1/M62 junction near Birmingham, England; the Chicago Circle, where eight highways

Figure 9.2. The Woodbridge cloverleaf interchange between Routes 4 and 25 in New Jersey in 1930. Neither of the routes were then freeways. The Jersey City to Trenton super highway was the upper route; the other was the shore route. *Photo from the Wesley Williams collection in Vic Roads.*

intersect; the Los Angeles Stack; the Spaghetti Bowl in Los Angeles; and the Mouse-trap in Denver.

The countries of Europe were not far behind the United States in postwar freeway construction. Italy and Germany already had operating systems. The British freeway system had been predicted with uncanny accuracy by Hilaire Belloc in 1923, but twenty-six years were to pass before the Special Roads Act introduced the legal concept of motorways, as freeways are called in that country.[7] The first British free-way was the 13 km long Preston Bypass, opened in 1958.

Not only was the time after World War II ripe technically for freeway con-struction, in terms of both network development and construction scale, but perhaps more importantly, such major civil engineering works provided an ideal mechanism for postwar reconstruction and employment. Most of the developed world now has an extensive freeway network, although the urban freeway is still largely an American phenomenon.

One significant feature of the freeway is that it usually represents a new road and a new route. Most roads have been upgrades of preexisting ways; rarely are new ways created. The freeway therefore represents a relatively unusual development. The incentive for this tendency to establish new routes was undoubtedly the fact that the freeway—on which vehicles travel at least 100 km/h—demanded completely new alignments. The progression from foot to animal to cart to coach to cycle could be handled in stages within the existing ways. But, as with rail in its halcyon days,

Figure 9.3. The intersection of the Santa Monica (*upper*) and Harbor freeways in Los Angeles, illustrating the land consumed by such a system. *National Geographic* photo by Bruce Dale, July 1983. *Courtesy of the National Geographic Society.*

the car could only be safely and efficiently accommodated by a completely new system.

Evidence can be seen in the decline in road safety between the 1920s and the 1950s as the capabilities of the car, both individually and collectively, began to require noticeably more attention than could be supplied by preexisting roads, or roads only marginally improved from the days of the horse and cart.

Managing the Car

The problems that are about to be discussed actually predate the car. The various and almost insurmountable environmental, economic, and traffic problems of urban transport also existed with the horse. The car-caused problems of traffic congestion, road accidents, air pollution, and fuel usage all had their nineteenth-century antecedents. In 1896 the residents of Brooklyn resisted street paving in their area because they believed it would make their street into a "thoroughfare for carts and vehicles of all kinds" (McShane 1979). The residents were concerned about the noise from passing horse carriages, which would be "so intolerable that it will make the street undesirable." They also feared that the lives of children would be in constant danger."

McShane in 1979 traced a change in perception of the street as a place, beginning in about 1880, to the use at the turn of the century of terms like *arteries* to describe those same streets. He described an active municipal engineering lobby encouraging this trend, armed largely with arguments about improved sanitation. In McShane's words, "These changes in streets literally paved the way for the automobiles."

Transportation in this century was ushered in by the invention of the car, an event welcomed by many as a universal panacea. Indeed, the car has subsequently fulfilled many of its promises and, along with the truck, has been a major cause of twentieth-century social and industrial development. However, the car and the truck have also brought with them some old and some major new problems.

Up to the 1930s the car was predominantly used for recreation, shopping, and social visits rather than for journeys to work. A 1934 survey showed that 45 percent of Pittsburgh wage earners owned cars, but only 20 percent of these car owners used their vehicles for the journey to work. Pleasure was the carrot; driving to work was a side benefit. This pattern was to change rapidly, for the car had many attractions, providing the following benefits (Lay 1990):

1. A common transport mode for all land-based trips of all lengths and to all land-linked origins and destinations
2. Reliable door-to-door service without the need for transport mode changes
3. Assured availability, for those with access to one
4. High operating speeds relative to other land modes
5. Ensured seating and privacy; for instance, in 1967, 40 percent of successful American marriage proposals were made in a car (Flink 1987).
6. Weather protection
7. Pride of ownership

It is little wonder that recent studies of car ownership levels—which have continued to increase—now commonly talk of saturation levels, that is, of the levels at which the practical demands for car ownership are effectively satisfied, with every potential user having access to a suitable car or fleet of cars. President Herbert Hoover had some inkling of this in the 1920s when he promised the American electorate "two chickens in every pot and a car in every garage." As early as 1923 Robert Lynd's classic Middletown study of an American small town showed that two out of every three families owned a car.

Increasing car ownership levels potentially provide the public with an increasing level of travel satisfaction, but have a number of serious drawbacks. For example, traffic facilities are soon unable to handle the demands placed on them, but they cannot be expanded indefinitely because there are usually social and economic constraints on providing more road space. An equilibrium must often be reached via traffic congestion, which is an inefficient, ugly, and unwanted solution. The traffic jam caused by visitors to Cape Canaveral to watch the 1981 space shuttle launch had not cleared by the time the shuttle had made its first orbit of the entire earth (Patton 1986). Traffic management measures have often delayed the need to attain this equilibrium. The creation of extra road facilities obviously also helps, but can be ineffective when travel demand significantly exceeds supply. Slowly the world has learned that travel demand management is as necessary as is the creation of road supply.

There was some perceptive early realization of the need for restraint. For example, the Melbourne Metropolitan Town Planning Commission (1929) commented:

> A street system has certain limits to the amount of traffic it can carry, and if the height and use of buildings are not properly regulated, the capacity of the street is likely to become insufficient. . . . Regulation and control of traffic must be a useful aid in the solution of the problem. . . . The claims of the motorist for greater facilities are persistent, and frequently unreasonable. . . . The use of the private motor vehicle for ordinary shopping purposes, or for merely personal transport to and from business, causes a most extravagant use of street space.

Restraint has not always been an easy lesson to learn, as potential followers of the American lead have found.

> When Nikita Khrushchev visited the United States in 1954 he was stunned by the enormous masses of American cars. . . . After his return to Russia he proclaimed that this was not what the people of the Soviet Union needed: instead of such masses of private cars the Soviet Union would devise new methods of public transportation in its cities. . . . This was one of the sources, perhaps even a principal source, of Khrushchev's unpopularity and of his demise. . . . His resignation was followed by the first construction of a modern giant car plant in the Soviet Union. His successor, Brezhnev, was a car addict, collecting them with . . . enthusiasm. (Lukacs 1984)

Parking has also developed into a major urban issue. Car ownership levels have approached one per eligible driver in some areas and the average car is parked for 96 percent of its life (Lay 1990). The parking lot has become a particularly unattractive component of many urban landscapes. Yet many merchants and landowners strongly, and probably correctly, see parking facilities as an essential urban provision. The first automatic parking meter operated in Oklahoma City in 1935.

Off the arterial road system, the car has had a major impact on the form of our suburbs. In addition, because it is dimensionally no larger than many of its transport predecessors, the car can be statically accommodated within the old streets and ways. However, its dynamic accommodation is much more difficult. The car's high speed, particularly relative to walking, creates an aggressiveness that must be constrained. Certainly it has not been possible to rely on the self-restraint of the individual motorist, whose motoring decision-making is usually singularly self-centered. For a while it was thought that simply providing more and more street space was the solution. But this has rarely proved effective, and one of the major advances in recent times has been the use of a restricted street system to forcibly civilize the car.

Such measures have been applied with verve and skill in many of the old towns of Europe and in the new suburbs of the New World. Based on restraint rather than on generosity, they represent a major maturing of our attitude toward the car. They indicate that the passionate love affair is over and that the marriage is now being made to work. In new developments, this approach has meant planning sensibly for the car. The seeds for this devolution of power away from the car were given in the earlier discussion of the garden city concept and the Radburn Plan. Perhaps not surprisingly, this revolutionary approach was initially adopted at a very slow pace.

Raymond Unwin was the first to emphasize clearly the need for an obvious hierarchy of ways—from freeways, through arterial roads, roads that collect and distribute local traffic, local streets, and culs-de-sac, to walking paths. His Hampstead Gardens in particular highlighted the usefulness of the cul-de-sac in the automobile age. The important ward or neighborhood concept of car access, but not car thoroughfare, came out of Ebenezer Howard's garden city, was firmly introduced by Clarence Perry in the 1920s, developed further by Alker Tripp, and then embraced by Colin Buchanan in his "environmental area" concept—epitomized in his landmark 1963 report *Traffic in Towns* with its specific intent of improving the quality of the local environment. This concept forms the basis of much current planning aimed at living with the car.

Reviewing the garden city era, and subsequent experience, it can be seen that successfully living with the car has involved such items as the following:

1. Pedestrian malls, where the car is either totally excluded or forced to travel at walking pace. The first conversion of a road used by cars to a pedestrian mall occurred in Germany in 1929.
2. Low-connectivity street networks, where most streets are places to be, rather than through routes
3. Traffic-calming measures such as small roundabouts to inhibit traffic movement and slow zones, where the car travels at speeds below 35 km/h
4. Careful siting of major car attracters, such as large shopping centers, away from residential areas

Such inhibitions on through traffic are not new. A Roman edict proclaimed: "The circulation of the people should not be hindered by numerous litters and noisy chariots."

Noise and air pollution from cars are significant problems in many urban areas. They are controllable, but users and producers have been slow to take the necessary steps because of the high costs involved. Nevertheless, those steps are occurring. Again, an equilibrium has occurred as rising noise and air pollution levels have forced action, despite the consequences to motorists.

Successful living with the car has also meant a resurgence of simple paths and ways. As we came to realize that the car could not meet all our transport needs, we also realized that it was usually incompatible with other desirable modes of travel such as walking and cycling. Malls, walking paths, and cycling paths are now the joys of many of our urban areas and riversides.

An extreme view was Le Corbusier's 1924 argument in his *City of Tomorrow* that the street would need to be made unnecessary if urban utopia was to be achieved. His *Ville Contemporaine* included a plan for redeveloping Paris that involved demolishing the core of the city. More realistically, perhaps we have finally taken the step that Mumford had in mind when he wrote, "The first lesson we have to learn is that a city exists, not for the constant passage of motorcars, but for the care and culture of men" (Mumford 1964).

Road-based Transport

The car has remained a relatively inefficient travel mode in terms of space, due to the desire of most motorists to travel separately and independently of their colleagues. So attractive is the individuality and privacy of the car that, in normal circumstances, few motorists can be convinced of the need to reduce that individuality by sharing their vehicle with another traveler. Consequently, the effective capacity of the average urban car is no greater than that of a motorcycle, but its fuel usage is ten times and its space usage five times as high as a motorcycle's.

Given this inherent inefficiency, there have been some initiatives based on extending the usefulness of the car. Car pooling and special lanes for high-occupancy vehicles have met with some success, provided the added incentives were sufficiently great. In 1914 some unemployed men in Los Angeles began operating Fords as freelance buses known as *jitneys,* picking up passengers in advance of the regular transport services and charging a standard flat fare. The operation spread widely, despite strenuous opposition from the regulators. Jitney service remains today as a popular transport mode in many Asian cities. Car rental began in a Chicago used car lot in 1918, using Model T Fords. In 1923 the business was bought by John Hertz, president of the Chicago Yellow Cab Company.

The car is not the only independent means of surface transport. The bicycle, motorcycle, truck, bus, and taxi have all been present since at least the birth of the car. They continue to flourish relative to other modes of travel, although the motorcycle's accident record and exposure to the weather have detracted from its wider use. Traveling at 100 km/h requires a degree of visible presence and occupant protection that has not yet been met within the confines of the motorcycle. Since the birth of the car there has been increasing ownership of specialized off-road vehicles for both recreation and business. For example, the modern cattle rancher frequently discards the horse for the motorcycle.

Trucks have captured much of the freight market, yielding to rail and canal only on long, bulk hauls and to air only on high-speed long-distance delivery. Private enterprise has flourished in the truck industry. Yet another equilibrium has occurred with truck sizes. It has usually been possible to justify increasingly larger and heavier trucks on the basis of economies of scale. However, finding financing for the road infrastructure costs associated with those increases in truck size has commonly proved a major constraint. Forty to 50 tonne truck limits are common, whereas 100 tonne capacities are both possible and desired by industry. At a human scale, the public fairly naturally finds sharing road space with huge high-speed trucks to be quite disturbing.

The tram or streetcar played a major role in urban transport from about 1880 on. In most cities the trams came to be replaced by buses in the period between the wars. Buses were more flexible in their operation and were thought to be cheaper. They were also encouraged by motorists and traffic authorities, who saw the removal of the tram as aiding the smooth flow of road traffic. In the United States at least, the changeover was also aided by public discontent with the attitudes of and service offered by the transit companies, mirroring the earlier discontent with the railroad companies (Bottles 1987). The physical turning point was the motorization of New York surface transport in 1935. San Francisco and Los Angeles also subsequently succumbed to this process, the latter in 1944.

Thus, the changeover was at least assisted by incentives from the car industry (Schaeffer and Sclar 1975). For example, between 1932 and 1956 General Motors, usually through its National City Lines, bought about a hundred near-bankrupt and poorly managed private tram systems in forty-five cities and encouraged their early conversion to bus operations (St. Clair 1986). National City Lines, formed in 1936 by General Motors in conjunction with Standard Oil of California and Firestone, was convicted of antitrust violations that occurred from 1946 on. GM pulled out in 1949. There has been strong debate over the years as to whether General Motors and its partners deliberately destroyed much of the nonmotorized urban transport in the United States or, alternatively, took reasonable commercial advantage of the situation in which they found themselves. The general consensus is that they merely aided and abetted and took advantage of a trend that was already in place: the trend toward low-density suburban living.

Buses and taxis have found their niches in urban transport, providing travel capacities below that of the train but above the car. Their effectiveness has largely been limited by the same congestion factors encountered by the urban car. In an attempt to solve the problem, some traffic lanes have been reserved for buses, thus providing a natural evolution through to trams (light rail) and on to heavy rail. Buses have also proved successful on low-volume, low-cost, long-distance operations.

Interestingly, no new modes of automotive travel have developed since the latter years of the last century. The vehicles of those halcyon days have become more reliable and more efficient and faster, but they are still noticeably the same vehicles.

Cars and trucks operate almost solely on petroleum, although a few other sources, such as natural gas and alcohol, have been used. Car operating efficiency has improved in recent times with increases in relative fuel prices and as governments have stepped in to demand increased efficiency levels to avoid excessive dependency on oil producers and too rapid a depletion of the world's petroleum resources. Nevertheless, the world remains wedded to petroleum as its main transport fuel. No technically and economically viable alternative is as yet on the horizon, although the electric vehicle may be just coming into sight. At the moment, the world's most effective response to another fuel crisis would be for more personal travel to be by road-based public transport and smaller gasoline-powered private vehicles and for more freight to be moved by rail.

On another tack, the modern road has also come to provide much more than a thoroughfare or street. Following Haussmann's nineteenth-century lead in Paris, most modern roads also provide access routes for public and private utility services such as power, water, sewage, and telecommunications, either above or below ground. In addition the road right of way has come to be shared with other transport modes and, occasionally, with property development of the space above the road. Of course, this trend has its historical precedents in the many medieval bridges—such as London Bridge—that provided both passage and a site for shops and dwellings.

Traffic Control

Perhaps it is shortsighted to say that as yet there looms on our transport horizon no alternative to the car, the truck, and the road. Rather than alternatives, the future seems rather to suggest that electronics and information technology may well see the car and the road evolve into an even more effective combination than they are today.

Nowhere has technological development been clearer than in the field of traffic control. Traffic control arose from the need to curb the otherwise unrestricted number of cars using the road (the supply), to find an acceptable equilibrium between car supply and road space, to civilize the car. Initially, methods such as speed restrictions and traffic signals at intersections arose in response to the need. Gradually, the control systems have grown more sophisticated as the technology has improved and the need increased. Motorists can now be automatically detected, identified, and fined for speeding.

The development of the technology for traffic signals at individual intersections was discussed in chapter 6. A major advance after World War II was in the application of scientific methods to determine appropriate signal settings and timings, which required an understanding of the delays and queues imposed on traffic waiting at signals, in order to use the signal settings more efficiently and to allocate the green time to the various roads approaching the intersection. This work was based on the application of queueing theory and operations research to the problem. These two methods had been developed during World War II and the subsequent cold war, particularly the Berlin airlift. Some benefit was also gained from their parallel application to telecommunications design, particularly for postwar telephone exchanges (Lay 1989). An important further breakthrough was in methods of linking together the timing of the traffic signal lights in an area in a coordinated manner. This electronic development began in Sydney, Australia, in the early 1960s, and that city's SCAT system is still the world leader in the field (Lay 1989).

People have also come to realize that controlling traffic at intersections is only a part of the total need. Giving priority to vehicles on one intersection leg or on one arterial road might not coincide with the community's overall urban or public transport strategies. Or, as another example, monitoring and inhibiting traffic entering a freeway via a ramp because the freeway is already at capacity might not coincide with the larger objective of favoring concentrated rather than dispersed land use developments. It might be preferable to give priority to the short-distance travelers entering the ramp and inhibit the long-distance travelers already on the freeway. A further example would be giving travelers moving circumferentially around a city center priority over radial car travelers with a public transport option.

Electronics and information technology have begun to make the bases of traffic control decisions much more transparent and to lay bare the principles actually being applied. Not surprisingly, a number of dilemmas involving traffic and transport planning philosophies are arising. At the same time, the new technology has provided improved traffic efficiency. This trend will increase rapidly as electronics provides the motorist with more guidance and advice as well as imposing increased control.

Traffic control and traffic information systems in cars have been segregated since the inception of motor vehicles. Hence, one-way communication has prevailed with drivers passively obtaining information about the status of the road system and guidance information about directions and the road environs from signs, signals, and markings. In more recent years, traffic information has also been conveyed to drivers via the car radio and by roadside variable-message signs.

The road system has received little reverse flow of information about the performance of the vehicles using it. Traffic engineers have collected information about the state of traffic operations from counting, measuring occupancy of carriageway zones, and visual surveillance of traffic incidents or congested facilities. Both historical and real time information are increasingly being used today to increase traffic flows, reduce delays, reduce fuel usage, improve safety, and met a number of other traffic efficiency criteria. But, by and large, the car and its driver still do not react actively with the road and traffic system.

Two-way communication between road vehicles and traffic control systems promises substantial improvements in traffic efficiency. A two-way system could lead to automated highway systems, which would retain individuality of travel but reduce some of the burdens of driving. Two-way communication could use speakers and VDUs (visual display units) to provide individualized information regarding best routes, ideal speeds, imminent traffic incidents, expected delay times, and so on. The same system could also inform traffic control systems of the current level of traffic on a road and predict traffic conditions in the immediate future. Such systems have the most potential value in high-density urban areas. One such system is the Siemens Euro-Scout route guidance technology currently under prototype testing in Berlin. In the Euro-Scout system, each participating vehicle supplies a central control room with details of its location and traffic speed. A computer then calculates the most efficacious routes through the road network and broadcasts the information to subscribers. Equipment in the car then converts this into a VDU map display showing the recommended route. At the same time, a synthesized voice in the car gives the driver specific instantaneous instructions on when and where to turn, diverge, or merge. Advice is also given on the availability of parking spaces.

Motorists pay for road use in an immediate sense largely through fuel taxes, which do not adequately represent in value or in incidence the costs of a car's use of a piece of road. For example, fuel tax does not necessarily reflect the sum of costs of road wear and tear, the traffic services provided, the congestion caused, or the environmental nuisance created. In addition, fuel tax is rarely perceived to apply when the costs are actually being incurred and so does not influence driver behavior. The fact that car usage is not properly priced has been one of the major impediments to its rational use. To overcome the problem, Arthur Pigou in his famous 1920 work *Economics of Welfare* proposed directly charging motorists for each piece of road that they used. However, no means of applying his road pricing theory then existed, although toll roads addressed the issue for specific road lengths. Current developmental efforts are using telecommunications and electronics to permit road use charges to be widely applied. Such experiments are destined to have a rocky path; no community will be happy about a new tax on an assumed human right—the right to drive unimpeded—nor accept the intrusion into the privacy of their travel.

The first serious road pricing proposal was put forward by Martin Beckman in the United States in 1956. A major British government report was produced in 1964, but was quickly put aside under political pressure. Paper studies ever since have come to naught. A trial of electronic road pricing was begun in Hong Kong in 1985. Vehicles carried an electronic tag and were automatically detected and their accounts charged each time they passed "readers" mounted beside the road. The trial was largely abandoned—again for political rather than technical reasons—in 1986. No road pricing application yet exists except for small schemes in Valetta, Malta, and—by stretching the point—a couple of cordon-restraint schemes. The Singapore cordon-restraint measure was introduced in 1975, and in 1986 Bergen, Norway, became the second city to adopt a cordon pricing scheme.

Nevertheless, because of the many advantages of the car and the truck, the two most effective techniques for traffic control have been road space restraints and the very high taxes imposed on car purchase in cities like Hong Kong and Singapore. This latter measure works because it suppresses car ownership and hence circumvents Lay's first Law of the Car: If you have a car, you will use it wherever and whenever possible. This law is a consequence of the various advantages of the car that were catalogued at the beginning of the chapter (Lay 1984).

Toll roads were discussed at some length in chapter 4. They have made some resurgence in this century through tolled freeways in France, Italy, and the United States. Modern smart-card methods of toll collection may increase their efficiency. The first of the new breed of American tollways were Merrit Parkway in Connecticut, which opened as a toll road in 1939, and the Pennsylvania Turnpike. But in many instances and countries, the road is generally regarded as a public good, to be funded and maintained for the public purpose. The electronic road pricing measures discussed earlier have the potential to make all roads toll roads.

The Future

Much has been said in this chapter about equilibrium being achieved between the demands of motor vehicle use, the supply of roads, the various community costs caused by motor vehicle use, and society's schizophrenic expectations. The realization that the ever-increasing demands for car and truck travel can be managed by controlling and redistributing those demands, as well as by continually increasing the supply of road space, has been one of the more important recent contributions to the history and evolution of the car. Importantly, it is a contribution whose potential has yet to be tapped, and so its impact will continue strongly into the future, where we will surely see the translation of this philosophy of restraint into a practical reality.

And what of the future? The community has clearly expressed its commitment to individual, private surface travel for personal transport, but the influence of burgeoning information technology on work patterns may mean that the long-term reasons for such travel may change from predominantly commercial to predominantly recreational.

In the shorter term we can expect the car to be civilized by the increasing input of electronics, control systems, and information technology into both the individual vehicle and the overall system. The road system will become less like a swarm of bullets, potentially lethal and intermittently maneuvered by distracted and less than fully alert drivers making apparently random choices and judgments. This is perhaps an exaggeration, but it must be how our current system would look to a visitor from outer space. Widespread gridlock and ever-extending parking lots are not parts of any rational agenda. The future road network will clearly be better organized, more systematic, safer, more interactive, and more efficient. Given that there are physical limits to how small a vehicle can be made and given our strong preferences for privacy, it could be predicted that the next major technological breakthrough will be a more efficient and effective power source than the IC engine.

Most important, it will be demonstrably obvious in the future that we have finally learned—undoubtedly just as the next transport invention appears on the scene—to civilize the car and to respect the human scale.

The ways of the world, after some twelve millennia of usefully serving society, have been given a great twentieth-century injection of mobility through the motorcar. Consequently, more than ever they are now vital, indispensable, and all-pervasive parts of our lives. Ways are key components of our society and, as such, are also true mirrors of that society. I hope this book has provided a few facets for such a mirror, for in that mirror we see our fortunate inheritance. As Colley Cibber said some three hundred years ago:

> The good of ancient times let others state,
> I think it lucky I was born so late.

Terminology for Road Pavements

The figure below shows the basic road terminology used in the book. The key element is the carriageway, which is the manufactured surface suitable for travel. The road is the entire right of way within which the carriageways are located. Primarily for safety reasons, there must be considerable space—ideally, 10 m on a high-speed road—between the edge of the carriageway and any obstruction (such as a property boundary). Of course, the separation distance may be as low as 2 m on a low-speed street.

The carriageway is ideally divided into traffic lanes about 3.5 m wide, with shoulders for emergency stopping and structural protection forming the edges of each carriageway (Lay 1990). The shoulders will be at least a meter wide and may be the same width as a traffic lane.

The pavement structure is in three parts. In the book, the local material is called the natural formation, but in pavement engineering it is commonly called the *subgrade*. If need be, any poor material is removed and replaced and the existing material enhanced. Its upper surface is then constructed to be parallel to the upper surface of the finished pavement. The more expensive structural course is a manufactured component, which distributes and reduces the traffic loads to a stress level that the natural formation can safely carry. It is usually called the *basecourse* in pavement engineering and may actually comprise a number of separate courses. The wearing course provides a good ride for the traffic using the road, protects the basecourse from local damage by tires, and keeps water from entering and damaging the underlying structural courses and natural formation. All these concepts are further explained in Lay (1990).

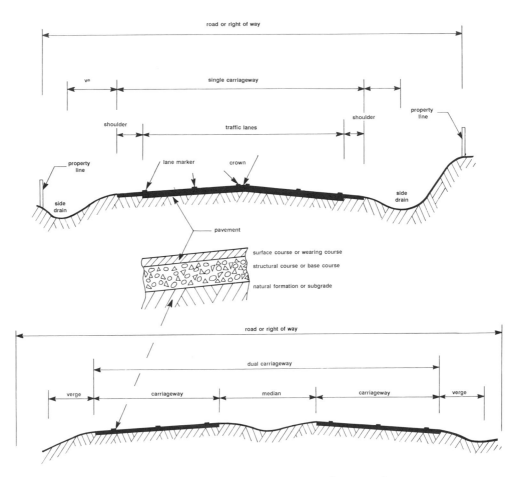

Figure A.1. Basic terminology, using a road cross section.

Chronology

The road is that physical sign or symbol by which you best understand any age or people.

H. BUSHNELL, 1864

10,000 B.C.	Pathways in use in some temperate zones, including much of today's inhabited world
9000 B.C.	Agricultural communities arise; springs of Elisha (Jericho) used by humans
8000 B.C.	Human travel becomes significant; first town at Jericho
7000 B.C.	Domestication of animals
6000 B.C.	Sleighs in use
5000 B.C.	First wheel; animal castration practiced; domesticated cattle used for hauling; ridgeways in use; sleds in use
4000 B.C.	Urban societies appear; stone paving in towns; Glastonbury site under development; keystone arch in use; regular trade routes established
3300 B.C.	First planned city; Sweet Path, earliest known extant manufactured path, built at Glastonbury
3000 B.C.	Ways maintained between towns; brick paving in Indian towns; bitumen in use as a mortar; small horses, donkeys, and carriages in use; development of wider ways
2800 B.C.	First chariots in use
2600 B.C.	Concrete used as a mortar for flagstone road in Crete; Egypt builds first manufactured bridge
2500 B.C.	Bitumen used for temple surfaces; first military vehicle; first metal tires; bitumen used as a mortar in brick roads; first four-wheeled wagon in use
2300 B.C.	Horses used for hauling
2200 B.C.	Oldest known map (of Lagash, in Mesopotamia)

335

2150 B.C.	First traffic tunnel
2000 B.C.	Advanced civilizations present; metal tools available; earliest extant stone road in use in Crete; major route from Iran to Mediterranean in use; oldest extant corduroy road in use (at Glastonbury); burnt clay bricks, mortared with bitumen, used for roadmaking; horse harnesses, spoked wheels, and copper nails in wooded tires in use; organized pack convoys in operation; fast military chariots in use in Middle East
1900 B.C.	Kahun, Egypt, laid out on a geometric basis
1700 B.C.	Barbarians begin using new chariots to defeat empires of Middle East
1500 B.C.	Earliest extant literature mentions Great Roads
1225 B.C.	Joshua burns wooden chariots
1200 B.C.	Iron Age begins; first pontoon bridge build, over Wei He River in China
1100 B.C.	Iron used in chariots; first road engineering corps established
1050 B.C.	First major pier and beam bridge built at Quzhou in China
1000 B.C.	Metals used in vehicle construction; cement used in roads in India; horse riding begins to take hold
900 B.C.	Baghdad reaches peak population for a walking city; Assyria starts first cavalry; plank road built at Glastonbury
850 B.C.	Oldest extant bridge built
800 B.C.	Isaiah proclaims the role of the road
780 B.C.	Covered bridge built in Babylon
700 B.C.	Bitumen and brick processional road built in Assur; iron tires used; corbell arch bridge built; first keystone arch bridge; parking bans introduced on Assyrian Royal Road
670 B.C.	King Esarhaddon orders roadmaking for trade and commerce
615 B.C.	Nabopolassar builds bitumen and brick processional road in Babylon
605 B.C.	115 m bridge built over the Euphrates River
600 B.C.	Bitumen used for road surfacing; Nebuchadnezzar makes roads in Lebanon
550 B.C.	King Cyrus builds roads in Jordan
539 B.C.	Virtues of suburban living extolled in Babylon
500 B.C.	Macadam-type construction in Crete; caravan route in use from Middle East to India; Persians build their Royal Road; amber roads in use in Europe; Hippodamus active in town planning; riding horses becomes more common than travel by chariot; swiveling axles introduced
450 B.C.	Roman legal code specifies path widths
400 B.C.	First Roman paving; double-shaft horse harnessing introduced; shrunk-on metal tires introduced; first recorded tolls
323 B.C.	Persian road system stops expanding
312 B.C.	Romans begin Via Appia
300 B.C.	Indian Royal Road in use; first distillation of petroleum; Silk Road from Middle East to China in use
285 B.C.	First recorded bamboo suspension bridge, in China
250 B.C.	Romans develop pozzolanic cements
244 B.C.	Via Appia reaches coast at Brindisi

219 B.C.	Longest extant Roman bridge built; use of bitumen in roadmaking disappears
200 B.C.	Chain-cable 120 m suspension bridge built at Shensi in China; central road administration established in China; Chinese drill for oil and gas
62 B.C.	Pons Fabricius built in Rome
55 B.C.	Romans enter Britain
50 B.C.	Romans introduce wheel load limits
45 B.C.	Julius Caesar bans vehicles from CBD (central business district) to alleviate congestion, introduces one-way streets and off-street parking requirements
30 B.C.	Romans build first major hard-rock tunnel near Naples
19 B.C.	Romans build Pont du Gard
A.D. 27	Vitruvius publishes *De Architectura*
A.D. 50	Claudius I extends traffic restraint to all Roman towns; Hero of Alexandria describes steam-powered vehicle
100	Wheelbarrow invented in China; first Chinese arch bridge
104	Trajan's truss bridge over Danube built
125	Hadrian limits number of vehicles entering Rome to limit congestion
134	Pons Aelius built in Rome
200	Roman road system at its peak; saddles introduced in Europe
300	Antonine itineraries (early road maps) written
399	First Chinese travels entire Silk Road
400	Iron suspension bridge built in India; rope suspension bridges in China; street lighting in Ephesus
406	Romans leave Britain
500	Roman influence disappears; Saxons settle in Britain
590	Zhaozhou Bridge built
635	First Westerner travels entire Silk Road
700	Stirrups introduced to Europe; horseshoes in common use
750	Harnesses introduced in Europe
800	Charlemagne constructs some roads in France and introduces corvée system; Norsemen build Varangian Road in Russia; travel increases
850	Saracens pave and light the streets of Córdoba
900	Population of Baghdad peaks and begins to decline; Slavs introduce vehicle suspension
950	Norse law defines road widths
1000	Most English road systems in place geographically
1066	Arrival of Normans sees travel begin to increase in Britain
1102	Henry I builds one of England's completely new roads
1135	Henry I decrees that English roads should be two carriages in width
1154	Paris passes street-cleaning edict; Chinese build longest bridge, Su Di
1184	Streets paved in Paris
1187	Pont d'Avignon built
1189	Murage tolls at city walls introduced in England; Lu Gou Bridge built in Beijing
1200	Use of saltpeter as an explosive noted
1209	London Bridge completed

1222	Roman roads restored in France; Mongols reopen Silk Roads
1235	First mention of Simplon Pass through Alps
1237	St. Gotthard Pass opened to foot traffic with opening of Stiebende suspended bridge
1245	First ceremonial accouterment added to wagon tray
1270	Bacon predicts high-speed travel
1271	Marco Polo begins journey to Beijing
1274	First English toll right granted; central road administration begins in Norway
1280	First formal street cleaners in England
1284	Northampton imposes first freight tax
1285	Statute of Winchester codifies right of way and manorial obligations for road maintenance
1286	Tolls introduced on London Bridge
1300	English Parliament orders London's Strand paved; efficient harnessing widely available; four-wheeled wagons reappear in Europe; Pope Boniface VIII suggests travelers keep to the left; timber blocks used for paving in Russia
1302	London paviors ordinance passed
1314	University of Paris begins courier service
1333	Oldest extant truss bridge built
1338	Septimer Pass opened through Alps
1345	First European flat arch bridge
1348	Paris requires landowners to pay for adjoining street paving
1353	England introduces a general tax for a particular road construction
1356	England introduces vehicle tax for road construction
1364	Scawageours noted in England
1377	Large stone arch bridge built over Adda in Italy
1390	Street surveyors noted in England
1391	First English town (Chester) paving act
1395	Paris bans emptying chamber pots into streets
1397	Trade of pavior (paveur) first mentioned
1400	Paris appoints a master inspector of pavements
1415	Stone setts first used (in Paris)
1416	London ordinance requires street lighting
1420	Tibetan iron-chain suspension bridge built with flat deck
1450	San Luis Rey suspension bridge built in South America
1464	Louis XI repairs French roads and introduces a courier service
1467	Earliest record of a causeway
1480	First solid-rock road tunnel since Romans
1485	First mention of road landscaping; arch buckling discussed in Alberti's book
1487	Paris bans trotting and galloping
1498	Columbus discovers Trinidad
1500	Leonardo da Vinci introduces concept of force
1501	Etzlaub produces pilgrims' route maps
1508	Louis XII has official control over all French roads
1509	Master of posts appointed in England
1525	First urban coach services, in Milan

1531	English statute defines bridge maintenance requirements
1533	English statute requires stone for paving to be brought from seashore to London
1540	Paris bans U-turns and requires horses to be led by hand
1546	Long-distance transport reappears in Europe
1550	Henry II embarks on major roadmaking in France
1552	Estienne publishes first book of road maps
1553	French make roadside tree planting compulsory; private coaches introduced into England
1555	English Parliament decrees that cartways should be 2.4 m wide; public weigh scales introduced in Dublin; English parishes given road responsibility; first general English road act passed; first English coach built
1556	Stone pavement constructed on segment between Paris and Orléans
1563	Statute labor made permanent in English Statute for Mending of Highways
1565	First manufactured road in the United States
1569	Ammanati introduces the elliptical arch
1570	Palladio builds truss bridge over the Cismon-Brenta River in Italy and publishes books on architecture
1573	First salt tax for roadmaking in France
1574	London expands beyond its city walls
1580	London bans building outside city walls
1585	Toglietta publishes book on pavements in Italy
1586	Stevin publishes on graphical statics
1587	Statute labor made perpetual
1590	Word *road* enters literature in place of *street*
1591	Tenda Pass packhorse route through the Alps
1595	Sir Walter Raleigh finds Trinidad Lake asphalt
1599	Regular cart-based freight services introduced in England
1600	Road centerline marked in Mexico City
1601	English Parliament restricts number of coaches; asphalt discovered in Germany
1602	Kent places toll on carts weighing over one tonne
1603	Wooden rails used in coal mines
1605	Hackney carriage introduced into England
1606	First North American cleared way
1607	Parishes responsible for American colonial roads
1609	Surveyor of the highways to the king appointed in England
1610	First English stagecoach; first English-language book on roads
1611	First U.S. bridge
1614	First use of separate footpaths beside urban streets
1616	Flagstone paving introduced into London
1620	First U.S. highway built
1622	James I of England decrees that only two-wheeled vehicles weighing less than one tonne can use roads; Bergier's book on Roman roads published
1625	First street paving in North America
1626	Jost studies effect of heat on asphalt

1629	English act prohibits more than five horses drawing any vehicle; English Post Office begins carrying private mail
1630	Covent Garden developed as London suburb
1632	First U.S. road legislation
1635	Regular highway mail services established in England with London–Edinburgh route; Charles I introduces measures to restrict hackney use
1638	Galileo introduces beam analysis
1652	Ban on wagons being galloped in New York City
1654	First U.S. toll bridge; Cromwell repeals corvée system, passes ineffective act for local taxation
1661	Glass windows first used in coaches
1662	English act requires street lighting and that wheel widths be at least 100 mm; London bans drains in center of street; first major U.S. bridge; Pascal's bus service operates in Paris
1663	Toll roads made legal in England
1664	France introduces state coach service using chaisses
1665	First tolls collected on Great North Road; first Berliner coach produced
1667	Construction of Champs-Elysées begins
1669	Horsepower limit in 1629 English act raised to seven
1670	Tempered-steel springs introduced for coaches
1672	Corps of Army Engineers established in France
1673	Huygens invents internal combustion; oldest extant timber arch completed (in Japan)
1675	Hooke discovers principle of the arch; Leibniz defines stress distribution in a beam; New York and Philadelphia linked by road
1676	Hooke discovers elasticity
1680	Newton sketches steam-powered vehicle; St. James's Square in London developed from a great estate
1681	Coal tar patented
1688	Human-drawn cart introduced into Europe
1690	Rim brakes introduced for coaches; bitumen found in Severn Valley in England
1691	Bernoulli produces beam moment-curvature relationship; first English parliamentary road statute confirms king's highway concept
1693	Gautier publishes book on roads and becomes inspector general of bridges and roads
1694	Bernoulli calculates beam deflections
1695	Toll barriers legalized in England; Hire introduces arch pressure-line concepts
1698	English law requires guideposts at crossroads
1699	Term *parkway* introduced in Virginia
1700	Flying coaches introduced into England
1703	First effective English toll act
1704	Route marking introduced in Maryland; New York and Boston linked by a bridle path

1705	French regulations specify road reservation widths
1712	Newcomen builds first steam engine; d'Eyrinys begins his asphalt discoveries
1714	Gautier publishes first book on bridges
1716	Central road authority established in France
1720	High-volume iron production begins (England); work commences on Moscow–St. Petersburg road; d'Eyrinys invents mastic
1721	First major manufactured road in North America
1722	"Keep left" rule introduced on London Bridge; Réaumur makes first malleable iron
1725	Book (Leopold's *Theatrum Machinarium* carries first drawing of a cast-iron road roller; General Wade begins roadmaking in Scotland
1726	First of the British toll riots
1729	Bélidor and Couplet apply theory to arch design
1734	Death penalty introduced in England for destroying turnpike facilities; first Western suspension bridge
1736	First national decree for driving on the right; lottery tried for funding Westminster Bridge
1737	Dissertation on roads presented to Royal Society by R. Phillips
1739	Coulomb applies pressure theory to arches
1741	Weighbridges legalized in England; first "classical" suspension bridge built in Europe
1743	Perronet active in building arch bridges
1747	Ecole nationale des ponts et chaussées established in France under Perronet; use of asphalt for roadmaking reported in Ireland
1750	Stagecoaches now dominant means of interurban transport
1751	Period of "turnpike mania" begins in England
1753	British act limits wheel spacing to 1.65 m and wheel width to at least 225 mm; Metcalf begins roadmaking career
1755	1753 act amended to allow 150 mm wheels for wagons pulled by fewer than seven horses; Braddock builds road to Pittsburgh
1756	First British law requiring driving on the left; Smeaton reintroduces cement
1757	Trésaguet begins working on pavements; Euler solves column buckling problem
1760	Industrial Revolution under way; golden era of canals begins
1761	Hand-cut stone setts first used in Britain; Philadelphia uses lottery to fund street construction
1764	Trésaguet begins pavement development in France; Grubenmanns builds large timber truss; Denmark begins road network development
1765	Blind Jack Metcalf builds turnpike in England; Irish presentment scheme initiated; Watt develops steam engine
1766	London adopts policy of installing footpaths
1767	First iron rails in use
1768	Smeaton builds road in Nottinghamshire
1769	Hunt develops iron hoop tire; Cugnot develops prototype steam vehicle

1770   First motorized vehicle; Young publishes famous commentary on
       turnpikes
1771   Cugnot steam vehicle operates
1772   First use of cast-iron road rollers; Neuilly Bridge completed
1773   British act permits wagons with wheels at least 400 mm wide to be
       drawn by any number of horses; Coulomb publishes *Memoir on
       Statics*
1775   Trésaguet publishes memoir on pavements; Perronet becomes
       inspector general of roads in France; American Revolution
1776   First attempt to abolish the corvée system of compulsory road labor
       in France; Coulomb publishes theory of beams
1778   First road and bridge built in Australia
1781   First iron bridge built; work begins on Siberian Highway
1782   Watt invents double-acting steam engine
1783   Rotary steam power available; iron structural shapes cast
1784   British mail changes from horseback to coach; American colonies
       form Union; National Road first discussed in the United States
1785   Reports of timber truss in the United States
1786   Somers Town, London, created as first planned suburb
1787   McAdam becomes involved in roads; end of corvée system in France
1789   French Revolution
1790   Lottery used to fund road construction in Virginia
1791   France abolishes central control of roads and permanently abolishes
       the corvée; first rural road built in Australia; first wrought-iron
       bridge
1792   Palmer begins truss-arches in the U.S.; first modern carriageway
       through the Alps; first keep-to-the-right rule in the United States
1795   First U.S. turnpike and paved arterial road opens (Lancaster)
1796   Mastic used to waterproof bridge deck; natural cement discovered in
       England; first Telford iron bridge built; first use of cement in a
       modern bridge
1800   Canadian law requires convicted drunks to work on roads
1801   Telford becomes involved in roads; first modern suspension bridge;
       Trevithick builds first British steam vehicle
1802   First "conceptual" U.S. turnpike opens in Virginia; advocacy for U.S.
       National Road begins; tax on land sales used to fund roads in Ohio;
       U.S. Corps of Engineers created
1803   Telford reports on Scottish roads; first British steam-powered vehicle
       on the roads
1804   First U.S. drive-on-the-right law; practical beam trusses used in the
       United States; elliptical spring invented; Ohio makes it illegal to
       leave tree stumps in roadways
1805   Irish act limits road grades; Napoleon completes Simplon Pass;
       Palmer completes first covered bridge, the Permanent in
       Philadelphia
1806   U.S. act specifies a 6 m width for the National Road; Napoleon
       reintroduces salt tax to pay for roads

1807	First (horse-drawn) passenger railway; Young publishes book on statics
1808	Gallatin defines procedures for the economic assessment of roads
1810	McAdam promulgates his work at a parliamentary inquiry; Telford given control of Holyhead Turnpike; mastic used for footpath construction in Paris; world's oldest extant suspension bridge built over Merrimack River
1811	Central road control reintroduced in France; construction of U.S. National Road begins; first streets lit by coal gas
1812	First bridge to span over 100 m built (in Philadelphia); Ellenborough ruling on right of passage; first horse-drawn low-cost bus
1813	Concord coaches introduced; first gas-powered internal-combustion engine; first pure beam-truss bridge
1814	Hackneys required to carry a number
1815	First steerable bike; Holyhead Road Commission established; largest timber arch bridge in the world built
1816	British stagecoaches allowed to use narrower wheels; McAdam appointed to Bristol Turnpike and publishes road pamphlet; first U.S. state road authority established in Virginia; first cable suspension bridge
1817	Macadam used in France; Rennie's Waterloo Bridge opened
1818	Dupuit publishes on tolls and on road maintenance; after-hours cycling banned in Milan; National Road reaches Wheeling, Virginia
1819	Largest cast-iron bridge built
1820	McAdam's name used as macadam to describe his method; Gourieff develops timber-block paving; lattice truss introduced; coaches provide faster travel than horseback; convicts used for road construction in Australia
1821	First structural analysis of a suspension bridge
1822	Macadam used in Australia; tar first used for blinding macadam
1823	Macadam used in the United States; McAdam family supervise 3.3 Mm of toll road
1824	Artificial cement invented; first asphalt blocks used for paving; first horse-drawn bus
1825	Macadam adopted for national standard in the United States; Stephenson's steam locomotive hauls first freight; steam-powered vehicles use roads; major cartway construction begins
1826	Menai Straits Bridge completed; Navier publishes structural textbook, including theory of arches and continuous beams
1827	First Howe truss
1828	Telford's Over Bridge opened; first use of modern concrete on roads (by Telford); steel first used in a bridge
1829	First stone plodder installed; Georgia repeals corvée in favor of use of slaves; Argentat bridge uses spun cables
1830	Macadam endorsed in France; first passenger steam railway; turnpike era at its peak; *Annales des ponts et chaussées* first published; travel speeds jump to 60 km/h

1831	Commercial steam coaches begin operating; safety fuse invented; Hodgkinson introduces practical column design; Broughton suspension bridge fails under troop traffic
1832	Horse-drawn trams in use; London Bridge demolished
1833	First bowstring truss bridge
1834	Hansom cabs in significant use; tar surface-dressing patent granted; Sarine Valley suspension bridge sets world span record; first electric vehicle
1835	British Highway Act requires cartways to be 6 m wide, specifies left-hand driving, introduces effective local-government taxation and ends corvée system; plank road invented; *Anti-Railway* journal launched
1836	Coaching reaches its peak in England; taxes placed on steam coaches; iron first used for street paving
1837	Irish conduct first traffic survey; London–Holyhead mail transfers to rail; native asphalt used as a surface dressing
1838	British act authorizes railways to carry mail; asphalt first used in the United States; major road experiment in London; Russian timber blocks first used in Britain; first steam-powered earthmoving equipment; Hughes publishes book on road drainage; federal funding for U.S. National Road ends; Paris adopts mastic for footpaths; London committee sees new urban roads as improving moral and physical health; carriage drivers required to carry a number
1839	First pedal bike; Goodyear vulcanizes rubber; first iron bridge in the United States
1840	Tarmacadam in use; rubber roads tested in London; iron used in trusses; first bridge in concrete
1841	U.S. National Road abandoned; mail carried on British railways; British act permits rates to be used to rescue turnpikes; iron plate girder introduced
1842	Mobile steam farm engine produced
1843	Rebecca Toll riots; the Brunels complete soft-ground tunnel under the Thames
1844	Marginal cost pricing introduced
1845	New York State specifies 6.5 m for road widths; turnpike era ends; pneumatic tire patented
1846	Ramming to rhythm introduced for pavement compaction; concrete in use as a structural course; Warren truss introduced
1847	First solid rubber tire; Zhurawski produces method for redundant truss analysis; Whipple publishes book on bridges
1849	First powdered-asphalt trials; Whipple publishes book on truss theory; natural asphalt method discovered at Neuchâtel; Wheeling suspension bridge sets world span record
1850	Dark ages of roads begins; work begins on Paris boulevards; road control reverts to local government in many countries; Britannia Rail Bridge opened; traffic congestion on Broadway (New York

City); wood-block street burns in San Francisco; first rolled wrought-iron beam

1851   Central road control abolished in Ireland; compressed air first used for caisson construction

1852   First tram with rails flush to the road; French keep-to-the-right decree

1853   First modern iron bridge; first water-filled road roller

1854   Hot powdered native asphalt successfully used in Paris to produce natural asphalt; Wheeling suspension bridge fails under wind resonance

1856   Nitrocellulose and nitroglycerin invented; Bessemer converter developed for steel

1857   First commercial oil refining

1858   Steam stone-crusher manufactured; gunpowder used for roadmaking; Carey produces gravity model for trip prediction; Central Park (New York City) opened

1859   First (steam) motorcycle; first practical internal-combustion engine; first oil strike

1860   Steam-driven road roller manufactured; population densities of New York and Philadelphia peak

1861   First commercial bicycle; de Smedt arrives in the United States; wooden streets burn in Chicago great fire; limits placed on coach speeds; open-hearth furnace developed for steelmaking

1862   First practical IC-powered vehicle; de Rochas publishes four-stroke paper; steel first used in major bridges

1864   Trinidad Lake asphalt commercially exploited; Culmann publishes book on bridge analysis; Maxwell publishes structural analysis method

1865   Red Flag Act introduced in Britain; concrete mix design first well understood; first full concrete road built; high-volume petroleum refining begins; Malo's work on asphalt published; Britain begins exporting steam vehicles; first penetration concrete; Culmann publishes work on graphical statics

1866   First concrete surface course; first concrete blocks used for paving

1867   Nobel develops dynamite; Paris licenses carriage drivers; largest drawbridge opened in Copenhagen

1868   Pedestrian crossing signals introduced in London; Winkler develops influence line method; Olmsted proposes Brooklyn parkway; first cycle race

1869   First successful use of precoated tarmacadam; France introduces abrasion test for macadam stone; first international bicycle race; first bicycle journal published; Crompton uses steam vehicles for road-trains in India; first road use of mastic

1870   Germany abandons central control of roads; mastic first used for road surfacing; rickshaw invented; gasoline first used for IC; first bicycle club

1871   Last London turnpike closed

1872   First trial of manufactured asphalt

1873	Bridge traffic loadings studied; Forder introduces improved hansom cab; first use of spray and chip seal; first trials of native asphalt paving by de Smedt
1874	Eads's St. Louis Bridge opened (longest fixed-arch bridge); Mohr publishes virtual displacement method
1875	Continuous concrete mixer developed; Public Health Act ushers in the British suburb; first reinforced-concrete bridge
1876	Otto develops first four-cycle IC engine; first all-steel bridge; blasting powder invented; de Smedt moves to Washington, D.C., to advance its pavements; wrought-iron Howe truss collapse shows need for reappraisal
1877	Safety bicycle invented; Washington, D.C., trials prove success of asphalt
1878	Blade grader introduced; Pennsylvania Avenue asphalted; Bicycle Touring Club formed in the United Kingdom; mild steel introduced
1879	Electric tram invented; Tay Bridge failure
1880	County councils established in England; asphalt-block compressing machine manufactured; French trial of bituminous sprays; subdivided trusses introduced; horse-drawn earthmoving equipment in use; League of American Wheelmen formed; Benz sells first engine
1883	Benefit-cost analysis applied to roads; Barber Company formed; Roeblings' Brooklyn suspension bridge opened, including centerline marking
1884	First modern road map; typical new suburban village developed at Chestnut Hill, Pennsylvania
1885	Leaning-wheel graders introduced; first modern bike; first practical IC vehicles produced by Benz (tricycle) and by Daimler and Maybach (motorbike); asphalt blocks successfully used; rolled steel joists produced
1886	Roads Improvement Association formed in Britain; Benz patents first vehicle and has first public trip in an IC car
1887	World's first car sale; Richardson succeeds de Smedt as inspector of pavements in Washington, D.C.; first use of pneumatic tires on a powered vehicle
1888	Dunlop reinvents the pneumatic tire; U.K. cyclists required to carry a continuously tinkling bell; Australian hardwood used for paving in Europe; penetration tester developed for bitumen; Barber Company controls Trinidad Lake asphalt; county councils given road powers in Britain; first long motor trip, by Bertha Benz; Selby sets stagecoach speed record of 23 km/h
1889	First IC bus; Firth of Forth Bridge opens
1890	American manufactured-asphalt technology exported to Australia; Good Roads Association formed in Australia; Benz introduces modern stub-axle steering
1891	New Jersey leads way back to centralized road control; first designed-for-the-purpose car; first IC truck

1892	National League for Good Roads formed in the United States; steam trucks introduced
1893	Diesel patented; first IC car in the United States; Massachusetts creates first U.S. highway department; U.S. Office of Road Inquiry formed
1894	Richardson joins the Barber Company; manufactured asphalt used in Britain; Richardson conducts trials in Britain; first oiled road; first motoring magazine; four new types of movable bridge introduced; first car race
1895	Last turnpike on Holyhead Road closes; electric cabs introduced in Paris; first car on modern tires; first use of unblended residual bitumen for roadmaking; first British IC car; Italian Touring Club erects cast-iron road signs; first cycle path; first conventional car race; the word *car* first used for a motor vehicle; first commercial motorbike; first IC bus; first motor show; first car clubs formed
1896	Red Flag Act relaxed in Britain; first electric car; first car accident in England, in France, and in the United States; first car accident insurance; steel culvert invented; first use of asphaltic concrete; Brooklyn residents resist street paving to stop through traffic
1897	Stanley Steamer introduced; French begin using motor vehicles in military maneuvers
1898	Rubber added to bitumen; first international motor show held in Paris; first independent car dealer; Howard's garden city movement begins
1899	First modern concrete arch bridge; first Australian IC car; first car fatality in the United States; driver licensing introduced in Chicago; London–Edinburgh car trip made
1900	Heated bitumen transport introduced; sheepsfoot roller invented; first *Michelin Guide,* regulations committee stops bicycle development; United States begins national pavement-material testing program
1901	First modern car; cyclists and motorists in the United Kingdom unite for better roads; Canadian and American railroads operate Good Roads trains; tarmac invented; vehicle registration introduced in New York
1902	Guglielminetti conducts spray trials at Monte Carlo; Swiss canton bans cars; American Road Makers formed; hand signals for drivers introduced (in Berlin); first mass-produced car (Oldsmobile)
1903	First car crosses the United States; emulsified bitumen introduced; Pont Adolphe in Luxemburg becomes largest stone arch; driver licensing introduced in Massachusetts; Ford Motor Company founded; first sports car; Letchworth new town established; seat belts first used (France)
1904	First crawler tractor; bitumen blowing in commercial use; glued, laminated timber produced; possible first electric traffic signal; traffic-counting stations set up in Maryland
1905	Richardson issues text on asphalt; last British stagecoach ceases

operation; first double-decker bus (Germany); first road safety conference; first traffic roundabout—Columbus Circle in New York City; New York borrows money for road construction; Plauen Bridge sets world arch span record; Hampstead Gardens suburban development introduces new standards

1906   IC buses capture market; cloverleaf intersection invented; pedestrian safety island introduced; Beverly Hills opened as transit company land speculation; Long Island Motor Parkway built; first filling station opened (United States)

1907   First car crosses Australia; U.K. tar spray trials; Quebec Bridge failure; Peking–Paris rally; French army subsidizes citizen purchase of trucks; stop line marking introduced in Virginia

1908   First controlled-access road opened (Long Island); Model T Ford begins production; General Motors founded; superelevation introduced; riveted joints used; Georgia legislates to use convicts for roadmaking; PIARC formed and suggests international traffic signs

1909   National Roads Board established in the United Kingdom with funds from a gasoline tax; first international convention on car regulations; first international convention on signs; car replaces horse as official transport for American presidents; portable asphalt mixer introduced

1910   Solid rubber truck tires widely available; American Road Builders Association established; U.S. trials of road primer; New Zealand begins spray and chip trials; PIARC conference concentrates on dust problem; Timoshenko begins his publication series; first symbolic traffic sign; soil stabilization patented in the United States

1911   British government replaces the horse with the truck in the British army; first centerline marking on bridges and road curves; first marked pedestrian crossing

1912   First pneumatic truck tire; the United States nominates specific truck for army use; Stanley Steamer sales peak; pavement test track built in England; first traffic control tower; probable first electric traffic signal; the United States introduces national road funding; international traffic priority rules recommended; the United States funds experimental study of rural roads

1913   First Australian state road authority established; Germany begins construction of Berlin dual carriageway; last corvée system abolished in Alabama; Ford begins mass production and introduces the assembly line

1914   First motorized vehicle used in war; first modern traffic signal; American Association of State Highway Officials (AASHO) formed; last horse-drawn bus in London; jitneys begin operation (in the United States); all inhabited continents crossed by car

1915   First stop sign; stone introduced into sand carpets in the United Kingdom

1916   Federal standards in the United States specify 5 m width for two-way roads; rubber blocks used for paving; federal road funding introduced in the United States; cord begins to replace canvas in common truck tires; first turn ban

1917   First coordinated traffic signals; first pavement use of soil stabilization; largest truss bridge built (in Quebec)

1918   Reflectors placed on railway crossing gates; U.S. 396 Bureau of Public Roads formed; first rental car business opens; first three-color traffic signal

1919   Welwyn new town established; Thomas MacDonald appointed commissioner of U.S. Bureau of Public Roads

1920   The United States has as many motor vehicles as horse-drawn ones; first four-directional traffic signa; Highway Research Board established in the United States; Bates Road Test begins; the United Kingdom begins bypass construction

1921   American states produce sign manuals; construction of first Autostrada begins; U.S. introduces Federal Aid Highway Act; first limited-access, divided-carriageway road opened

1922   Le Tourneau introduces tracked scraper; U.S. Army produces Pershing road plan; first motel; Eno Foundation and Harvard traffic research bureau established; joints used to control concrete slab cracking

1923   First bulldozer; traffic counting promoted by U.S. government; Maine makes first modern traffic predictions; bitumen blends in the United States officially reduced to nine; first vehicle-based roughometer; peak year for production of Model T Fords; Bronx River Parkway opened; bullseye reflectors developed

1924   First road safety conference; Moses becomes commissioner of Long Island Parkway Commission; Le Corbusier produces plan for new Paris

1925   Hydraulic controls for bulldozer blades; glass beads used for signs; last Australian stagecoach ceases operation; the word *motel* first used; first autostrada opened; Hubbard and Field develop stability test for asphalt

1926   League of Nations Sign Convention; British car ownership per capita passes peak for horse-drawn carriages; Westergaard introduces rational concrete-slab design

1927   First overhead lane controls; Hamlin predicts lane capacity; Holland Tunnel under Hudson completed; gravity model for travel applied in Boston

1928   First vehicle-activated traffic signal; word *freeway* coined; prestressed arch built in Germany; Pigou proposes optimum congestion level

1929   Contruction of first autobahn begins; last horse tram closes, in Budapest; Ambassador suspension bridge sets world span record; planners firmly introduce concept of car access, but not car thoroughfare; first pedestrian mall introduced (in Germany); Radburn new town established

1930   CBR test introduced for natural formations

1931   First diesel-powered equipment; second League of Nations Traffic

Sign Convention; Bayonne Bridge completed and sets world arch span record; George Washington Bridge opened and sets world span record

1932   First section of George Washington Memorial Parkway opened; first drive-in movie; discomfort criterion used in curve design; first traffic engineering codes; Cross produces moment distribution method; first freeway opened (Bonn–Cologne)

1933   Parisian taxi drivers strike over slippery asphalt; Proctor compaction test introduced

1934   Design-speed concept firmly in place; U.S. Congress funds road planning; first urban parkway opened (Henry Hudson)

1935   All U.S. Army equipment to be able to travel on roads; Greenshields produces linear speed–density link; textiles used in pavements; first automatic parking meter; catseye reflectors developed

1936   Freyssinet builds first modern prestressed-concrete bridge

1937   Golden Gate Bridge sets world span record; President Franklin D. Roosevelt recommends a national highway system

1939   New York Worlds Fair highlights roads of the future; first new-generation U.S. toll road (Merrit) opened

1940   Menai Straits Bridge rebuilt; Tacoma Narrows Bridge fails in wind; first radial freeway (Pasadena) opens in Los Angeles; Pennsylvania Turnpike opens

1941   The United States passes defense act for roads

1942   CBR-thickness curves introduced by the U.S. Corps of Engineers

1944   Interstate system recommended by a U.S. presidential committee; first urban travel studies

1945   Highway system advocated for emergency evacuations in the United States; interstate system mentioned in federal act

1947   Last horse-drawn cab license surrendered in London

1949   United Nations Sign Convention approved; Britain passes legislation for motorways

1950   Britain builds last steam truck; Germany reintroduces cable-stayed bridge

1952   First money spent on U.S. interstate system; WASHO Road Test conducted

1953   Wilson of General Motors announces that what's good for GM is good for America; MacDonald ends thirty-four years in charge of Bureau of Public Roads

1956   Eisenhower authorizes U.S. interstate system; Beckman argues for road pricing

1958   AASHO Road Test begins; first British motorway opened; first major cable-stayed bridge (T. Heuss)

1961   Duplessis Bridge collapses due to brittle fracture

1962   Kings Bridge collapses due to brittle fracture; U.S. road funding act adds planning mandate

1964   Verrazano Narrows Bridge in New York establishes world span record

1965   Area traffic control introduced in Sydney

1967	Silver Bridge collapses in Virginia
1968	Vienna UN convention on road signs
1969	Buchanan publishes *Traffic in Towns*; Fourth Danube and Milford Haven bridges collapse
1970	Seat belts made compulsory for the first time (in Victoria); Westgate Bridge collapses due to design problems
1975	Singapore cordon-restraint system introduced; Tasman Bridge collapses due to ship impact
1977	New River Gorge Bridge completed and sets world span record for arch bridge
1981	Humber Bridge in England establishes world span record
1983	Mianus Bridge collapses in Connecticut
1985	Electronic road pricing trial in Hong Kong
1991	Akashi Straits Bridge in Japan establishes world span record

Notes

Chapter 1: The First Ways

1. The original link between the two ways was probably where the route crossed the Thames near Goring (Taylor 1979). The name is sometimes spelled Ickneild or Ikenild. Icknield is its newer Anglo-Saxon name and refers to the Iceni people through whose Norfolk land it passed.

2. It has also been called the Royal, Pilgrims', Host, and Ox roads. In Danish the name means "host road."

3. Queen Bodicea's name is also spelled Boudicca, Boadicia, or Bonduca. The name meant "victory." Her husband was king of the Iceni.

4. In his poem "The Rolling English Road," published by Methuen of London in *The Flying Inn*.

5. One such road may have been the Fosse or Foss Way. *Foss* is an Anglo-Saxon word that means "ditch," and the way may have served as a boundary between peoples in post-Roman times. The maximum extent of the pre-Roman way was claimed in about 1940, and the name Jurassic Way applied to it at that very recent time referred to the assumed use of the way to carry quarried Jurassic limestone found along parts of its length. There is much unresolved doubt about the pre-Roman existence of the Fosse Way (Taylor 1979). In part it may have been linked with Wan's or Woden's Dyke in Wiltshire. *Dyke* had the same meaning as ditch and was used as a means of defense. Wan or Woden was the chief Norse god.

6. The underlying factors influencing the growth of cities—their morphogenesis—are covered in Vance ([1977] 1986).

7. Herodotus was a Greek author who lived in Asia Minor in the fifth century B.C. and wrote a comprehensive *History of the Persian Wars*. He has been called the Father of History and the world's first travel writer (Casson 1974).

8. Aristotle considered Hippodamus eccentric because he wore his hair long and favored expensive jewelery and cheap clothes. Miletus and Priene were adjacent cities on the estuary of the Meander River in Turkey (see fig. 3.1). Miletus had been a leading center for geographic studies since 600 B.C.

9. St. Augustine, settled in 1564, was the first permanent white settlement in the United States.

Chapter 2: The Demands of Transport

1. The word *wagon* (or waggon) is commonly used for four-wheeled vehicles and *cart* for two-wheelers. A *carriage* is any vehicle for carrying people, and a *dray* is a cart used for carrying freight.
2. Emperor Elagabalus extended the list in A.D. 200 by variously using naked girls, tigers, lions, and leopards to haul his personal carriage.
3. Hence the origin of the word *post* now applied to mail deliveries. It was of this post service that Herodotus wrote: "Neither snow, nor heat, nor darkness of the night stays these messengers from completing their task as swiftly as possible." This quotation was later used as the motto of the U.S. Postal Service.
4. The word *diploma* comes from the pass the Romans issued to permit free passage of the postriders and their other official messages.
5. The donkey, ass, and onager are all members of the horse family. The donkey—or burro—is the domesticated version of the wild ass, and breeding produced a sturdier animal than either the ass or the onager. The mule is a horse-donkey hybrid, which, although sterile, is tougher than the horse and stronger than the ass. Figure 2.3 shows that dogs were also used as pack animals.
6. Packhorse paths were also called packways, pack and prime ways, causeways, and causeys. The word *causeway* comes from the same Norman root, *cauce,* as the French word for road, *chaussée.* As befits its origin, *causeway* is now often used for a raised embankment-like way crossing wet or marshy ground. In the nineteenth century, *causeway* was also used in Scotland to describe a street paved with stone. An 1857 report to the city of Edinburgh read: "The stones in many places . . . should be lifted and recausewayed. . . . The Ratho [a stone type] causeway would cost about . . ." (Paxton 1977).
7. In 1769 Hunt developed the modern hoop tire and was once thought to have been the first to invent the technique (Severy 1977).
8. Cicero noted that some Roman chariots were patterned on those used by Queen Bodicea (Strong 1956).
9. It was once believed, on the basis of artists' drawings that did not show the smaller front wheels characteristic of a steerable vehicle, that medieval vehicles were not steerable. It now appears that many such drawings were not accurate; artists used artistic license to show all wheels of uniform size.
10. Technically, wheel loads and tire dimensions determine the contact pressures between the wheel and the running surface of the path, the stresses in the pavement structure, and the stresses in the underlying natural formation. The contact pressure largely determines both the ability of the wheel to traverse poor ground and the damaging effect of the wheel on the running surface. The magnitude and frequency of the stresses on the pavement structure and underlying natural formation determine the life of the pavement.
11. The number of axles rather than the number of wheels is usually counted. Legal axle loads around the world range between 8 and 14 tonnes. A design axle load is often called an equivalent standard axle (ESA), and pavements are usually designed for at least a million ESA. A large truck might cause three ESA of damage (Lay 1990).
12. From the twelfth century, Beverley was a major market town involved in an active cloth trade with Europe.
13. Chapter 7 gives a similar criticism of legal edicts in terms of four hundred years of street cleaning laws.

Chapter 3: Roadways

1. Ur and Lagash in Mesopotamia were the two famous Sumerian cities founded in about 4000 B.C. and 5500 B.C. respectively. Ur, the home of Abraham, was the Sumerian capital for four hundred years from 2800 B.C. Despite various successful invasions, Sumerian influence remained strong in the region until about 600 B.C.

2. Taklimakan is also spelled Takla Makan; the name means "desert of no return." A third middle route went from Kuqa through the desert along the Tarim River to Ruoqiang.

3. There was an alternative northern route between Lanzhou and Xian that used the Jing He rather than the Wei He River.

4. Note that Woolley's dates and deductions are rather ambitious. A more restrained view, leading to the dates given in the text is found in Moorey (1982).

5. An extensive discussion of street paving in the Middle East in the millennia before Christ is given in Forbes (1934).

6. The name Hermes probably derived from the Greek word for a pile of stones. His commonest symbol was that of a phallus. Interestingly, Japan also had a tradition of placing phallic symbols at crossroads.

7. This construction is said to have given rise to the term *highway* (O'Flaherty 1967), although the word more probably derived from the use of embankments to provide passage through marshy ground. An alternative explanation is that *highway* derived from *ridgeway*.

8. There is some question about the terminology used here, which is drawn from the somewhat vague descriptions of Vitruvius. It may refer more to floor than to road construction (Leighton 1972). For example, *pavimentum* was also the word for a floor of compacted earth.

9. A portion of Watling Street (also called Gwethelin Street) subsequently served as the boundary between Warwickshire and Leicestershire. Another portion closely paralleled the London end of the Holyhead Road (see fig. 1.1).

10. For the millennium after the Romans, no permanent bridge crossed the Rhine. Charlemagne did build one wooded bridge at Mainz. Construction took ten years, but the bridge was soon destroyed by fire. Some believe that a stone arch bridge—the Constantine Bridge—existed at Cologne from the fourth to the tenth centuries, when it was destroyed by the order of the archbishop of Cologne.

11. Or Abd ar-Rahman II. His public works program was one of the highlights of his spectacular reign from 822 to 852. He also installed extensive street lighting (chapter 7).

12. There are graphic descriptions of just how bad the roads of those times were. An excellent review of roads of this period is given in Wilkinson (1934).

13. Gautier described himself as "architecte, ingénieur, and inspector des grands chemins, ponts et chaussées du Royaume." He also published the first book on bridge engineering in 1714 (see chapter 8).

14. Chapter 4 discusses the close links between Trésaguet, Turgot, and events leading to the French Revolution.

15. As was McAdam (Manton 1956); the sobriquet must have appealed to nineteenth-century humor. Other nicknames were Pontifex Maximus (see chapter 8 for its real meaning) for Telford and McAdam the Magician for McAdam.

16. Edgeworth had been funded by the Dublin Society in 1816 to carry out research leading to the book. His inventions related to carriage wheels and springs. His daughter Maria became a leading novelist of her time.

17. This does not quite excuse Seely (1987) for describing macadam as "packed stone bound together with earth."

18. See chapter 8. Navier's book on the subject was published in 1823. An abridged, updated, and more accessible version of this work is available as Navier (1831). Another book translating McAdam's work into French was published as Cordier (1823).

19. An extensive set of resistance calculations for various circumstances is given in Baker (1903).

Chapter 4: Motives and Management

1. Culloden was the battle at which the English crushed the 1745 Scottish rebellion. The sentence is chronologically misleading; as shown in chapter 3, Wade had built his Scottish roads from 1725–1737, and Telford did not begin his until 1803.

2. Napoleon III had another civil engineering interest. While in exile in London he attended Institution of Civil Engineers meetings, had a published discussion (*Proceedings* 6:427) and was made an honorary member in 1869 (Institution of Civil Engineers 1873).

3. Excellent background to the subject is given in Webb and Webb ([1913] 1963) and Jackman (1916), and so only a summary of the tortuous history is given here.

4. The gild was established in the thirteenth century to support poor priests. Its charter was strengthened by Richard II. One of its other roles until 1553 was to manage Stratford-upon-Avon.

5. Edward Clarke was an associate of commentator Arthur Young, whom we encounter later in the chapter.

6. The "ancient highway and post road" from London to Scotland via York. The Great North Road, or Great Road North, went from London via the Angel Inn at Islington, Stevenage, and Biggleswade to Huntingdon (Alconbury). It followed old Roman roads leading to Stilton, Lincoln, Doncaster, and thence to York, and eventually via Newcastle and Berwick to Edinburgh (Harper 1901). At its southern end the Roman road was known by its later Saxon name of Ermine (or Hermin or Irmine) Way or Street. Ermin was a Saxon god. Much of the route is now the A1. The southern portion of the road actually has two parts. The older second part is called the Old North Road and reaches Huntingdon via Tottenham, Ware, Wadesmill, and Royston. It is shown as a solid line in figure 1.1 and is the route now followed by the A10 and A14. The newer route is shown dashed.

7. Extensive original documents related to English tolls and turnpikes are reproduced in Searle (1930).

8. Young was a prominent activist of the time. Although he played a significant role in road reform he is better known for his role in agricultural reform, leading to his appointment in 1793 as the first secretary of the British Board of Agriculture. The quotation is from Law and Clark (1907).

9. The post office had been expanded to carry private mail in 1629 and had established regular postal services in 1635.

10. John Palmer's idea led to his appointment as controller general of the post office in 1786. He was "ignominiously dismissed" in 1792 for administrative malpractice.

11. A fine review of the French road system during this time is given in Vance (1986).

12. Or Corps imperial du génie. Colbert had previously established an Academy of Science in 1666 and an Academy of Architecture in 1671. He had also combined the grand voyer's role with his other official duties.

13. The political history of the bureau is covered in Seely (1987). A number of its early research initiatives are discussed in chapter 7.

14. Seely (1987) offers the counterview that AASHO was formed by eastern states to counter the strong rural focus advocated by Page.

15. We meet Agg in chapter 6, and his text is referred to frequently in chapter 7. In 1923 one member of the committee was W. E. Lay, an assistant professor of mechanical engineering at the University of Michigan (Hatt 1923).

16. The Jeu de Paume is memorable to many for its subsequent role for many years as a gallery of French impressionist paintings.

Chapter 5: A Surge of Power

1. These dates are confused. Some say that the carriage arrived in 1560 or 1564 and not in 1571.

2. The name of the town, also spelled Kocsi, Kocz, Kozo, or Coki, is pronounced "koch." The town is near Budapest.

3. Diane de Poitiers's residence at Chenonceaux was one of those served by the famous 1556 Toury to Artenay road built by Henry II and discussed in chapter 4. Some accounts suggest that the coach owner might have been the king's daughter, Diane de France.

4. Chapter 4 noted how the Royal Mail in 1780 was still taking forty hours on horseback to reach nearby Bath, and Vale (1960) reports that the standard mail trip in 1782 took thirty-eight hours.

5. Horse trams were also called horse cars, street railways, street railroads, or streetcars. Electric trams were also called trolleys. A tram running in its own right of way is sometimes called light rail.

6. An early Telford biographer (Rickman 1838) saw it differently, noting that "Telford was not generally adverse to railways."

7. An alternative view is that no practical two-wheelers existed prior to Drais von Sauerbronn's invention (Wilson 1986).

8. A good history of the tire is given in Tompkins (1981).

9. A detailed technical history of the internal-combustion engine is given in Newcomb and Spurr (1989).

10. This Samuel Brown is not to be confused with the naval officer of the same name who was making major contributions to bridge technology at the same time (chapter 8).

11. Some texts give him as a Luxemburger, but Duncan (1928) is adamant that Lenoir was born in Mussy la Ville in Belgium and was naturalized as a Frenchman after 1871.

12. A good review of the history of the development of the IC car in Germany is given in Nuebel and Seherr-Thoss (1983).

Chapter 6: Power in the Road

1. It is arguable whether it was the first because the steam cars of the 1830s had brought forth *The Journal of Elemental Locomotion* in 1832.

2. From Graham's *Ruthless Rhymes,* republished by Edward Arnold, London, in 1984.

3. The various impacts of the car on American life between 1895 and 1910 are admirably reviewed in Flink (1974).

4. A comprehensive review of this period of autocamping is given in Belasco (1979).

5. Schlereth (1985) claims that electrically powered red and green lights were operating on Euclid Avenue in Cleveland, in 1904.

6. A comprehensive review of American signal development is given in Sessions (1976).

7. Rømer was also a successful astronomer who made the first realistic measurement of the speed of light.

8. Early signing practice is reviewed in Sessions (1976).

9. However, Stenton (1951) reports a late fourteenth-century guide to English roads in the Bodleian Library.

10. A fine historical collection of French road maps is available in Reverdy (1986).

11. The fascinating and confused history of the mistransfer of this particular technology over some fifty years has been explored by Good (1978).

Chapter 7: Pavements

1. From the author's experience in England in the summer of 1985—it rained on forty consecutive days—the watering task might not have been demanding. In 1726 Daniel Defoe wrote in his *Tour through the Whole Island of Great Britain,* "Sometimes a whole summer is not dry enough to make the roads passable."

2. *Viscous* means that, given time, the material does flow under force. A rigid material with infinitely high viscosity does not flow at all, whereas a material with low viscosity, such as water, flows almost instantly.

3. The publication date for Jost's study has been given as 1626, but it has not been possible to reconcile this date with his 1711 meeting with d'Eyrinys. Although d'Eyrinys described himself as a professor of Greek and a doctor of medicine, Forbes referred to him as a "Greek doctor" (1936) and later (1958c) observed, "He claims to be a Greek by birth." Others have called him a "Russian doctor."

4. Robb (1947) and, by citation, James (1964b) have claimed that asphalt was used in 1747 by the earl of Kildare (he became the first duke of Leinster in 1766) and his German architect Richardi Castle for carriageways at Leinster House in Dublin (later the seat of both the Irish Houses of Parliament). Castle, who came to Ireland in 1729, is claimed to have copied a method used on the estate of Prussian noblemen using an asphaltic mixture of ship's bitumen and lime on a lime-concrete structural course. The evidence given is a 1747 record in a manuscript in the duke of Leinster's library. The Prussian origins are consistent with the asphalt discoveries noted above. However, careful investigations by P. J. O'Keeffe (private communication, 1988) of the available evidence in Dublin indicates no substance to the claim. Sketches and subsequent references to men raking the gravel carriageways of Leinster House confirm this view.

5. Electric lighting began in the 1880s.

6. The French suggest that Vicat may have preceded Aspidin by some six years.

7. The Committee on Science and the Arts (1843) claimed that creosoting had been proposed by the German, M. Moll, prior to 1843. Creosote is made by distilling coal tar.

8. Cartways were also known as wheelers, tramways, and plodders.

9. U.S. 727505, issued in 1903.

10. Equipment used in asphalt construction is comprehensively reviewed in Tunnicliff, Beaty, and Holt (1974).

11. An unfortunate but common terminology—rigidity is absolute and so all natural materials by definition lack rigidity.

12. A review of bitumen test development in the United States during this century is given in Hveem (1971).

13. "Road surfacing" does not necessarily mean a bound surface, but could also cover sand-clays and macadam. Indeed, Rose (1953) indicates 200 Mm of surfaced road in the United States in 1900, whereas the data in chapter 6 indicate that less than 10 Mm, or 5 percent, was paved with a coherent surface.

Chapter 8: Bridges

1. In addition, an erudite classical and social history of the bridge is given in Robins [1950].

2. Izmir was then called Smyrna (see fig. 3.1).

3. Other writers describe the bridge as a multispan crossing of one of the two Min Jiang rivers, built in about 1050 B.C. with an overall length of 400 meters.

4. The word *Euphrates* meant "well bridged."

5. The word *Bosphorus* meant "cow ford."

6. For example, there are doubtful records of iron chains in a 120 m span bridge in Shaanxi in China in 206 B.C. (Knox 1974), and of a 75 m iron chain bridge built in A.D. 65 near Jinghong, over the Lan Cang—or Mekong—River in Yunnan (Needham 1971; Wittfoht 1984). Needham first indicated doubt, which was subsequently confirmed, about the A.D. 65 date. The Ji Hong Bridge appears to be the one that many writers have incorrectly claimed to have been built in A.D. 65 (Tang 1987).

7. Or Tang Dong Gan Po or Tong Tang gyal-po in Chinese.

8. The Bramaputra was also called the Zangbo River.

9. Marc Brunel was a naval engineer who had fled from France during the Revolution. He later sent his son, Isambard, back to France for an engineering education.

10. Olaf later became King Olaf II of Norway, brought Christianity to that country, and was canonized as St. Olaf.

11. There is a suggestion that the Pont-St.-Esprit, which was built over the Rhône near Bollene (about 40 km upstream from Avignon) in 1305 might qualify as the first.

12. Coulomb is best known in civil engineering for the development of the earth pressure theories still used for retaining-wall design, but is perhaps most widely known for his work on electricity and magnetism. An outstanding theoretician, he began his career as an engineer with the Corps des ponts et chaussées, a background that ensured his research remained of great practical relevance. Indeed, he was moved to his earth pressure theory by dissatisfaction with the applicability of Bélidor's work to military fortifications he was building in Martinique. Coulomb's contributions are gathered together in his famous *Memoir on Statics,* published in 1773 (Coulomb 1773; Heyman 1972).

13. Twenty-two years later Barlow became famous as the designer of the iron framework for the glass and iron St. Pancras Station in London.

14. Good reviews of early United States bridges are given in ASCE (1976) and Fletcher and Snow (1932).

15. Howe's nephew Elias invented the sewing machine.

16. A fine and comprehensive general review of the history of bridge analysis is available in Hopkins (1970).

17. *Stress* is force divided by the area over which it is applied. *Strain* is deformation divided by original length.

18. The Seguin family lived in Annonay and was related to the balloon Montgolfiers. Some authors misspell the name Sequin.

19. The poem, called "The Tay Bridge Disaster," appeared in his *Poetic Gems,* published in 1890. The verse quoted is the first of nine.

20. Its proper name was the Point Pleasant Bridge. It was called Silver Bridge locally because it was the first bridge painted with aluminum paint.

Chapter 9: From the Past into the Future

1. The role of the U.S. auto industry lobby has been critically reviewed in St. Clair (1986). The role of the British road lobby, particularly the Roads Improvement Association and the British Road Federation, is discussed in Hamer (1987).

2. In the 1830s Robert Owen coined the word *socialism*.

3. The cul-de-sac is called a *blind lane* in earlier English and a *court* or *close* in some countries.

4. In 1901 the Pacific Electric had grown out of the Southern Pacific Railroad, which had arrived from San Francisco in 1876 (Bottles 1987). However, a local railway had been operating since 1869.

5. From his 1933 book, *Happy Days*.

6. Chinese records from the Western Zhou dynasty (1066–771 B.C.) mention the construction of ring roads (Jiang 1986).

7. A comprehensive history of the development of British motorways is given in Charlesworth (1984). Important acts to control access from adjoining properties and to control ribbon development had been introduced in 1909 and 1935, respectively.

References

Abraham, H. 1960. *Historical Review and Natural Raw Materials*. Vol. 1 of *Asphalts and Allied Substances*. New York: Van Nostrand.

Addison, W. 1980. *The Old Roads of England*. London: Batsford.

Agg, T. R. 1916. *The Construction of Roads and Pavements*. New York: McGraw-Hill.

Aitken, T. 1907. *Roadmaking and Maintenance*. London: Charles Griffin.

Albert, W. 1972. *The Turnpike Road System in England, 1663–1840*. Cambridge: Cambridge Univ. Press.

Anderson, R. M. 1932. *The Roads of England*. London: Benn.

ARBA (American Road Builders' Association). 1977. "History of ARBA, 1902–1937." *American Road Builder* 5413–5414:1–79.

Armstrong, E. L. 1976. *History of Public Works in the United States, 1776–1976*. Chicago: American Public Works Association.

ASCE. 1976. *American Wooden Bridges*. American Society of Civil Engineers, (ASCE), Historical Publication no. 4. New York: ASCE.

Attwooll, A. W. 1955. "Rolled Asphalt Road Surfacing Materials." *Journal of the Institution of Highway Engineers* 3 (6): 59–75.

Aveling-Barford Co. 1965. *A Hundred Years of Road Rollers: A Pictorial Record*. Surrey: Oakwood.

Bagshawe, R. W. 1979. *Roman Roads*. Aylesbury, Eng.: Shire Archaeology.

Bagwell, P. S. 1974. *The Transport Revolution from 1770*. London: Batsford.

Baker, I. O. 1903. *A Treatise on Roads and Pavements*. New York: Wiley.

Baker, J. N. 1960. "Medieval Trade Routes." In *Social Life in Early England*, ed. G. Barraclough. London: Routledge and Kegan Paul.

Ballen, D. 1914. *Bibliography of Road-Making and Roads in the United Kingdom*. London: King.

Balogun, J. A.; Robertson, R. J.; Goff, F. L.; Edwards, M. A.; Cox, P. C.; and Metz, K. F. 1986. "Metabolic and Perceptual Responses whilst Carrying External Loads by Head and by Yoke." *Ergonomics* 29 (12): 1623–1637.

Barker, T. C., and Robbins, N. 1963. *A History of London Transport*. London: George Allen and Unwin.

Barth, D. O. 1980. "America's Covered Bridges." *Civil Engineering* (ASCE) 50 (2): 51–54.

Beasley, D. R. 1988. *The Suppression of the Automobile*. New York: Greenwood.

Beckman, M., McGuire, C. B., and Winsten, C. B. 1956. *Studies in the Economics of Transportation*. New Haven: Yale Univ. Press.

Belasco, W. J. 1979. *Americans on the Road: From Autocamp to Motel, 1910–1945*. Cambridge, Mass.: MIT Press.

Belloc, H. 1923. *The Road*. Manchester: C. Hobson for the British Reinforced Concrete Engineering Co. Ltd.

———. 1926. *The Highway and Its Vehicles*. London: Studio.

Benvenuto, E. 1991. *An Introduction to the History of Structural Mechanics*. New York: Springer-Verlag.

Better Roads. 1980. "History and Progress: Georgia." *Better Roads*. November, 16–18.

Beuscher, J. H. 1956. "Roadside Protection through Nuisance and Property Law." In *Land Acquisition, 1955*. Highway Research Board Bulletin no. 113.

Bickmore, A. S. 1865. *Travels in the East Indian Archipelago*. London: Murray.

Billington, D. P. 1981. "Bridge Design and Regional Aesthetics." *Proceedings of the American Society of Civil Engineers* 107 (ST3): 473–486.

———. 1983. *The Tower and the Bridge*. New York: Basic Books.

Birch, E. L. 1980. "Radburn and the American Planning Movement: The Persistence of an Idea." *Journal of the American Planning Association* 46 (4): 424–439.

Bird, A. 1960. *The Motor Car, 1765–1914*. London: Batsford.

———. 1969. *Roads and Vehicles*. London: Longmans.

Blake, P. 1963. *God's Own Junkyard*. New York: Holt, Rinehart, and Winston.

Blamey, T. A. 1930. "Traffic Control from the Point of View of the Traffic Man." *Safety News*, January, 39–41.

Blumenfeld, H. 1965. "The Modern Metropolis." *Scientific American* 213 (3): 64–74.

Bodey, H. 1971. *Roads*. London: Batsford.

Bokonyi, S. 1980. "The Importance of Horse Domestication in Economy and Transport." Symp. *Transport and Social Change*, no. 2, Stockholm, 15–21.

Bone, F. J. 1952. "Modern Road Construction—1851." *Quarry Managers' Journal* 36 (3): 158–159.

Borth, C. 1969. *Mankind on the Move*. Washington, D.C.: Automotive Safety Foundation.

Bottles, S. L. 1987. *Los Angeles and the Automobile*. Berkeley and Los Angeles: Univ. of California Press.

Boulnois, H. P. 1919. *Modern Roads*. London: Edward Arnold.

Boulton, W. H. [1931] 1969. *The Pageant of Transport through the Ages*. Reprint. New York: Blom.

Boumphrey, G. 1939. *British Roads*. London: Thomas Nelson.

Boyer, M. N. 1959. "Medieval Suspended Carriages." *Speculum* 34 (July): 359–366.

———. 1960. "Medieval Pivoted Axle." *Technology and Culture* 1 (2): 128–138.

Bridenbaugh, C. 1968. *Vexed and Troubled Englishmen: 1590–1642*. New York: Oxford Univ. Press.

Brindle, R. E. 1983. "Major Routes and Networks in New Towns and Growth Centres." Chap. 5 in *Town Planning and Road Safety Review*. Australian Road Research Board Internal Rept. AIR 319-7. Vermont South, Victoria.

———. 1984. *Town Planning and Road Safety: A Review of the Literature and Practice*. Australian Road Research Board Special Rept. SR 28, May. Vermont South, Victoria.

British Parliamentary Papers. 1765. *Report from the Committee Appointed . . . for Repairing any Particular High-way*. 2:466–467. London.

Broome, D. C. 1963. "The Development of the Modern Asphalt Road." *Surveyor and Municipal Engineer* 122 (3728): 1437–1440 and 122 (3729): 1472–1475.

Bruckberger, F. 1959. *Image of America*. New York: Viking.

Buchanan, C. D. 1958. *Mixed Blessing: The Motor Car in Britain*. London: Hill.

Burden, V. 1985. *Discovering the Ridgeway*. 4th ed. Aylesbury, England: Shire Publications.

Burgoyne, J. F. 1861. *Rudiments of the Art of Constructing and Repairing Common Roads*. 3d ed. London: John Weale.

Burt, O. W. 1968. *The National Road.* New York: John Day.

Buttrick, G. A. 1962. *The Interpreter's Dictionary of the Bible.* New York: Abingdon.

Cannon, M. M. 1967. *The Land Boomers.* Melbourne: Melbourne Univ. Press.

Carey, A. E. 1914. *The Making of Highroads.* London: Crosby Lockwood.

Carey, H. C. 1858. *Principles of Social Science.* Philadelphia: Lippincott.

Casson, L. 1974. *Travel in the Ancient World.* London: George Allen and Unwin.

Catton, M. D. 1959. "Early Soil-Cement Research and Development." *Proceedings of the American Society of Civil Engineers* 85 (HW1): 1–16.

Central Road Research Institute. *See* CRRI.

Chabrier, E. 1875. "The Applications of Asphalt." *Proceedings of the Institution of Civil Engineers* 43 (1): 276–295.

Chandler, T., and Fox, G. 1974. *Three Thousand Years of Urban Growth.* New York: Academic Press.

Charlesworth, G. 1984. *A History of British Motorways.* London: Thomas Telford.

———. 1987. *A History of the Transport and Road Research Laboratory: 1933–1983.* London: Avebury.

Chevallier, R. 1976. *Roman Roads.* London: Batsford.

Chrichton, M. 1976. *Eaters of the Dead.* London: Cape.

Civil Engineering. 1981. "Notes of the Month." *Civil Engineering* (U.K.). Anniversary Number Supplement, November, 11–15.

Clark, C. 1977. *Population Growth and Land Use.* 2d ed. London: Macmillan.

Clark, D. 1965. *The Ingenious Mr. Edgeworth.* London: Oldbourne.

Clarke, Edward. 1793. *A Tour through the South of England, Wales, and Part of Ireland Made during the Summer of 1791.* N.p.

Clifford, J. M. 1977. "History of Compaction and Development of Early Equipment." Part 1 of *Compaction of Road Materials.* National Institute of Road Research. Rept. RC/5/77. Pretoria.

Coane, J. M.; Coane, H. G.; and Coane, J. M. 1908. *Australian Roads.* Melbourne: Robertson, and 4th ed., 1927.

Coke, E. 1628. *Institutes of the Laws of England, or a Commentary upon Littleton, Not the Name of the Author Only, but of the Law Itself.* Vol. 1 (commonly called "Coke upon Littleton"). London: Societie of Stationers.

Coles, B., and Coles, J. M. 1986. *Sweet Track to Glastonbury: The Somerset Levels in Prehistory.* London: Thames and Hudson.

Coles, J. M. 1984. "Prehistoric Roads and Trackways in Britain." In *Loads and Roads in Scotland and Beyond.* Edited by A. Fenton and G. Stell. Edinburgh: John Donald.

Collingwood, F. 1891. "Street Paving." *Engineering Record,* 7 March, 223–224.

Committee on Science and the Arts. 1843. "Report on the Best Modes of Paving Highways." *Journal of the Franklin Institute* 6, ser. 3, no. 3 (Sept.): 145–168 and (Oct.): 217–233.

Copeland, J. 1968. *Roads and Their Traffic: 1750–1850.* Newton Abbott, Eng.: David and Charles.

Cordier, M. J. 1823. *Ponts et chaussées: Essais sur la construction des routes.* Lille: Reboux-LeRoy.

Coulomb, C-A. 1773. "Essai sur une application des règles de maximis and minimis a quelques problemes de statique, relatifs à l'architecture." In *Coulomb's Memoir on Statics.* *See* Heyman 1972.

Crofts, J. 1967. *Packhorse, Waggon, and Post.* London: Routledge and Kegan.

Cron, F. W. 1974. "Beginnings of Traffic Measurement." Part 1 of "Highway Design for Motor Vehicles—a Historic Review." *Public Roads* 38 (3): 85–95.

———. 1975a. "The Beginnings of Traffic Research." Part 2 of "Highway Design for Motor Vehicles—a Historic Review." *Public Roads* 38 (4): 163–174.

————. 1975b. "The Interaction of the Driver, the Vehicle, and the Highway." Part 3 of "Highway Design for Motor Vehicles—a Historic Review." *Public Roads* 39 (2): 68–79.

————. 1975c. "The Vehicle-Carrying Capacity of the Highway." Part 4 of "Highway Design for Motor Vehicles—a Historic Review." *Public Roads* 39 (3): 96–108.

————. 1976a. "The Dynamics of Highway Curvature." Part 5 of "Highway Design for Motor Vehicles—a Historic Review." *Public Roads* 39 (4): 163–171.

————. 1976b. "The Evolution of Highway Grade Design." Part 7 of "Highway Design for Motor Vehicles—a Historic Review." *Public Roads* 40 (2): 78–86.

————. 1976c. "The Evolution of Highway Standards." Part 8 of "Highway Design for Motor Vehicles—a Historic Review." *Public Roads* 40 (3): 93–100.

CRRI. 1963. *History of Road Development in India.* Delhi: Central Road Research Institute.

Cummins, C. L. 1976. "Early IC and Automotive Engines: A History of the Automobile Internal Combustion Engine." *Society of Automobile Engineers,* SP-409, paper no. 760604.

Dalgeish, A. 1980. "Telford and Steam Carriages." In *Thomas Telford: Engineer,* ed. A. Penfold, 117–128. London: Thomas Telford.

Daniels, G. H., and Rose, M. H. 1982. *Energy and Transport: Historical Perspectives on Policy Issues.* Beverly Hills: Sage.

Darcy, H. P. 1850. "Sur le pavage et le macadamisage des chaussées de Londres et de Paris." *Annales des Ponts et Chaussées.* 2d ser., 20 (233): 1–58.

Davison, C. S. 1961. "Bridges of Historical Importance." *Engineer* 211 (5481): 196–198.

Dawson, D. G. 1986. "Short History of Human-powered Vehicles." *American Scientist* 74 (14): 350–357 and 74 (6): 578.

Dawson, G. J. 1876. "Street Pavements." *Journal of the Liverpool Polytechnic Society,* 15–75. Available in Institution of Civil Engineers Tracts, 8vo, vol. 270.

Day, M. 1984. Letter to the editor. *Current Archaeology* 92: 287.

Deacon, F. 1879. "Sheet Carriageway Pavements." *Proceedings of the Institution of Civil Engineers* 58 (4): 1–30.

Deacon, W. 1810. *Remarks on Conical and Cylindrical Wheels, Public Roads, Wheel Carriages, etc.* 2d ed. London: W. Deacon.

de Camp, L. S. 1960. *The Ancient Engineers.* New York: Doubleday.

de Coulaine, Q. 1850. "Sur l'emploi des substances bitumineuses dans la construction des chaussées, sur la nature, la composition, les propriêtés de ces substances, et leurs diverses applications." *Annales des Ponts et Chaussées,* Mem. et Documents, 2d cahier, 19: 240–308.

Delano, W. H. 1880. "On the Use of Asphalt and Mineral Bitumen in Engineering." *Proceedings of the Institution of Civil Engineers* 60 (1673): 249–303.

Deloche, J. 1968. "Recherches sur les routes de l'Inde du temps des Mogols." *Publication l'Ecole Française d'Extreme Orient,* 67.

————. 1973. "Les ponts anciens de l'Inde." *Publication l'Ecole Française d'Extreme Orient,* 93.

Dennison, G., and Tomlinson, K. Y. 1969. "Let's Put Brakes on the Highway Lobby." *Reader's Digest,* May, 97–102.

de Saint-Hardouin, T. 1877. "On the Wear of Macadamised Roads." *Annales des Ponts et Chaussées,* 226. As reported in *Proceedings of the Institution of Civil Engineers* 49 (3): 303–304.

de Villefosse, R. H. 1975. *Histoire des grandes routes de France.* Paris: Libraire Academique Perrin.

d'Eyrinys, E. 1721. *Dissertation sur l'asphalte ou ciment naturel.* Paris: Lottin. A copy is given as Appendix B in Malo 1866.

Dickinson, R. E. 1961. *The West European City.* London: Routledge and Kegan Paul.

Dore, L.A.G., and Jerrold B. [1872] 1968. *London: A Pilgrimage.* London: Grant. Facsimile edition. New York: Blom.

"Dr. Tar." 1989. *Routes/Roads* 268 (2): 89–91.

Duncan, H. O. 1928. *The World on Wheels*. Paris: H. O. Duncan. Copy in Royal Automobile Club Library in London.

Dupuit, A. J. 1844. "De la mesure de l'utilité des traveaux publics." *Annales des Ponts et Chaussées,* 2d ser., 8:332–375.

Dyos, H. J., and Aldcroft, D. H. 1969. *British Transport: An Economic Survey from the Seventeenth Century to the Twentieth*. Leicester: Leicester Univ. Press.

Earle, J. B. 1971. *A Century of Roadmaking Materials*. Oxford: Blackwell.

———. 1974. *Black Top*. Oxford: Blackwell.

Edge, R. 1990. "Old Wirral Highways and the Ancient Birkenhead Bridge." *Highways and Transportation* 37 (4): 31–33.

Edgeworth, R. L. 1817. *Essay on the Construction of Roads and Carriages*. . . . . 2d ed. London: Hunter.

Edwards, P. J. 1898. *History of London Street Improvements 1865–1897*. London: County Council.

Ellice-Clark, E. B. 1880. "Asphalte and Its Application to Street Paving." *Engineering News,* 1 May, 145–157.

Engineering Record. 1890. "Slipperiness of Pavements." *Engineering Record* 23 (25 January): 113–128.

———. 1891a. "Details of Asphalt Pavement Work at Washington, D.C." *Engineering Record* 25 (14 February): 178–180.

———. 1891b. "European Streets and Highways." *Engineering Record* 26 (5 December): 3.

———. 1891c. "Recent Experience with Asphalt in Washington, D.C." *Engineering Record* 26 (31 October): 348–349.

———. 1907. "Recent Progress in the Asphalt Paving Industry." *Engineering Record* 56 (1 June): 653–655.

———. 1910. "A Noteworthy Boulevard." *Engineering Record* 61, no. 9 (26 February): 252.

Everett, S. 1980. *World War I: An Illustrated History*. London: Bison.

FHA (U.S. Federal Highway Administration). 1976. *America's Highways, 1776–1976*. Washington: U.S. Department of Transportation.

Field, D. C. 1958. "Mechanical Road-Vehicles." In *A History of Technology*. Edited by C. Singer et al., 5:414–437. Oxford: Oxford Univ. Press.

Findley, R. 1980. "The Pony Express." *National Geographic* 158 (1): 45–71.

Fletcher, R., and Snow, J. P. 1932. "A History of the Development of Wooden Bridges." *Proceedings of the American Society of Civil Engineers* 11 (9): 1455–1498.

Fletcher, W. [1891] 1972. *The History and Development of Steam Locomotion on Common Roads*. London: Span. Reprint. David and Charles.

Flink, J. J. 1974. *America Adopts the Automobile, 1895–1940*. Cambridge, Mass.: MIT Press.

———. 1985. "Innovation in Automotive Technology." *American Scientist* 73 (2): 151–161.

———. 1987. *The Automobile Age*. Cambridge, Mass.: MIT Press.

Flower, R., and Jones, M. W. 1981. *One Hundred Years on the Road*. New York: McGraw-Hill.

Forbes, R. J. 1934. *Notes on the History of Ancient Roads and Their Construction*. Vol. 3. Amsterdam: N. V. Noord-Hollandsche. Archaeologisch-historische Bijdragen, University of Amsterdam.

———. 1936. *Bitumen and Petroleum in Antiquity*. Leiden: Brill.

———. 1938. "Bibliography of road building." *Roads and Road Construction* 16 (186): 189–196.

———. 1953. "Roads of the Past." *Chemistry and Industry* 4 (24 Jan.): 70–74.

———. 1958a. "Roads to c1900." In *A History of Technology*. Edited by C. Singer, E. J. Holmyard, A. R. Hall, and T. I. Williams, 4:520–547. Oxford: Oxford Univ. Press.

———. 1958b. "Petroleum." In *A History of Technology*. Edited by C. Singer et al., 5:102–123. Oxford: Oxford Univ. Press.

———. 1958c. *Studies in Early Petroleum History*. Leiden: Brill.

Ford, Henry. 1924. *My Life and Work*. 2d ed. London: Heinemann.

Fordham, H. G. 1924. *The Road-Books and Itineraries of Great Britain, 1570–1850*. Cambridge: Cambridge Univ. Press.

Foster, M. S. 1981. *From Streetcar to Superhighway*. Philadelphia: Temple Univ. Press.

Fox, C. 1931. "Sleds, Carts, and Waggons." *Antiquity* 5:185–199.

Frazer, J. G. 1914. *The Golden Bough: A Study in Magic and Religion*. 3d ed. London: Macmillan.

Fugl-Meyer, H. 1937. *Chinese Bridges*. Shanghai: Kelly and Walsh.

Gallatin, A. 1808. "Roads and Canals." *American State Papers: Miscellaneous*. Vol. 1:724–741. Washington, D.C.

Gautier, H. 1693. *Traité de la construction des chemins*. Paris: Cailleau.

Geddes, N. B. 1940. *Magic Motorways*. New York: Random House.

Georgano, G. N., ed. 1972. *A History of Transport*. London: J. M. Dent.

Gibbons, F., and Evans, G. 1978. *The Pictorial History of Trucks*. London: Orbis.

Gies, J. 1962. *Adventure Underground*. London: Hale.

————. 1963. *Bridges and Men*. New York: Doubleday.

Gillespie, A. K., and Rockland, M. A. 1989. *Looking for America on the New Jersey Turnpike*. New Brunswick, N. J.: Rutgers Univ. Press.

Gillespie, W. M. 1856. *A Manual of the Principles and Practice of Road-Making*. New York: Barnes.

Gillmore, Q. A. 1876. *A Practical Treatise on Roads, Streets, and Pavements*. New York: Van Nostrand.

Ginzrot, J. C. 1817. *Der Wagen und Fahrwerke der Griechen und Romer und anderer altern Volker: nebst der Bespannung, Zaumung und Verzierung ihrer Zug-Reit-und-Last Thiere*. 2 vols. Munich: J. Lentner.

Good, M. C. 1978. *Road Curve Geometry and Driver Behaviour*. Australian Road Research Board Special Rept. no. 15. Vermont South, Victoria.

Gordon, A. 1836. *A Treatise upon Elemental Locomotion*. London: Tegg.

Gramsborg, P. 1986. "London Bridge Is Falling Down." *Norseman* 26 (2): 11.

Grattesat, G. 1982. *Ponts de France*. Paris: Presses de l'Ecole Nationale des Ponts et Chaussées.

Greene, F. V. 1885. "Pavements Here and Abroad." *Engineering News and American Contracting Journal,* 19 September, 186.

Gregory, J. W. 1931. *The Story of the Road*. London: Maclehose.

Griffin, E. 1887. "The Streets of Washington, D.C., in 1886–1887." *Engineering News,* 22 October, 297–298.

Hadfield, W. J. 1934. *Highways and Their Maintenance*. London: Contractors Record.

Haldane, A. R. 1962. *New Ways through the Glens*. London: Thomas Nelson.

Hall, G. D., ed. 1965. *The Treatise on the Laws and Customs of the Realm of England Commonly Called Glanvill*. Book 9, 11. London: Thomas Nelson.

Hall, P. 1973. "Urban Growth in England." In *The Containment of Urban England,* ed. P. Hall. London: George Allen and Unwin.

————. 1988. *Cities of Tomorrow*. London: Basil Blackwell.

Halstead, W. J., and Welborn, J. Y. 1974. "History of the Development of Asphalt Testing Apparatus and Asphalt Specifications." *Proceedings of the Association of Asphalt Paving Technologists* 43A:89–120.

Hamer, M. 1987. *Wheels within Wheels*. London: Routledge and Kegan Paul.

Hamilton, S. B. 1958. "Building and Civil Engineering Construction." In *A History of Technology*. Edited by C. Singer, E. J. Holmyard, A. R. Hall, and T. I. Williams, 4:442–488. Oxford: Oxford Univ. Press.

Harding, A., ed. 1980. *The Guinness Book of Car Facts and Feats*. 3d ed. Enfield, U.K.: Guinness Superlatives.

Harlow, A. F. 1928. *Old Post Bags*. New York: Appleton.

Harper, C. G. 1901. *The Great Road North.* 2 vols. London: Chapman and Hall.

Hartmann, C. H. 1927. *The Story of the Roads.* London: Routledge.

Haskins, C. H. 1929. *Studies in Mediaeval Culture.* New York: Ungar.

Hatt, W. K., ed. 1923. *Proceedings of the Second Annual Meeting of the Advisory Board on Highway Research, Division of Engineering, National Research Council.* Bulletin of the National Research Council 6 (1), no. 32, May. Washington, D.C.

Haverfield, F. 1913. *Ancient Town-Planning.* Oxford: Oxford Univ. Press.

Haywood, W. 1871. *Report to the Streets Committee of the Hon, the Commissioner of Sewers of the City of London, upon Granite and Asphalte Pavements.* London: Judd.

HEB. 1930. *Tours of the International Highway Engineers: October 1930.* Washington, D.C.: Highway Education Board.

Heightchew, R. E. 1979. "TSM: Revolution or Repetition?" *Institute of Transport Engineers Journal* 48 (9): 22–30.

Hey, D. 1980. *Packmen, Carriers, and Packhorse Roads.* Leicester: Leicester Univ. Press.

Heyman, J. 1972. *Coulomb's Memoir on Statics.* Cambridge: Cambridge Univ. Press.

———. 1988. "Poleni's Problem." *Proceedings of the Institution of Civil Engineers* 84 (1): 737–759.

Heyman, J., and Threlfall, B. D. 1972. "Telford's Bridge at Over." Part 2 of "Two Masonry Bridges." *Proceedings of the Institution of Civil Engineers* 52:319–330.

Highway Education Board. *See* HEB.

Hilf, J. W. 1957. "Compacting Earth Dams with Heavy Tamping Rollers." *Proceedings of the American Society of Civil Engineers* 83 (SM2): 1205–1228.

Hill, F. G. [1957] 1977. *Roads, Rails, and Waterways.* Norman: Univ. of Oklahoma Press. Reprint. Westport, Conn.: Greenwood.

Hill, P. 1980. "The League's Earliest Years." *American Wheelmen* 16 (5): 8–14.

Hilton, G. W. 1969. "Transport Technology and the Urban Pattern." *Journal of Contemporary History* 4:123–135.

Hindley, G. 1971. *A History of Roads.* London: Peter Davies.

Hogentogler, C. A. 1923. "Apparatus Used in Highway Research Projects in the United States." *Bulletin of the National Research Council* 6 (4): no. 35. Washington, D.C.

Hogsbro, K-E. 1986. *To Bridge the Danish Way.* Copenhagen: Danish Road Directorate.

Hoiberg, A. J. 1965. *Bituminous Materials: Asphalts, Tars, and Pitches.* Vol. 2, Pt. 1, *Asphalts.* New York: Interscience, Wiley.

Holmes, F. H. 1972. "The State-of-the-Art in Urban Transportation—and How We Got There." *Transportation* 1 (4): 379–401.

Holmstrom, J. F. 1934. *Railways and Roads in Pioneer Development Overseas.* London: P. S. King.

Home, G. 1931. *Old London Bridge.* London: John Lane the Bodley Head.

Hopkins, H. J. 1970. *A Span of Bridges.* Newton Abbott, Eng.: David and Charles.

Hopper, R. H. 1982. "Left-Right: Why Driving Rules Differ." *Transportation Quarterly* 36 (4): 541–48.

HRB. 1971. *Ideas and Actions: A History of the Highway Research Board, 1920–1970.* Washington: Highway Research Board.

Hughes, A. C.; Adam, W. G.; and China, F. J. 1938. *Tar Roads.* London: Arnold.

Hughes, S.; Law, H.; and Burgoyne, J. F. 1868. *Rudimentary Papers on the Art of Constructing and Repairing Common Roads.* London: Virtue.

Hughill, P. J. 1981. *The Elite, the Automobile and the Good Roads Movement in New York: The Development and Transformation of a Technological Complex, 1904–1913.* Syracuse University, Department of Geography Discussion Paper no. 70, March.

Hulbert, A. B. [1904] 1971. *Paths of the Mound-Building Indians and Great Game Animals.* Vol. 1 of *Historic Highways of America.* 16 vols.; early ones appeared in 1902. New York: AMS.

Humber, W. 1864. *A Complete Treatise on Cast and Wrought Iron Bridge Construction.* London: Lockwood.

Hutchins, J. G. 1977. *Transportation and the Environment.* London: Elek.

Hveem, F. N. 1960. "Devices for Recording and Evaluating Pavement Roughness." *Highway Research Board Bulletin* 264:1–26.

———. 1971. "Asphalt Pavements from the Ancient East to the Modern West." *Highway Research News* 42 (Winter): 21–39.

Hyslop, J. 1984. *The Inka Road System.* Orlando: Academic Press.

Institution of Civil Engineers. 1873. "Memoir: Emperor C. L. Napoleon Bonaparte, 1808–1873." *Proceedings of the Institution of Civil Engineers* 38 (1): 284–286.

———. 1878. "Apparatus for Measuring the Comparative Strength of Broken Stones." *Proceedings of the Institution of Civil Engineers* 57 (3): 331.

Institute of Transportation Engineers. 1980. *A History of Traffic Control Devices.* Washington: Institute of Transportation Engineers.

Ironbridge Gorge Museum Trust. 1979. *The Tar Tunnel: Telford, England.* Ironbridge: Ironbridge Gorge Museum Trust.

ITE. *See* Institute of Transportation Engineers.

Jackman, W. T. 1916. *The Development of Transportation in Modern England.* Cambridge: Cambridge Univ. Press.

Jackson, J. B. 1980. *The Necessity for Ruins and Other Topics.* Amherst: Massachusetts Univ. Press.

Jackson, K. T. 1985. *Crabgrass Frontier: The Suburbanization of the United States.* New York: Oxford Univ. Press.

Jackson, W. T. 1982. *Wagon Roads West.* Berkeley and Los Angeles: Univ. of California Press.

James, J. G. 1964a. "Fifty Years of White Lines." *Roads and Road Construction,* Dec., 409–414.

———. 1964b. "History of Bituminous Surfacings." Letter. *Surveyor and Municipal Engineer* 123 (3736): 29, 25–26.

James, P. 1984. *London from the Air.* London: Weidenfeld and Nicholson.

Jeffreys, W. R. 1949. *The King's Highway.* London: Batchworth.

Jiang, C. F., ed. 1986. *China Encyclopaedia: Volume on Transport* (in Chinese). Beijing: China Encyclopaedia Publishing.

Jones, B. O. 1985. *Australia as a Post-Industrial Society: Questioning the Future.* Melbourne Commission for the Future Occasional Paper no. 2, December.

Jordan, R. P. 1985. "Viking Trail East." *National Geographic* 167 (3): 278–317.

Keats, J. 1958. *The Insolent Chariots.* Philadelphia: Lippincott.

Kellett, J. R. 1969. *The Impact of Railways on Victorian Cities.* London: Routledge and Kegan Paul.

Kelley, B. 1971. *The Pavers and the Paved.* New York: Scribner's.

Kemp, E. L. 1979. "Links in a Chain." *Structural Engineer* 57A (8): 255–263.

Kemp, J. R., and Crawford, D. A. 1930. *Weight and Speed Regulation of Motor Vehicles.* Brisbane: Queensland Government Printer.

Kennerell, E. J. 1958. "Roads from the Beginning." *Journal of the Institute of Highway Engineers* 5 (3): 176–205.

Kincaid, P. 1986. *The Rule of the Road.* New York: Greenwood.

Kirby, R. S.; Withington, S.; Darling, A. B.; and Kilgour, F. G. 1956. *Engineering in History.* New York: McGraw-Hill.

Kirkaldy, A. W., and Evans, A. D. [1915]. *The History and Economics of Transport.* London: Sir Isaac Pitman.

Klindt-Jensen, O. 1949. "Foreign Influences in Denmark's Early Iron Age." *Acta Archaeologica* (Copenhagen) 20:87–108.

Knightly, C., and Cyprien, M. 1985. *A Traveller's Guide to Royal Roads.* London: Routledge and Kegan Paul.

Knox, H. S. 1974. "The Suspension Bridge: Its History and Development." *Society of Engineering Journal and Transactions* 65 (1): 55–71.

Krampen, M. 1983. "Icons of the Road." *Semiotica* 43½: 1–204.

Krchma, L. C., and Gagle, D. W. 1974. "A U.S.A. History of Asphalt Refined from Crude Oil and Its Distribution." *Proceedings of the Association of Asphalt Paving Technologists* 43A: 25–88.

Kuzmanovic, B. O. 1977. "History of the Theory of Bridge Structures." *Proceedings of the American Society of Civil Engineers* 103 (ST5): 1095–1111.

Labatut, J., and Lane, W. J. 1950. *Highways in Our National Life*. Princeton: Princeton Univ. Press.

Lampard, E. 1955. "The History of Cities in the Economically Advanced Areas." *Economic Change and Development* 3: 81–136.

Lane, R. H. 1935. "Waggons and Their Ancestors." *Antiquity* 9: 140–150.

Larkin, F. J., and Wood, S. 1975. "Past and Future of Construction Equipment." *Proceedings of the American Society of Civil Engineers* 101 (CO2): 309–315, 101 (CO4): 689–698.

Laurence, C. J. 1980. "Roundabouts: Evolution, Revolution, and the Future." *Highway Engineering* 27 (5): 2–10.

Law, H., and Clark, D. K. 1907. *The Construction of Roads and Streets*. London: Crosby Lockwood and Sons.

Law, W. M. 1962. The Development and Use of Coated Macadam. Institute of Works and Highways Superintendents, Annual Conference brochures, Hastings.

Lay, C. D. 1920. "Notes on the Influence of Automobiles on Town, Country, and Estate Planning." *Landscape Architecture* 10 (2): 89–95.

Lay, M. G. 1964a. "Bridge Design." Chap. 4 in *Structural Steel Design*. Edited by L. Tall. New York: Ronald Press.

———. 1964b. "The Experimental Bases for Plastic Design: A Survey of the Literature." *Welding Research Council Bulletin*, 99.

———. 1982. *Structural Steel Fundamentals*. Melbourne. Australian Road Research Board. Vermont South, Victoria.

———. 1983. "The Future Role of the Motor Car." *Proceedings of the Third World Conference on Transport Research* (Hamburg) 2: 1257–1263.

———. 1984. *History of Australian Roads*. Australian Road Research Board Special Rept. no. 29, December. Melbourne: Australian Road Research Board. Vermont South, Victoria.

———. 1985. *Source Book for Australian Roads*. 3d ed. Melbourne: Australian Road Research Board. Vermont South, Victoria.

———. 1989. *A Citation Analysis of Australian Road Technology*. Australian Road Research Board Research Rept. ARR 156, April. Melbourne: Australian Road Research Board. Vermont South, Victoria.

———. 1990. *Handbook of Road Technology*. 2d ed. London: Gordon and Breach.

Lee, C. E. 1971. "One Hundred Years of Railway-associated Omnibus Services." In *Omnibus: Readings in the History of Road Passenger Transport*. Edited by J. Hibbs, 147–180. Newton Abbott, Eng.: David and Charles.

Lee, G. E. 1947. "The Highways of Antiquity." *Railway Gazette* (pamphlet). Westminster.

Leeming, J. J. 1965. "Famous Road Engineers: 1—Thomas Telford." *Journal of the Institute of Highway Engineers*, July, 16–17.

Leibbrand, K. 1970. *Transportation and Town Planning*. Cambridge, Mass.: MIT Press.

Leighton, A. C. 1972. *Transport and Communications in Early Medieval Europe, A.D. 500–1100*. Newton Abbott, Eng.: David and Charles.

Leonhardt, F. 1982. *Brucken/Bridges*. London: Architectural Press.

Lochhead, W. 1878. Letter to the editor. *Engineer* 46 (15 Nov.): 358.

Lock, J. B.; Gelling, M. J.; and Colquhoun Kerr, D. J. 1982. "Limitation of Access onto

Arterial Roads." *Proceedings of the Eleventh Australian Road Research Board Conference* 11 (4): 76–84.

Lopez, R. S. 1956. "The Evolution of Land Transport in the Middle Ages." *Past and Present* 9:17–29.

Lukacs, J. 1984. *Outgrowing Democracy: A History of the United States in the Twentieth Century.* New York: Doubleday.

Luton Museum. 1970. *The Turnpike Age.* Luton: Luton Museum and Art Gallery.

McAdam, J. L. 1816. *Remarks (or Observations) on the Present System of Roadmaking.* 1st ed. (9th ed. 1827). London: Longman, Rees, Orme, Brown and Green.

MacDonald, T. H. 1928. "The History and Development of Road Building in the United States." *Transactions of the American Society of Civil Engineers* 92 (1685): 1181–1206.

Mace, T. 1675. *Profit, Conveniency, and Pleasure to the Whole Nation . . . Concerning the Highways of England.* In British Museum Library, 1391.c.20.

McKay, J. P. 1976. *Tramways and Trolleys.* Princeton: Princeton Univ. Press.

McLean, J. R. 1980. "Two-lane Road Traffic Flow and Capacity." Australian Road Research Board Internal Report AIR 359-1. Vermont South, Victoria.

McShane, C. 1979. "Transforming the Use of Urban Space: A Look at the Revolution in Street Pavements, 1880–1924." *Journal of Urban History* 5 (3): 297–307.

Malcolm, L. W. 1934. "Early History of the Streets and Paving of London." *Transactions of the Newcomen Society for the Study of the History of Engineering and Technology* 14:83–95.

Malo, L. 1861. "Note sur l'asphalte: son origine, sa preparation, ses applications." *Annales des Ponts et Chaussées* 1, 4th ser., 69–100. Mem. et Documents. Paris: Dunod.

———. 1866. *Guide practique pour la fabrication et l'application de l'asphalte et des bitumes.* Paris: Lacroix.

———. 1885. "The Asphalt Pavements of Berlin." *Proceedings of the Institution of Civil Engineers* (minutes). 81:376–377.

Manton, B. G. 1956. "John Loudon McAdam: Born September 21st, 1756." *Highways and Bridges and Engineering Works* 24 (1157): 6–10.

Margary, I. D. 1955. *Roman Roads in Britain.* Vol. 1. London: Phoenix House.

Margetson, S. 1967. *Journey by Stages.* London: Cassell.

Martyn, N. 1987. *The Silk Road.* Sydney: Methuen.

Matthiessen, H. 1961. *Haervejen, en tusindarig vej.* Copenhagen: Glydendalske.

Maudsley, A. 1888. *Highways and Horses.* London: Chapman and Hall.

Mayhew, H. [1850] 1968. *London Labour and London Poor.* Reprint. New York: Dover.

Meigs, M. 1898. "Oil as a Road Material." Letter to editor. *Scientific American,* 24 December, 407.

Melbourne Metropolitan Town Planning Commission. 1929. *Plan of General Development.* Melbourne: Government Printer.

Merdinger, C. J. 1952. "Roads through the Ages: I and II, Military Engineer." *Journal of the Society of American Military Engineers* 44 (300): 268–273 and 44 (301): 340–344.

Mesqui, J. 1979. "Les routes dan le Brie et la Champagne occidentale." *Revue Générale des routes et des Aérodromes* 557 (Oct.): 65–79.

———. 1980. "Techniques et ouvrages routiers avant le XVIIIe siecle. Part VII." *Revue Générale des routes et des Aérodromes* 562 (March): 21–35.

———. 1986. *Le pont en France avant le temps des Ingénieurs.* Paris: Picard.

Meyer, J. R., and Gomez-Ibanez, J. A. 1981. *Autos, Transit, and Cities.* Cambridge, Mass.: Harvard Univ. Press.

[Meyn, L.?] 1873. "The Asphalts." *Van Nostrand's Eclectic Engineering Magazine,* 74–82 and 123–126.

Mills, F. 1981. "Effects of Beltways on the Location of Residences and Selected Workplaces." *Transport Research Board Records* 812:26–33.

Mitchell, C. G. 1972. "New Technology in Urban Transport." *Proceedings of the Institution of Civil Engineers,* 52 (Aug.): 127–147.

Mitchell, J. 1867. "On a New Method of Constructing the Surface of Streets and Thoroughfares." *Engineer* 2 (30): 186.

Moaligou, C. 1982. "Le pave de Paris." *Revue Générale des routes et des Aérodromes* 583:39–43.

Mogridge, M. J. 1986. "Road Pricing: The Right Solution for the Right Problem." *Transport Research Annual* 20A (2): 157–167.

Mongait, A. L. 1961. *Archaeology in the U.S.S.R.* Middlesex: Pelican/Penguin.

Montgomery, J. A. 1988. *Eno: The Man and the Foundation.* Westport, Conn.: Eno Foundation for Transportation.

Moore, H. C. 1902. *Omnibuses and Cabs: Their Origin and History.* London: Chapman and Hall.

Moore, R. 1876. "Street Pavements: An Exhaustive Review." *Engineering News,* 6 May, 146; 20 May, 163 and 170; 3 June, 179.

Moorey, P. R. 1982. *Ur "of the Chaldees."* Ithaca: Cornell Univ. Press.

Morris, A. E. 1972. *History of Urban Form.* London: Godwin.

Mountain, A. C. 1894. "The Evolution of the Modern Road." *Proceedings of the Vic Institute of Engineers,* November.

———. 1903. *Wood Paving and Road Making in Australia.* 2d ed. Melbourne: Modern Printing.

Mueller, E. A. 1970. "Aspects of the History of Traffic Signals." *Transactions of the IEEE (Vehicle Technology),* VT 19-1, February, 6–16.

Mumford, L. 1961. *The City in History.* London: Martin, Secker and Warburg.

———. 1964. *The Highway and the City.* New York: Harcourt Brace Jovanovich.

Musgrave, R. A. 1959. *The Theory of Public Finance.* New York: McGraw-Hill.

Navier, C. L. 1831. "Considerations sur les traveaux d'entretion des routes en Angleterre." *Annales des Ponts et Chaussées,* 1st ser., 2 (18): 32–156.

Needham, N. J. 1971. *Science and Civilisation in China.* Vol. 4 (3), *Civil Engineering and Nautics.* Cambridge: Cambridge Univ. Press.

Nelson, L. H. 1990. *The Colossus of 1812: An American Engineering Superlative.* New York: American Society of Civil Engineers.

Netherton, R. D. 1963. *Control of Highway Access.* Madison: Univ. of Wisconsin Press.

Newcomb, T. P., and Spurr, R. T. 1989. *A Technical History of the Motor Car.* Bristol: Adam Hilger.

Newcomen, T. G. 1951. "Pre-Roman Roads in Britain." *Contractors' Record and Municipal Engineering* 14 and 21 (Feb.): 11–16, 28; also in *Journal of the Institute of Highway Engineers* 2 (3): 14–39.

Newlon, H., Jr. 1987. "Private-Sector Involvement in Virginia's Nineteenth-Century Transportation Improvement Program." *Transport Research Record* 1107:3–11.

Newlon, H., Jr., Pawlet, N. M., et al. 1985. *Backsights.* Virginia: Virginia Department of Highways and Transportation.

Nicholson, T. R. 1970. *Passenger Cars: 1863–1904.* London: Blandford.

———. 1971. *Passenger Cars: 1905–1912.* London: Blandford.

Nixon, S. C. 1936. *The Invention of the Automobile.* London: Country Life.

North, E. P. 1879. "The Construction and Maintenance of Roads." *Transactions of the American Society of Civil Engineers* 8 (May): 95–147 and 8 (Dec.): 333–380.

Nuebel, O., and Seherr-Thoss, H. C. 1983. "The Development of Automobilism in Germany." *Automobilism,* Historical Issue: The Automobile Story, vol. 11, January (Also dated December 1982).

O'Connor, C. 1975. "The Design of Bridges: An Historical Study." Inaugural lectures, 13 June 1974. Brisbane: Univ. of Queensland Press. Typescript.

O'Flaherty, C. A. 1967. *Highways*. London: Edward Arnold.

Ogilby, J. [1675] 1970. *Britannia*. Facsimile ed. Holland: Theatrum Orbis Terrarum.

O'Keeffe, P. J. 1973. "The Development of Ireland's Road Network." Paper presented to the Civil Division of the Institute of Engineers of Ireland, 5 November, Dublin.

————. 1985. "Roads in Ireland: Past and Ahead." National Institute for Physical Planning and Construction Research, Roads Division, report, February, Dublin, Ireland.

Ontario Ministry of Transport and Communications. 1984. *Footpaths to Freeways: The Story of Ontario's Roads*. Downsview: Ontario Ministry of Transport and Communications.

Orski, C. K. 1979. "Transportation Planning As If People Mattered." *Practicing Planner* 9 (1), March.

Page, L. W. 1910. "Effect of Motor and Horse-drawn Vehicles on Roads." *Engineering Record* 61 (9): 251–252.

Pannell, J. P. 1964. *An Illustrated History of Civil Engineering*. London: Thames and Hudson.

Parkes, J. 1925. *Travel in England in the Seventeenth Century*. London: Oxford Univ. Press.

Parnell, H. 1838. *A Treatise on Roads*. London: Longman, Orme, Brown, Green and Longmans.

Parsons, W. B. 1939. *Engineers and Engineering in the Renaissance*. Cambridge, Mass.: MIT Press.

Paterson, J. 1825. *M'Adam and Roads*. Pamphlet, 4 pages. Copy in the Institution of Civil Engineers Library in London.

Paterson, J. 1927. *The History and Development of Road Transport*. London: Pitman.

Patton, P. 1986. *Open Road*. New York: Simon and Schuster.

Paulett, N. M., and Lay, K. E. 1980. *Historic Roads of Virginia: Early Road Location. Key to Discovering Historic Resources?* Charlottesville: Virginia Highway and Transportation Research Council.

Pawson, E. 1977. *Transport and Economy: The Turnpike Roads of Eighteenth Century Britain*. London: Academic Press.

Paxson, F. L. 1946. "The Highway Movement: 1916–1935." *American Historical Review* 51: 236–253.

Paxton, R. R. 1977. *Early Concrete Roads in Edinburgh*. Concrete, U.K.: 11 (9): 20–22.

Peckham, S. F. 1909. *Solid Bitumens*. New York: Myron Clark.

Penfold, A., ed. 1980. *Thomas Telford: Engineer*. London: Thomas Telford.

Pennant, T. 1813. *Some Account of London*. 5th ed. London: Faulder, White, Cockrane.

Pennybacker, J.; Fairbank, H. S.; and Draper, W. F. 1916. *Convict Labour for Roadwork*. U.S. Department of Agriculture Bulletin no. 414, 15 December Washington, D.C.: GPO.

Permanent International Association of Road Congresses. *See* PIARC.

Perronet, J. R. 1782. *Descriptions des projects et de la construction des ponts*. Paris: N.p.

Peters, T. F.; Billington, D. P.; Dubas, P.; Epprecht, W.; Hauri, H. H.; Menn, C.; and Muheim, H. 1981. *The Development of Long Span Bridge Building*. Zurich: Federal Institute of Technology.

Pettifer, T., and Turner, N. 1984. *Automania*. London: Collins.

Phillips, R. 1737. "A Dissertation Concerning the Present State of the High Roads of England." Paper presented before the Royal Society. London: Giliver and Clark. In ICE (Institution of Civil Engineers) tracts, 8vo, vol. 74.

PIARC. 1910. General Report of the Second Congress, Brussels. Paris: Permanent International Association of Road Congresses.

————. 1969. *AIPCR-PIARC, 1909–1969*. Paris: Permanent International Association of Road Congresses.

Piggott, S. 1968. "The Beginnings of Wheeled Transport." *Scientific American* 219 (1): 82–90.

————. 1983. *The Earliest Wheeled Transport*. London: Thames and Hudson.

Plowden, D. 1974. *Bridges: The Spans of North America*. New York: Norton.

Plowden, W. 1971. *The Motor Car and Politics: 1896–1970*. London: Bodley Head.

Poldolny, W. 1976. "Cable-stayed versus Classical Suspension Bridge." *Transport Engineering Journal* (proceedings of the American Society of Civil Engineers) 102 (TE2): 291–311.

Porter, O. J. 1949. "Development of the Original Method for Highway Design." *Proceedings of the American Society of Civil Engineers* 75 : 11–17.

Postan, M., and Rich, E., eds. 1952. *The Cambridge Economic History of Europe*. Vol. 2, *Trade and Industry in the Middle Ages*. Cambridge: Cambridge Univ. Press.

Powell, W. D.; Potter, J. F.; Mayhew, H. C.; and Nun, M. E. 1984. "The Structural Design of Bituminous Roads." Transport and Road Research Laboratory Rept. LR1132. Crowthorne.

Pratt, E. A. [1912] 1970. *A History of Inland Transport and Communication*. London: Kegan Paul. Reprint. New York: Kelley.

Prebble, J. 1956. *The High Girders*. London: Secker and Warburg.

Pritchett, W. K. 1980. *Roads*. Part 3 of *Studies in Ancient Greek Topography*. Berkeley and Los Angeles: Univ. of California Press.

Procter, T. [1610] 1977. *A Worthy Worke Profitable to This Whole Kingdome Concerning the Mending of All High-waies*. London. Facsimile ed. W. Johnson (An earlier edition was printed by E. Allde of London in 1607.)

Public Roads. 1926. "An Instrument for the Measurement of Relative Road Roughness." *Public Roads* 7 (7): 144–151.

Pyne, W. H. 1806. *Microcosm, or Picturesque Groups for the Embellishment of Landscape*. 2 vols. London: Nattes. There were many subsequent editions. The facsimile edition by Blom in New York in 1971 is based on a collected version published by Nattali of London in 1845.

Rae, J. B. 1971. *The Road and the Car in American Life*. Cambridge, Mass.: MIT Press.

Ray, G. K. 1964. "History and Development of Concrete Pavement Design." *Proceedings of the American Society of Civil Engineers* 90 (HW1): 79–101.

Reader, W. J. 1980. *McAdam: The McAdam Family and the Turnpike Roads, 1798–1861*. London: Heinemann.

Reverdy, G. 1980. "Les chemins de Paris en Languedoc." Part 5 of "Histoire des grandes liaisons françaises." *Revue Générale des Routes et des Aérodromes* 583 : 39–43.

———. 1982. *Histoire des grande liaisons françaises*. 2 vols. Paris: Revue Générale des Routes et des Aérodromes.

———. 1986. *Atlas historique des routes de France*. Paris: Presses de l'Ecole Nationale des Ponts et Chaussées.

Richardson, C. 1905. *The Modern Asphalt Pavement*. New York: John Wiley.

Rickman, J., ed. 1838. *Life of Thomas Telford, Civil Engineer: Written by Himself*. London: Hansard.

Rimmer, P. J. 1986. *Riksha to Rapid Transit*. Sydney: Pergamon.

Ristow, W. W., ed. 1961. *A Survey of the Roads of the United States of America: 1789 by Christopher Coles*. Cambridge, Mass.: Harvard Univ. Press.

Road Research Laboratory. 1962. *Bituminous Materials in Road Construction*. London: HMSO.

Roads. 1839. *The Roads and Railroads, Vehicles and Modes of Travelling of Ancient and Modern Countries*. London: Parker.

Robb, C. J. 1947. "Asphalt Roads in History." *Kings Highway*, 20–21.

Robbins, R. M. 1985. *A Public Transport Century*. Brussels: International Union of Public Transport.

Robertson, J. N. 1958. "The Pavement of Presidents." *Asphalt Institute Quarterly* 10 (2): 4–7.

Robertson, R. B. 1985. "The Evolution of Transportation Planning." *Transport Research Record* 1014 : 1–10.

Robins, F. W. [1950]. *The Story of the Bridge*. Birmingham: Cornish Bros.

Robinson, C. 1901. *Modern Civic Art*. New York: Putnam.

Robinson, J. 1971. *Highways and Our Environment*. New York: McGraw-Hill.

Rochester Executive Board. 1886. "The Report of the Executive Board of Rochester, New York." *Engineering News and American Contractors Journal* 16 (20 Nov.): 328–330.

Rock, J. F. 1926. "Through the Great River Trenches of Asia." *National Geographic* 50 (2): 133–186.

Roe, F. G. 1929. "The 'Wild Animal Path' Origin of Ancient Roads." *Antiquity* 3 (11): 299–311.

———. 1939. "The winding road." *Antiquity* 13 (49): 191–406.

Rolt, L. C. 1976. *Red for Danger: A History of Railway Accidents and Railway Safety*. Newton Abbott, Eng.: David and Charles.

Ronan, C. A. 1983. *The Cambridge Illustrated History of the World's Science*. Cambridge: Cambridge/Newnes.

Roots, I. 1972. "Cromwell's Ordinances: The Early Legislation of the Protectorate." In *The Interregnum*. Edited by G. Aylmer. London: Macmillan.

Rose, A. C. 1950a. "1611—The First American Bridge." *American Highways* 29 (2): 24.

———. 1950b. "1755—Braddock's Road." *American Highways* 29 (4): 28–31.

———. 1950c. "1700—The Iroquois Trail" and "1751—The Pennsylvania Road." *American Highways* 29 (3).

———. 1951a. "1795—The Philadelphia and Lancaster Turnpike Road." *American Highways* 30 (1): 34.

———. 1951b. "1809—The Natchez Trace." *American Highways* 30 (3).

———. 1952a. *Public Roads of the Past*. Washington, D.C.: American Association of State Highway Officials.

———. 1952b. "1823—First American Macadam Road." *American Highways* 31 (1): Jan.

———. 1952c. "1839—Our First Iron Bridge" and "1840—The National Pike." *American Highways* 31 (4): Oct.

———. 1953. *Historic American Highways*. Washington, D.C.: American Association of State Highway Officials.

Rose, M. H. 1979. *Interstate: Express Highway Politics, 1941–1956*. Lawrence: Univ. Press of Kansas.

Rosengarten, T. 1986. *Tombee, Portrait of a Cotton Planter*. New York: Morrow.

Ruddock, T. 1979. *Arch Bridges and Their Builders: 1735–1835*. Cambridge: Cambridge Univ. Press.

Rush, P. 1960. *How Roads Have Grown*. London: Routledge and Kegan Paul.

St. Clair, D. J. 1986. *The Motorization of American Cities*. New York: Praeger.

St. George, P. W. 1879. "On the Street and Footwalk Pavements of Montreal, Canada, from the Year 1842 to 1878." *Proceedings of the Institution of Civil Engineers* 58 (4): 287–309.

Salkfield, T. 1953. *Road Making and Road Using*. 4th ed. London: Sir Isaac Pitman.

Salusbury, G. T. 1948. "Street Life in Medieval England." 2d ed. London: Pen-in-hand. The first edition was issued in 1939 and the book was reissued in 1975 by Harvester Press.

Saville, M. H. 1935. "The Ancient Maya Causeways of Yucatan." *Antiquity* 9: 67–73.

Sayenga, D. 1983. *Ellet and Roebling*. York, Pa.: American Canal and Transportation Center.

Schaeffer, K. H., and Sclar, E. 1975. *Access for All*. Middlesex: Penguin.

Scherocman, J. A., and Martenson, E. D. 1984. "Placement of Asphalt Concrete Mixes: Placement and Compaction of Asphalt Mixtures." American Society for Testing Materials (ASTM) STP 829: 3–27.

Schlereth, T. J. 1985. *U.S. 40: A Roadscape of the American Experience*. Indianapolis: Indianapolis Historical Society.

Schreiber, H. 1961. *The History of Roads*. London: Barrie and Rockliff.

Scientific American. 1918. "Extracting Stories from Stones." *Scientific American* 118: 186–187, 200–202.

Scott-Giles, C. W. 1946. *The Road Goes On*. London: Epworth.

Searle, M. [1930]. *Turnpikes and Toll-bars*. London: Hutchinson.

Seely, B. E. 1987. *Building the American Highway System*. Philadelphia: Temple Univ. Press.

Sessions, G. M. 1976. *Traffic Devices: Historical Aspects Thereof*. Washington, D.C.: Institute of Traffic Engineers.

Severy, M. 1977. "The Celts." *National Geographic* 151 (5): 582–634.

Shank, W. H. 1973. *Vanderbilt's Folly: A History of the Pennsylvania Turnpike*. 2d ed. York, Pa.: American Canal and Transportation Center.

———. 1974. *Indian Trails to Super Highway*. 2d ed. York, Pa.: American Canal and Transportation Center.

———. 1976. *Three Hundred Years with the Pennsylvania Traveller*. York, Pa.: American Canal and Transportation Center.

———. 1980. *Historic Bridges in Pennsylvania*. 3d ed. York, Pa.: American Canal and Transportation Center.

Sherrard, H. M. 1930. *Report on Observations of Highway Work in the United States*. ARRB Library:625.7/SHE.

Shirley Smith, H. 1953. *The World's Great Bridges*. London: Phoenix House.

Siegel, J. 1982. *Traditional Bridges of Papua New Guinea*. Papua New Guinea University of Technology, Appropriate Technology Development Institute, Traditional Technology Series no. 1. Lae.

Silk, G. 1984. *Automobile and Culture*. New York: Abrams.

Simms, F. W. 1838. "Practical Observations on the Asphaltic Mastic." Private treatise. ICE (Institution of Civil Engineers) Tracts, 8vo, Vol. 201.

Simon, L. L. 1970. "The History of Highway Engineering as Related to Materials Science and Pavement Design." Ph.D. diss., University of New South Wales, Sydney.

Simpson, J. 1967. *Everyday Life in the Viking Age*. London: Batsford.

Simsa, P. 1986. "From the Engine to the Automobile." *Mercedes Benz In aller Welt* 199 (1): 4–25.

Sitwell, O. 1948. *Great Morning*. London: Macmillan.

Smiles, S. 1874. *Lives of the Engineers*. Vol. 3, *History of Roads*. London: Murray.

Stanley, C. C. 1979. *Highlights in the History of Concrete*. Fulmer: Cement and Concrete Association.

Stave, B. M., ed. 1981. *Modern Industrial Cities: History, Policy, and Survival*. Beverly Hills: Sage.

Steinman, D. B., and Watson, S. R. 1941. *Bridges and Their Builders*. New York: Putnam.

Stenton, D. M. 1951. *English Society in the Middle Ages, 1066–1307*. Harmondsworth: Penguin.

Stenton, F. M. 1936. "The Road System of Medieval England." *Economic History Review* 7 (1): Nov.

Stern, R. A. 1981. "La Ville Bourgeoise." *Architectural Design* 51 (10–11): 4–12.

Stevenson, R. 1824. "Roads and Highways." *Edinburgh Encyclopedia*. Edinburgh: Balfour.

Straub, H. 1952. *A History of Civil Engineering*. London: Leonard Hill.

Strong, L. A. 1956. *The Rolling Road*. London: Hutchinson.

Syme, R. 1952. *The Story of Britain's Highways*. London: Pitman.

Tafton-Brown, T. 1986. "The Topography of Anglo-Saxon London." *Antiquity* 60 (228): 21–28.

Tang, H. C. 1987. *Ancient Chinese Bridges*. Beijing: Historical Relic Publishers.

Taylor, C. 1979. *Roads and Tracks of Britain*. London: Dent.

Taylor, G. W. 1951. *The Transportation Revolution*. New York: Rinehart.

Taylor, I. 1823. *Scenes of British Wealth*. London: Harris.

Taylor, W. 1976. *The Military Roads in Scotland*. Newton Abbott, Eng.: David and Charles.

Telford, T. 1802. "A Survey and Report of the Coasts and Central Highlands of Scotland." London: House of Lords. Printed 1803 in *Reports on Highland Roads and Bridges*.

Thomas, J. M. 1970. *History at Source: Roads before the Railways, 1700–1851*. London: Evans.

Thompson, F. M. 1970. *Victorian England: The Horse-Drawn Society*. London: Bedford College, Univ. of London.

Thompson, J., and Smith, A. 1877. *Street Life in London*. London: Low, Marston, Searle and Rivington.

Tillson, G. W. 1897. "Asphalt and Asphalt Pavements." *Transactions of the American Society of Civil Engineers* 38, no. 814 (December): 215–235.

———. 1901. *Street Pavements and Paving Materials*. New York: Wiley.

Tillson, G.; Haylow, J. H.; and Richardson, C. 1907. "Will the Paving Materials of the Present Be Used in the Construction of the Pavements of the Future?" *Papers and Discussion, American Society of Civil Engineers* 33 (6): 711–721.

Tobias, J. J. 1967. *Crime and Industrial Society in the Nineteenth Century*. London: Batsford.

Todd, K. 1989. "The Roundabout: Its History and Development in Britain." *Journal of the Institute of Highways and Transportation* 36 (1): 25–31.

Tompkins, E. 1981. *The History of the Pneumatic Tyre*. London: Eastland.

Toynbee, A. 1970. *Cities on the Move*. London: Oxford Univ. Press.

Trésaguet, P. M.-J. 1831. "Memoire sur le entretier des chemins de la generalité des Limoges." *Annales des Ponts et Chaussées,* 1st ser., 2: 243–258.

Trinder, B. 1980. "The Holyhead Road: An Engineering Project in Its Social Context." In *Thomas Telford: Engineer,* ed. A. Penfold, 41–61. London: Thomas Telford.

Tronquoy, M. 1868. *Instruction sur l'execution des levés et des dessins au net*. Paris: Ecole Imperiale des Ponts et Chaussées.

Tunnicliff, D. G.; Beaty, R. W.; and Holt, E. H. 1974. "A History of Plants, Equipment and Methods in Bituminous Paving." *Proceedings of the Association of Asphalt Paving Technologists* 43A: 159–296.

Tuttle, A. S. 1924. "Increasing the Capacity of Existing Streets." *Proceedings of the American Society of Civil Engineers* 50 (5): 593–605 (discussion in *Transactions,* 1925, 88: 223–224).

Upton, N. 1975. *An Illustrated History of Civil Engineering*. London: Heinneman.

U.S. Congress. House. 1939. "Toll Roads and Fee Roads." 76th Cong., 1st sess., H. Doc. 272. Washington, D.C.: GPO.

U.S. Federal Highway Administration. *See* FHA.

Vaknalli, S. 1980. "Indian Roads from Rig Veda to Raj to Republic." *Highway Engineering* 27 (6): 40–41.

Vale, E. 1960. *The Mail Coach Men of the Late Eighteenth Century*. London: Cassell.

Van Beek, G. W. 1987. "Arches and Vaults in the Ancient Near East." *American Scientist* 257 (1): 78–85.

Vance, J. E. [1977] 1986. *This Scene of Man*. New York: Harper's College Press. Reprint entitled *The Continuing City*. Baltimore: Johns Hopkins Univ. Press.

———. 1986. *Capturing the Horizon: The Historical Geography of Transportation Since the Sixteenth Century*. Baltimore: Johns Hopkins Univ. Press.

Van Nostrand. 1876. "A Decade of Steam Road-rolling in Paris." *Van Nostrand's Eclectic Engineering Magazine* 6: 298–302.

Vitruvius, M. 1960. *De Architectura*. Translated by M. H. Morgan. New York: Dover.

Waddell, J. A. 1916. *Bridge Engineering*. Vol. 1. New York: Wiley.

Warner, S. B. 1962. *Streetcar Suburbs*. Cambridge, Mass.: Harvard Univ. Press.

Warren, F. J. 1901. "The Development of Bituminous Pavements." *Engineering Record* 44 (9): 198–199.

Watson, H. R. 1950. "Bitumen Refining." *Quarry Managers' Journal* 34 (4): 209–213.

Watson, W. J., and Watson, S. R. 1937. *Bridges in History and Legend*. Cleveland: Jansen.

Webb, S., and Webb, B. [1913] 1963. *The Story of the King's Highway*. London: Longmans. Reprint. London: F. Cass.

Weber, A. F. 1899. *The Growth of Cities in the Nineteenth Century*. London: Macmillan.

Weiner, E. 1986. "Urban Transportation Planning in the United States: An Historical Overview." U.S. Department of Transportation, DOT-1-86-09 (February) Washington, D.C.: GPO.

Wellborn, A. S. 1969. "History of Asphalt and the Asphalt Institute." *Civil Engineering* (ASCE) 39 (5): 62–69.

Welty, E. M., and Taylor, F. J. 1958. *The Black Bonanza.* New York: McGraw-Hill.

Wheeler, W. H. 1876. "Roads and Highways: Their History, Construction, Cost, Repair and Management." *Journal of the Royal Agriculture Society (England)* 2d ser., 12:382–406.

White. 1939. "White Centre Line in Ancient Highway in Mexico." *Concrete* 47 (5): 31.

White, L. 1962. *Medieval Technology and Social Change.* Oxford: Oxford Univ. Press.

Whitehead, R. A. 1975. *A Century of Steam-Rollers.* London: Ian Allen.

Whitehill, W. M. 1966. *Boston in the Age of John F. Kennedy.* Norman: Univ. of Oklahoma Press.

Whitelock, D. 1952. *The Beginnings of English Society.* Harmondsworth: Pelican.

Whitt, F. R., and Wilson, D. G. 1982. *Bicycle Science.* Cambridge, Mass.: MIT Press.

Wilkin, F. T. 1979. "The Highland Transport System: Genesis, Growth and Survival." *Journal of the Institute of Highway Engineers* 26 (2): 9–19.

Wilkinson, T. W. 1934. *From Track to By-pass.* London: Methuen.

Williams, R. I. 1986. *Cement-Treated Pavements.* London: Elsevier.

Wilson, D. G. 1986. "A Short History of Human-powered Vehicles." *American Scientist* 74 (4): 350–357.

Wilson, G. 1855. *Researches on Colour-Blindness: With a Supplement on the Danger Attending the Present System of Railway and Marine Coloured Signals.* Edinburgh: Sutherland and Knox.

Windeyer, T. W. 1958–1959. 102, *Commonwealth Law Reports* (Australian), 280, 303. Melbourne.

Wiseman, A., and Coles, P., translators. 1980. *Julius Caesar: The Battle for Gaul.* London: Chatto and Windus.

Wittfoht, H. 1984. *Building Bridges.* Dusseldorf: Beton-Verlag.

Wixom, L. W. 1975. *ARBA Pictorial History of Road Building.* Washington: American Road Builders' Association.

Wood. 1923. "Wood Paving Experiments." *Roads and Road Construction* 1 (1 January): 18.

Woolley, L. C. 1954. *Excavations at Ur.* London: Ernest Benn.

Yago, G. 1984. *The Decline of Transit.* Cambridge: Cambridge Univ. Press.

Yvon, M. 1985. "Pierre-Marie-Jérôme Trésaguet, Ingénieur des Ponts-et-Chaussées, 1716–1796." In *Les Routes de Sud de la France.* Proceedings 10th Nationale Congres des Sociétés Savantes. Montpellier. April, 295–329.

Zammit, T. 1928. "Prehistoric Cart Tracks in Malta." *Antiquity* 2:18–25.